MW00814584

Springer Proceedings in Physics

Volume 230

Indexed by Scopus

The series Springer Proceedings in Physics, founded in 1984, is devoted to timely reports of state-of-the-art developments in physics and related sciences. Typically based on material presented at conferences, workshops and similar scientific meetings, volumes published in this series will constitute a comprehensive up-to-date source of reference on a field or subfield of relevance in contemporary physics. Proposals must include the following:

- name, place and date of the scientific meeting
- a link to the committees (local organization, international advisors etc.)
- scientific description of the meeting
- list of invited/plenary speakers
- an estimate of the planned proceedings book parameters (number of pages/ articles, requested number of bulk copies, submission deadline).

More information about this series at http://www.springer.com/series/361

P. C. Deshmukh · E. Krishnakumar ·
Stephan Fritzsche · M. Krishnamurthy ·
Sonjoy Majumder
Editors

Quantum Collisions and Confinement of Atomic and Molecular Species, and Photons

Select Proceedings of the 7th Topical
Conference of ISAMP 2018

 Springer

Editors
P. C. Deshmukh
Indian Institute of Technology Tirupati
Tirupati, India

E. Krishnakumar
Raman Research Institute
Bangalore, India

Stephan Fritzsche
Helmholtz Institute Jena
Jena, Germany

M. Krishnamurthy
Tata Institute of Fundamental Research
Mumbai, India

Sonjoy Majumder
Indian Institute of Technology Kharagpur
Kharagpur, India

ISSN 0930-8989   ISSN 1867-4941   (electronic)
Springer Proceedings in Physics
ISBN 978-981-13-9968-8   ISBN 978-981-13-9969-5   (eBook)
https://doi.org/10.1007/978-981-13-9969-5

This Springer imprint is published by the registered company Springer Nature Singapore Pte Ltd.
The registered company address is: 152 Beach Road, #21-01/04 Gateway East, Singapore 189721, Singapore

Foreword

The 7th Topical Conference of the Indian Society of Atomic and Molecular Physics was held at Tirupati on January 6–8, 2017. It was jointly hosted by the Indian Institute of Technology Tirupati (IITT) and the Indian Institute of Science Education and Research Tirupati (IISERT), with generous support from the Director of the IITT (Prof. K. N. Satyanarayana) and the Director of the IISERT (Prof. K. N. Ganesh). While Prof. P. C. Deshmukh and Prof. Bhas Bapat served as the Joint-Conveners of this conference, Dr. S. Sunil Kumar served as the Conference Secretary.

Tirupati is the only place in India which houses an IIT and also an IISER. Both of these institutions have taken a major initiative in the field of Atomic, Molecular and Optical Physics. The ISAMP-TC7 is the *first* conference jointly convened by the two institutions, and early enough just within a few years that the institutions started operating since August 2015.

Apart from 67 posters that were presented at this meeting, 50 talks were delivered. The delegates came from near and far places in India (Chennai, Bangalore, Thiruvananthapuram, Mumbai, Delhi, Kolkata, Ahmedabad, Hyderabad, Patna, Bhopal, Vellore, Dhanbad, Indore, etc.), and also from 11 other countries (UK, Japan, USA, France, Czech Republic, Germany, Australia, Spain, Argentina, Singapore, Switzerland). The keynote addresses at the inaugural session were delivered by Prof. Anatoli Kheifets, ANU-Canberra (Theory), Prof. E. Krishnakumar, TIFR-Mumbai/RRI-Bengaluru (Experiment).

The 26 selected articles presented in this special volume of the Springer Proceedings in Physics represent the high quality of work that was discussed at the ISAMP TC7. All the contributions at the conference could not be included in this special volume, but the editors are happy to compile the articles in this volume to be left behind as the contribution of the ISAMP TC7 to primary scientific literature.

Tirupati, India   P. C. Deshmukh
Bangalore, India   E. Krishnakumar
Jena, Germany   Stephan Fritzsche
Mumbai, India   M. Krishnamurthy
Kharagpur, India   Sonjoy Majumder

Contents

Editors and Contributors

About the Editors

Dr. P. C. Deshmukh is Professor at the Department of Physics, and Dean, Sponsored Research and Consultancy, at the Indian Institute of Technology Tirupati. He obtained his Ph.D. from Nagpur University, followed by post doctoral work at the University of Aarhus, the University of Notre Dame, and Georgia State University. Prior to joining IIT Tirupati, Dr. Deshmukh was Professor and Head at the Department of Physics, IIT Madras. His primary areas of interest include photoabsorption processes in free/confined atoms, molecules and ions. He is a former President of the Indian Society of Atomic and Molecular Physics (ISAMP), and is a member of several national and international scientific advisory committees. He has published several articles in international peer-reviewed journals, and has also edited a previous proceedings of ISAMP, and also of AISAMP.

Dr. E. Krishnakumar retired as Senior Professor from Tata Institute of Fundamental Research (TIFR) Mumbai in 2017 and currently a Raja Ramanna Fellow and Emeritus Scientist at the Raman Research Institute, Bangalore. He obtained his Ph.D. from Physical Research Laboratory, Ahmedabad following which he worked as a scientist at the Space Applications Centre, Ahmedabad and the Jet Propulsion Laboratory at Caltech, Pasadena. His research interests include electron-molecule collisions in gas and condensed phase, molecular dynamics, electron and ion momentum imaging and spectroscopy, photoionization and photodetachment. He has published more than 100 papers in international peer-reviewed journals and delivered around 50 invited talks. Dr. Krishnakumar was President of the Indian Society of Atomic and Molecular Physics (ISAMP), has been a member of several national and international advisory committees and a Fellow of Indian Academy of Sciences.

Dr. Stephan Fritzsche is Professor and Theory-Chair for 'Correlated Quantum Systems' at the Helmholtz Institute Jena. He obtained his Ph.D. from the University of Kassel. Prior to joining the Helmholtz Institute, he worked at Oulu University (Finland) and the Frankfurt Institute of Advanced Studies. His research focuses mainly on the structure and dynamics of finite quantum systems with applications in atomic, optical and nuclear physics. Dr. Fritzsche and his group is working to develop new many-body techniques for describing the electron dynamics of ions, atoms and plasma in strong fields.

Dr. M. Krishnamurthy is Professor at the Department of Nuclear and Atomic Physics, TIFR Mumbai. After completing his Ph.D. from TIFR, he was a post-doctoral researcher at the Joint Institute for Laboratory Astrophysics, University of Colorado Boulder. His research interests include non-linear optics, intense lasers, plasma physics, mass spectrometry, nanoclusters, and X-ray VUV generation. He has published over 100 articles in international peer-reviewed journals, and also delivered over 75 invited talks at national and international conferences. Dr. Krishnamurthy has received several awards and distinctions such as INSA Young Scientist Medal, and the DAE-SRC Outstanding Investigator Award. He is also an elected Fellow of the Indian Academy of Sciences.

Dr. Sonjoy Majumder is Associate Professor at the Physics department, IIT Kharagpur. He obtained his Ph.D. from the Indian Institute of Astrophysics Bangalore. His major research areas include physics of cold and ultra-cold atoms, light-matter interaction, computational many-body physics, and astrophysics. He has published over 50 articles in international journals of repute. He is currently the principal investigator on two DST and DAE sponsored projects.

Contributors

Minori Abe Department of Chemistry, Tokyo Metropolitan University, Hachioji, Tokyo, Japan

Mario E. Alcocer-Ávila Université de Bordeaux, CNRS, CEA, CELIA (Centre Lasers Intenses et Applications), UMR 5107, Talence, France

D. Angom Physical Research Laboratory, Ahmedabad, Gujarat, India

Bobby Antony Atomic and Molecular Physics Lab, Department of Physics, Indian Institute of Technology (Indian School of Mines), Dhanbad, Jharkhand, India

Rukmani Bai Physical Research Laboratory, Ahmedabad, Gujarat, India; Indian Institute of Technology Gandhinagar, Gandhinagar, Gujarat, India

Renu Bala Department of Physics, Indian Institute of Technology Roorkee, Roorkee, India

Soumik Bandyopadhyay Physical Research Laboratory, Ahmedabad, Gujarat, India;
Indian Institute of Technology Gandhinagar, Gandhinagar, Gujarat, India

Bhas Bapat Indian Institute of Science Education and Research Pune, Pune, Maharashtra, India

Dolan Krishna Bayen Department of Physics, Visva-Bharati, Santiniketan, India

Jayanta Bera Indian Institute of Technology of Patna, Bihta, Patna, Bihar, India

Swati Bharti Indian Institute of Technology Roorkee, Roorkee, India

Soumen Bhattacharyya Atomic and Molecular Physics Division, Bhabha Atomic Research Centre, Mumbai, India;
Homi Bhabha National Institute, Anushaktinagar, Mumbai, India

Anal Bhowmik Department of Physics, Indian Institute of Technology Kharagpur, Kharagpur, India

Birger Böning Theoretisch Physikalisches Institut, Friedrich Schiller Universität Jena, Jena, Germany

Kajol Chakraborty Department of Physics and Astrophysics, Kalindi College, University of Delhi, New Delhi, India

Christophe Champion Université de Bordeaux, CNRS, CEA, CELIA (Centre Lasers Intenses et Applications), UMR 5107, Talence, France

Marcello Coreno Consiglio Nazionale delle Ricerche – Istituto di Struttura della Materia, Trieste, Italy

Alessandro D'Elia Department of Physics, University of Trieste, Trieste, Italy

P. C. Deshmukh Indian Institute of Science Education and Research Tirupati, Tirupati, India;
Indian Institute of Technology Tirupati, Tirupati, India

Amar Dora Department of Chemistry, North Orissa University, Baripada, Odisha, India

Stephan Fritzsche Theoretisch Physikalisches Institut, Friedrich Schiller Universität Jena, Jena, Germany;
Helmholtz Institut, Jena, Germany

R. K. Gangwar Department of Physics and Astronomy, National Institute of Technology Rourkela, Rourkela, India;
Department of Physics, Visvesvaraya National Institute of Technology, Nagpur, India

Chirashree Ghosh Department of Environmental Studies, University of Delhi, New Delhi, India

Suranjana Ghosh Amity University Patna, Rupaspur, Patna, Bihar, India

Ram Gopal TIFR Centre for Interdisciplinary Sciences, Hyderabad, Telangana, India

Ruchika Gupta Department of Physics and Astrophysics, Kalindi College, University of Delhi, New Delhi, India

S. Gupta Department of Physics, Indian Institute of Technology Roorkee, Roorkee, India

Barun Halder Indian Institute of Technology of Patna, Bihta, Patna, Bihar, India

Anatoli Kheifets Research School of Physics, The Australian National University, Canberra, ACT, Australia

E. Krishnakumar Raman Research Institute, Bangalore, India

M. Krishnamurthy TIFR Centre for Interdisciplinary Sciences, Hyderabad, India

S. R. Krishnan Indian Institute of Technology Madras, Chennai, Tamil Nadu, India

N. Kundu Department of Physics, Indian Institute of Technology Patna, Bihta, Patna, India

Chetan Limbachiya Department of Physics, The M. S. University Baroda, Vadodara, Gujarat, India

Sonjoy Majumder Department of Physics, Indian Institute of Technology Kharagpur, Kharagpur, India

Ankur Mandal Indian Institute of Science Education and Research Tirupati, Tirupati, India

Suddhasattwa Mandal Indian Institute of Science Education and Research Pune, Pune, Maharashtra, India

Swapan Mandal Department of Physics, Visva-Bharati, Santiniketan, India

Laura K. McKemmish Department of Physics and Astronomy, University College London, London, UK;
School of Chemistry, University of New South Wales, Kensington, Sydney, Australia

Paresh Modak Atomic and Molecular Physics Lab, Department of Physics, Indian Institute of Technology (Indian School of Mines), Dhanbad, Jharkhand, India

Juan M. Monti Instituto de Física Rosario, CONICET – Universidad Nacional de Rosario, EKF Rosario, Argentina

Marcel Mudrich Department of Physics and Astronomy, Aarhus University, Aarhus, Denmark

Sheo Mukund Atomic and Molecular Physics Division, Bhabha Atomic Research Centre, Mumbai, India;
Homi Bhabha National Institute, Anushaktinagar, Mumbai, India

S. G. Nakhate Atomic and Molecular Physics Division, Bhabha Atomic Research Centre, Mumbai, India;
Homi Bhabha National Institute, Anushaktinagar, Mumbai, India

H. S. Nataraj Department of Physics, Indian Institute of Technology Roorkee, Roorkee, India

L. Natarajan Department of Physics, University of Mumbai, Mumbai, India

Sukla Pal Physical Research Laboratory, Ahmedabad, Gujarat, India

Willi Paufler Theoretisch Physikalisches Institut, Friedrich Schiller Universität Jena, Jena, Germany

Vaibhav S. Prabhudesai Tata Institute of Fundamental Research, Mumbai, India

Dineshkumar Prajapati Shree M. R. Arts & Science College, Rajpipla, Gujarat, India

Priti Department of Physics, Indian Institute of Technology Roorkee, Roorkee, India

Michele A. Quinto Instituto de Física Rosario, CONICET – Universidad Nacional de Rosario, EKF Rosario, Argentina

R. Rajeev TIFR Centre for Interdisciplinary Sciences, Hyderabad, India

Robert Richter Elettra-Sincrotrone Trieste, Basovizza, Trieste, Italy

Roberto D. Rivarola Instituto de Física Rosario, CONICET – Universidad Nacional de Rosario, EKF Rosario, Argentina

Tom Rivlin Department of Physics and Astronomy, University College London, London, UK

Utpal Roy Department of Physics, Indian Institute of Technology of Patna, Bihta, Patna, Bihar, India

Soumyajit Saha Indian Institute of Technology Madras, Chennai, India

Lalita Sharma Indian Institute of Technology Roorkee, Roorkee, India

Vandana Sharma Indian Institute of Technology Hyderabad, Sangareddy, Telangana, India

Suvam Singh Atomic and Molecular Physics Lab, Department of Physics, Indian Institute of Technology (Indian School of Mines), Dhanbad, Jharkhand, India

Vishwanath Singh Atomic and Molecular Physics Lab, Department of Physics, Indian Institute of Technology (Indian School of Mines), Dhanbad, Jharkhand, India

Nidhi Sinha Atomic and Molecular Physics Lab, Department of Physics, Indian Institute of Technology (Indian School of Mines), Dhanbad, Jharkhand, India

Hemkumar Srinivas Max-Planck-Institut für Kernphysik, Heidelberg, Germany

Rajesh Srivastava Department of Physics, Indian Institute of Technology Roorkee, Roorkee, India

K. Suthar Physical Research Laboratory, Ahmedabad, Gujarat, India

Jonathan Tennyson Department of Physics and Astronomy, University College London, London, UK

Himani Tomer Atomic and Molecular Physics Lab, Department of Physics, Indian Institute of Technology (Indian School of Mines), Dhanbad, Jharkhand, India

Nafees Uddin Atomic and Molecular Physics Lab, Department of Physics, Indian Institute of Technology (Indian School of Mines), Dhanbad, Jharkhand, India

Pankaj Verma Atomic and Molecular Physics Lab, Department of Physics, Indian Institute of Technology (Indian School of Mines), Dhanbad, Jharkhand, India

Punita Verma Department of Physics, Kalindi College, University of Delhi, New Delhi, India

Ch. Vikar Ahmad Department of Physics and Astrophysics, Kalindi College, University of Delhi, New Delhi, India

Minaxi Vinodkumar Electronics Department, V. P. & R. P. T. P. Science College, Vallabh Vidyangar, Gujarat, India

P. C. Vinodkumar Department of Physics, Sardar Patel University, Vallabh Vidyanagar, Gujarat, India

Hitesh Yadav Department of Physics, Sardar Patel University, Vallabh Vidyanagar, Gujarat, India

Time-Resolved Theory of Atomic and Molecular Photoionization for RABBITT and Attoclock

Anatoli Kheifets[(✉)]

Research School of Physics, The Australian National University, Canberra,
ACT 0200, Australia
A.Kheifets@anu.edu.au
http://people.physics.anu.edu.au/~ask107/

Abstract. We outline a theoretical framework and present its numerical implementation for modeling of recent experiments on time-resolved atomic and molecular photoionization. We focus on RABBITT measurements of the Wigner time delay and the attoclock determination of the tunneling time. Our theoretical modeling is based on a numerical solution of the time-dependent Schrödinger equation driven by weak XUV probe and IR pump pulses in RABBITT experiments as well as by intense IR pulses in self-referencing attoclock measurements.

1 Introduction

Time-resolved studies of atomic photoionization with various pump–probe techniques such as attosecond streaking [1] or RABBITT [2], and self-referencing techniques like attoclock [3] opened up a new and rapidly developing area of research collectively termed attosecond chronoscopy [4]. The attosecond streaking and RABBITT measurements determine the photoelectron group delay which is related to the photoelectron phase and its energy derivative known as the Wigner time delay [5]. These studies bring one step closer what had been thought of as a complete photoionization experiment. The attoclock measurement can be related to the tunneling time, i.e., the time a photoelectron spends under the barrier in a classically inaccessible region. The new measurements reopened decade-long debate about a finite tunneling time [6].

In this presentation, the recent theoretical advances in evaluation of the Wigner time and tunneling time in atoms and molecules will be reviewed and connection with ongoing experimental activities will be made. The following topics will be highlighted:

1. Wigner time delay in photoionization of free and encapsulated noble gas atoms. This topic includes the relativistic effects and angular-dependent time delay. Connection with the recent measurements in heavy noble gas atoms will be made [7].

© Springer Nature Singapore Pte Ltd. 2019
P. C. Deshmukh et al. (eds.), *Quantum Collisions and Confinement of Atomic and Molecular Species, and Photons*, Springer Proceedings in Physics 230,
https://doi.org/10.1007/978-981-13-9969-5_1

2. Wigner time delay in molecular photoionization including complex heteronuclear molecules.
3. Tunneling time measurements [8] and calculations [9] in atomic hydrogen and their implications for the finite tunneling time problem.

2   Wigner Time Delay and RABBITT

RABBITT (Reconstruction of Attosecond Beating By Interference of Two-photon Transitions) technique has been used widely to study time-resolved dynamics of atomic [2, 7, 10–14] and molecular [15–17] photoemission. Recently, this technique has become angular resolved [18–20]. While the initial studies have been confined to a narrow photon energy range not exceeding 40 eV (twenty-fourth harmonic of the fundamental radiation at $\lambda \simeq 800$ nm), the most recent experiments [14, 21, 22] extended this range above 100 eV. This has enabled the study of photoemission dynamics beyond the outer atomic shells and to probe the subvalent $4d$ shell of the Xe atom [21].

The RABBITT technique builds on the interference of two ionization processes leading to the same photoelectron state by (i) absorption of the XUV frequency ω_{2q-1} of an odd $2q - 1$ harmonic and an IR quantum ω or (ii) absorption of ω_{2q+1} and stimulated emission of ω. Both ionization processes lead to the appearance of a sideband (SB), in between the one-photon harmonic peaks in the photoelectron spectrum. The sideband magnitude oscillates with the relative phase between the XUV and IR pulses [23, 24]

$$S_{2q}(\tau) = A + B \cos[2\omega\tau - C], \quad C = \Delta\phi_{2q} + \Delta\theta_{2q}, \tag{1}$$

where $\tau = \varphi/\omega$ denotes the phase delay of the IR field relative to the XUV. The term $\Delta\phi_{2q} = \phi_{2q+1} - \phi_{2q-1}$ denotes the phase difference between two neighboring odd harmonics $2q \pm 1$ that is related to the finite-difference group delay of the attosecond pulse as $\tau_{2q}^{(\mathrm{GD})} = \Delta\phi_{2q}/2\omega$.

2.1   Atomic RABBITT Simulations

Computational Technique Our atomic RABBITT modeling is based on solution of the one-electron time-dependent Schrödinger equation (TDSE) for a target atom [25],

$$i\partial\Psi(\boldsymbol{r})/\partial t = \left[\hat{H}_{\mathrm{atom}} + \hat{H}_{\mathrm{int}}(t)\right]\Psi(\boldsymbol{r}). \tag{2}$$

In the following, we adopt the system of atomic units and set the charge and mass of the electron as well as the Planck constant to unity $e = m = \hbar = 1$. The radial part of the atomic Hamiltonian

$$\hat{H}_{\mathrm{atom}}(r) = -\frac{1}{2}\frac{d^2}{dr^2} + \frac{l(l+1)}{2r^2} + V(r) \tag{3}$$

contains an effective one-electron potential $V(r)$. Several choises exist for this potential. We use either an optimized effective potentials (OEP) [26] or a localized Hartree–Fock potential (LHP) determined as prescribed in [27]. In the case of an encapsulated atom trapped inside a fullerene C_{60} cage, an attractive spherical square well potential is added

$$\Delta V(r) = \begin{cases} -U_0 < 0 \text{ if } R_{\text{inner}} \le r \le R_{\text{inner}} + \Delta \\ 0 \text{ otherwise.} \end{cases} \quad (4)$$

Here $R_{\text{inner}} = 5.8$ a.u., $\Delta = 1.9$ a.u., and $U_0 = 0.302$ a.u. [28].

The Hamiltonian $\hat{H}_{\text{int}}(t)$ describes interaction with the external field and is written in the velocity gage

$$\hat{H}_{\text{int}}(t) = \boldsymbol{A}(t) \cdot \hat{\boldsymbol{p}}, \boldsymbol{A}(t) = - \int_0^t \boldsymbol{E}(t') \, dt'. \quad (5)$$

This external field is comprised of both XUV and IR pulses. The XUV field is modeled by an attosecond pulse train (APT) with the vector potential

$$A_x(t) = \sum_{n=-5}^{5} (-1)^n A_n \exp\left(-2\ln 2 \frac{(t - nT/2)^2}{\tau_x^2}\right)$$
$$\times \cos\left[\omega_x(t - nT/2)\right], \quad (6)$$

where

$$A_n = A_0 \exp\left(-2\ln 2 \frac{(nT/2)^2}{\tau_T^2}\right).$$

Here A_0 is the vector potential peak value and $T = 2\pi/\omega$ is the period of the IR field. The XUV central frequency is ω_x and the time constants τ_x, τ_T are chosen to span a sufficient number of harmonics in the range of photon frequencies of interest for a given atom.

The vector potential of the IR pulse is modeled by the cosine squared envelope

$$A(t) = A_0 \cos^2\left(\frac{\pi(t - \tau)}{2\tau_{\text{IR}}}\right) \cos[\omega(t - \tau)].$$

The IR pulse is shifted relative to the APT by a variable delay τ such that the RABBITT signal of the even SB $2q$ oscillates as

$$S_{2q}(\tau) = A + B \cos[2\omega\tau - C]. \quad (7)$$

Solution of the TDSE (19) is found using the iSURF method as given in [29]. A typical calculation with XUV and IR field intensities of 5×10^9 and 3×10^{10} W/cm^2, respectively, would take up to 35 CPU hours for each τ.

The RABBITT parameters A, B, and C entering (7) can be expressed via the absorption and emission amplitudes

$$A = |\mathcal{M}_{\boldsymbol{k}}^{(-)}|^2 + |\mathcal{M}_{\boldsymbol{k}}^{*(+)}|^2, \ B = 2\text{Re}\left[\mathcal{M}_{\boldsymbol{k}}^{(-)} \mathcal{M}_{\boldsymbol{k}}^{*(+)}\right]$$
$$C = \arg\left[M_{\boldsymbol{k}}^{(-)} M_{\boldsymbol{k}}^{*(+)}\right] = 2\omega\tau_a. \quad (8)$$

Here $\mathcal{M}_{\boldsymbol{k}}^{(\pm)}$ are complex amplitudes for the angle-resolved photoelectron produced by adding or subtracting an IR photon, respectively. By adopting the soft photon approximation (SPA) [30], we can write

$$A, B \propto |J_1(\boldsymbol{\alpha}_0 \cdot \boldsymbol{k})|^2 |\langle f|z|i\rangle|^2 \tag{9}$$
$$\propto [1 + \beta P_2(\cos\theta_k)] \cos^2\theta_k.$$

Here we made a linear approximation to the Bessel function as the parameter $\alpha_0 = F_0/\omega^2$ is small in a weak IR field (see Appendix of [31] for derivation). In (8), θ_k is the angle between the photoelectron emission direction $\hat{\boldsymbol{k}}$ and the electric field vector of the linearly polarized light. By fitting the calculated angular dependence of the A and B parameters with the SFA expression (9) we can obtain the two sets of the angular anisotropy parameters β_A^{SB} and β_B^{SB} and compare them with the values calculated in the random phase approximation with exchange (RPAE), the latter including inter-shell correlation and exchange of the photoelectron with the remaining ionic core. These effects are not included in the single active electron TDSE model. At the same time, we derive the angular dependence from the odd high harmonic (HH) peaks by fitting angular variation of their amplitude with $1 + \beta^{\mathrm{HH}} P_2(\cos\theta_k)$. Thus, for each target atom, three sets of β parameters are extracted and analyzed over a wide photon energy range.

By using (8), the C parameter is converted to the atomic time delay

$$\tau_a = C/(2\omega) = \tau_W + \tau_{cc}, \tag{10}$$

which contains the two distinct components [32]. Here τ_W is the Wigner-like time delay associated with the XUV absorption and τ_{cc} is a correction due to the continuum–continuum (CC) transitions in the IR field. The latter term, τ_{cc}, can also be understood as a coupling of the long-range Coulomb ionic potential and the laser field in the context of streaking [33,34].

Numerical Results In Fig. 1, we display the angular anisotropy β parameters for the Ne $2p$ (left), Ar $3p$ (center), and Xe $4d$ (right) extracted from the TDSE calculations. The β^{HH} parameters extracted from the angular dependence of the high harmonic peaks are plotted along with the β^{SB} parameters extracted from the angular variation of the RABBITT A and B parameters in (8). The RPAE calculation is shown with the solid line. This calculation is known to reproduce accurately the experimental β parameters across the studied photon energy range [35].

We see that the harmonics and sidebands TDSE calculations of β parameters are consistent between each other and are fairly close to the XUV-only RPAE calculation. In Ne $2p$ and Ar $3p$, they are close to the experimental values. In the case of Xe $4d$, the experimental data are somewhat off due to the $4d/4p$ inter-shell correlation which is not included in the present calculation. Agreement with experiment can be significantly improved in a fully relativistic RRPA calculation with inclusion of all interacting subshells [39].

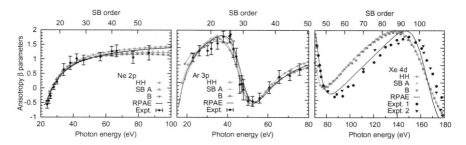

Fig. 1. Angular anisotropy β parameters for Ne $2p$ (left), Ar $3p$ (center), and Xe $4d$ (right) extracted from the TDSE calculations with the localized Hartree–Fock potential (LHF). The β^{HH} parameters extracted from the angular dependence of the high harmonic peaks are plotted with (red) filled circles. Same parameters β^{SB} extracted from the angular variation of the RABBITT A and B coefficients in (8) are plotted with (orange) triangles and (blue) asterisks, respectively. The RPAE calculation is shown with the solid line. The experiment [36] for Ne, [37] for Ar, and [38] for Xe are given by points with error bars (where available)

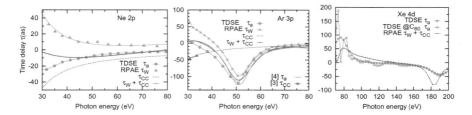

Fig. 2. Time delay in the polarization axis direction for Ne $2p$ (left), Ar $3p$ (center), and Xe $4d$ (right). The atomic time delay τ_a from the TDSE calculation (red filled circles) is compared with the Wigner time delay (orange triangle) from the RPAE calculation. The CC correction τ_{CC} is shown with the thin dotted line, whereas the sum $\tau_W + \tau_{CC}$ is displayed with the (blue) dotted line. On the right panel, the TDSE calculation of the atomic time delay τ_a for $4d$ shell of the encapsulated Xe@C$_{60}$ is shown with the open purple circles

The time delay in the polarization axis direction is shown in Fig. 2 for the Ne $2p$ (left), Ar $3p$ (center), and Xe $4d$ (right). We compare the atomic time delay τ_a extracted from the TDSE calculation with the Wigner time delay τ_W from the RPAE calculation. The hydrogenic CC correction τ_{CC}, which is shown separately, is then added to the Wigner time delay. This correction, as a function of the photoelectron energy, is represented by an analytic expression

$$\tau_{CC}(E) = NE^{-3/2}[a\log(E) + b], \tag{11}$$

where the coefficients N, a, and b are found from fitting the regularized continuum–continuum delay shown in Fig. 7 of [5]. We see that except for the near-threshold region where the photoelectron energy is very small and where

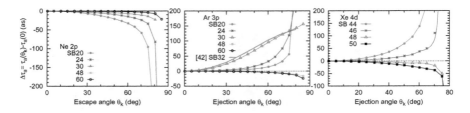

Fig. 3. Angular variation of the atomic time delay $\Delta\tau_a = \tau_a(\theta_k) - \tau_a(0)$ with the photoelectron emission direction for various side bands. Left: Ne $2p$, center: Ar $3p$, and right: Xe $4d$. In the case of Ar $3p$, an angular variation of time delay for SB32 from [43] is shown for comparison

the regularization of τ_{CC} may not be applicable, the identity $\tau_a \simeq \tau_W + \tau_{cc}$ holds well for Ne $2p$ but less so for Ar $3p$. In the case of Xe $4d$ (right panel), we also show the atomic time delay τ_a from the TDSE calculation on the encapsulated Xe@C$_{60}$ atom. This calculation shows multiple confinement resonances which are particularly prominent when the photoelectron de Broglie wavelength is comparable with the size of the cage [40,41].

Angular dependence of the atomic time delay $\tau_a(\theta_k)$ as a function of the escape angle is shown in Fig. 3 for Ne $2p$ (left), Ar $3p$ (center), and Xe $4d$ (right). This angular dependence is weakest in Ne where it is only prominent at large emission angles relative to the polarization axis of light. This is so because the $2p \rightarrow Ed$ photoemission channel is strongly dominant over $2p \rightarrow Es$ due to the Fano propensity rule [42]. However, at the magic angle close to $54°$, the spherical harmonic Y_{20} which determines the angular dependence of the stronger channel has a kinematic node and a normally weaker channel becomes competitive. This rings the angular dependence of the time delay shown in the left panel of Fig. 1. In other atoms, the competitions of the two major photoemission channels are strong because of a Cooper minimum (Ar $3p$) or a centrifugal barrier (Xe $4d$). Hence, the angular dependence of the time delay is considerably stronger.

Because of a very fine energy resolution, the spin–orbit split components of the valence shells of heavier noble gas atoms, Kr and Xe, could be resolved in a recent RABBITT measurement [7]. When the accessible photon energy range was extended to 90 eV, a similar measurement could be performed on the $4d$ subshell of Xe [21]. To reproduce these measurements within the present TDSE model, the one-electron potential in (19) was adjusted to obtain the experimental threshold energies of the corresponding spin–orbit split bound states. The corresponding TDSE results are compared with the experiment in Fig. 4. On the left panel, we make a comparison for the atomic time delay difference $\tau_a^{3/2} - \tau_a^{1/2}$ of the $5p$ subshell of Xe. In addition to the TDSE results, we display the relativistic RRPA result amended by the CC correction [44]. On the right panel, a similar comparison is made for the atomic time delay difference $\tau_a^{5/2} - \tau_a^{4/2}$ of the $4d$ subshell. The shown RRPA + CC result is from [39].

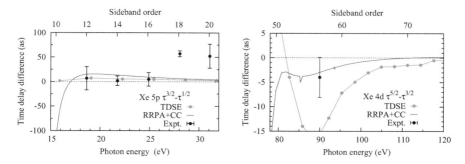

Fig. 4. The atomic time delay difference $\tau_a^{3/2} - \tau_a^{1/2}$ for the $5p$ subshell (left) and $\tau_a^{5/2} - \tau_a^{3/2}$ for the $4d$ subshell of Xe. The TDSE results are visualized with the filled red circles. The RRPA calculations amended by the CC correction for $5p$ [44] and $4d$ [39] are shown with the solid blue line. The experimental data for the $5p$ [7] and $4d$ [21] subshells are shown with error bars

2.2 Molecular RABBITT Simulations

Computational Technique Our molecular RABBITT simulations [45, 46] are based on the TDSE solution in an expanding coordinate system [47]. This method is termed time-dependent coordinate scaling (TDCS). In this method, the following variable transformation is made:

$$\mathbf{r} = a(t)\boldsymbol{\xi}. \tag{12}$$

Here $a(t)$ is a scaling factor with an asymptotically linear time dependence $a(t \to \infty) = \dot{a}_\infty t$ and $\boldsymbol{\xi}$ is a coordinate vector. Such a transformation makes the coordinate frame to expand along with the wave packet. In addition, the following transformation is applied to the wave function:

$$\Psi(a(t)\boldsymbol{\xi}, t) = \frac{1}{[a(t)]^{3/2}} \exp\left(\frac{i}{2} a(t)\dot{a}(t)\xi^2\right) \psi(\boldsymbol{\xi}, t). \tag{13}$$

Such a transformation removes a rapidly oscillating phase factor from the wave function in the asymptotic region [47]. Thus, transformed wave function satisfies the equation

$$i\frac{\partial \psi(\boldsymbol{\xi}, t)}{\partial t} = \left[\frac{\hat{p}_\xi^2}{2[a(t)]^2} - \frac{\mathbf{A}(t)\hat{\mathbf{p}}_\xi}{a(t)} + U[a(t)\boldsymbol{\xi}]\right] \psi(\boldsymbol{\xi}, t) , \tag{14}$$

where $\hat{\mathbf{p}}_\xi = -i\nabla_\xi = -i\left(\frac{\partial}{\partial \xi_x}, \frac{\partial}{\partial \xi_y}, \frac{\partial}{\partial \xi_z}\right)$. We note that if the spectrum of the operator $\hat{p}_\xi^2$ is upper limited, which is the case for any numerical approximation of a differential operator, then the first term in the RHS of (14) tends to zero as $[a(t)]^{-2}$ for $t \to \infty$. In the meantime, the potential term with a long-range Coulomb asymptotic $U(\mathbf{r} \to \infty) \sim 1/r$ is transformed to $U[a(t)\boldsymbol{\xi}] \sim Z/a(t)\xi$.

This means that both the Coulomb term and the vector potential term are decreasing in time as $1/a(t)$. Therefore, when solving (14), we can increase the time propagation step $\Delta t = a(t)\Delta t_0$ which accelerates the solution quite considerably [47].

For complex molecules, we employ the density functional theory (DFT) with the self-interaction correction (SIC) [48]. This correction is necessary to restore the Coulomb asymptotics of the photoelectron interaction with the residual ion which is essential for time delay calculations. The density functional with the SIC [48] contains the Hartree $E_H\{\rho\}$ and the exchange-correlation $E_{XC}\{\rho_\uparrow, \rho_\downarrow\}$ components:

$$E_{SIC} = E_H\{\rho\} + E_{XC}\{\rho_\uparrow, \rho_\downarrow\} - [E_H\{\rho_i\} + E_{XC}\{\rho_i, 0\}], \tag{15}$$

where the electron density $\rho(\mathbf{r}) = \sum_{i=1}^{N_e} \rho_i(\mathbf{r})$ is the sum of the particle densities of the i-th electron orbital $\rho_i(\mathbf{r}) = |\varphi_i(\mathbf{r})|^2$. The exchange-correlation functional is expressed in the local density approximation (LDA)

$$E_{XC}\{\rho_\uparrow, \rho_\downarrow\} = \int [\rho_\uparrow(\mathbf{r}) + \rho_\downarrow(\mathbf{r})]\varepsilon_{XC}[\rho_\uparrow(\mathbf{r}), \rho_\downarrow(\mathbf{r})]d\mathbf{r}, \tag{16}$$

where $\varepsilon_{XC}[\rho_\uparrow(\mathbf{r}), \rho_\downarrow(\mathbf{r})]$ is the exchange-correlation energy per electron. The effective potential acting upon an i-th electron by the rest of the many-electron ensemble is expressed as a functional derivative

$$u_i(\mathbf{r}) = \frac{\delta E_{SIC}}{\delta \rho_i(\mathbf{r})}. \tag{17}$$

The Kohn–Sham effective potential is the sum of the nuclear and electron components:

$$U(\mathbf{r}) = u_{nucl}(\mathbf{r}) + u_i(\mathbf{r}). \tag{18}$$

The SIC can be applied to any density functional. In the present application, we used two different correlation-exchange functionals: the one proposed in [49] (GL76-SIC) and a simpler pure exchange functional (X-SIC).

Numerical Results The energy and angular variation of the time delay in H_2^+ are displayed on the left and right panels of Fig. 5, respectively. Both these dependencies are very different from that of atomic H and He. The energy variation of τ_a with E_e for H_2^+ is non-monotonous. The angular dependence of H_2^+ displays an additional strong variation in the range of emission angles $\theta = 30°$–$50°$. To visualize clearly this molecular effect, we make a comparison of the angular-dependent time delay in H_2^+ with the spherically symmetric He^+ ion. To account for different ionization potentials, we carried out the He^+ calculation at the central frequency $\omega_{XUV} = 43\omega$. It is clearly seen that the atomic and molecular ions display the angular-dependent time delay which differs considerably not only by additional strong angular variation but also the magnitude of

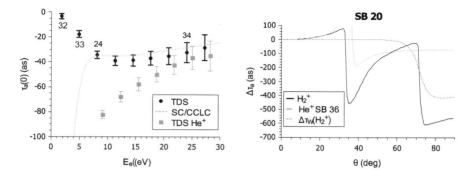

Fig. 5. Left: The time delay τ_a for H_2^+ as a function of the photoelectron ejection energy E_e at a fixed ejection angle $\theta = 0$. The SC/CCLC results (green dotted line) and dots for He^+ (red squares) are also shown for comparison. The corresponding SB indices are marked. Right: The angular variation of the time delay $\Delta\tau_a = \tau_a(E_e, \theta) - \tau_a(E_e, 0)$ (black solid line) and Wigner's time delay $\Delta\tau_W = \tau_W(E_e, \theta) - \tau_W(E_e, 0)$ (green dotted line) of H_2^+ for several fixed photoelectron energies E_e. The results for He^+ for SB with close energies are also shown (red dashed line)

the sharp drop of the time delay near the 90° emission angle. We note that the asymptotic field of the ion remainder is the same in both cases. Hence, should be the same the CC term of the atomic time delay (10). Therefore, the difference of the atomic time delay in the H_2^+ and He^+ ions should be attributed largely to the Wigner component τ_W of the time delay.

In Fig. 6, we display the atomic time delay of H_2 calculated by the TDCS method. Every energy point on the graph corresponds to a given SB. The error bars indicate the accuracy of the cosine fit to (7). For comparison, another calculation is shown in which the Wigner time delay is calculated by the prolate spheroidal exterior complex scaling (PSECS) [50] and the CLC correction is introduced analytically [51]. The PSECS is an ab initio technique and it returns the exact Wigner time delay for diatomic molecules.

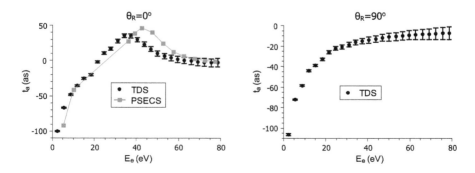

Fig. 6. Atomic time delay of H_2 as a function of the photoelectron energy E_e for emission in the polarization direction. The molecular axis is aligned along (left) and perpendicular (right) to the polarization direction

The Wigner time delay was also estimated by a classical approximation to CLC (CCLC) derived in [51]. The TDCS and PSECS + CCLC results agree very well close to the threshold and are qualitatively similar at large excess energies. In the parallel molecular orientation, both set of calculations display a peak in the atomic time delay. However, in the TDCS calculation this peak is shifted by 7 eV toward lower photoelectron energies ($E_e = 35$ eV in TDCS versus 42 eV in PSECS + CCLC). Such a large difference can be explained by poor performance of the DFT for such few-electron systems like H_2. We note that the peak displacement by 7 eV far exceeds an error of 1 eV in the ionization potential. This indicates that such a dynamic quantity as the atomic time delay is much more sensitive to inter-electron correlation than the static ionization potential.

On the right panel of Fig. 7, we display the time delay difference between the $3a_1$ and $1b_1$ orbitals of the H_2O molecule (visualized on the left panel) at the same photon energy. It may seem surprising that the time delay difference is nearly vanishing in the whole studied energy range. This, however, may be explained by the fact that the randomly oriented water molecule may look like the neon atom and these two states in Ne differ only by the nodal plane orientation. In the same figure, we plot the experimental results [17] which show a sign variation of the time delay difference. The frozen-core Hartree–Fock (FCHF) calculation by the same authors [17,52] is also overplotted. Neither calculation reproduces the experimental results within the stated error bars.

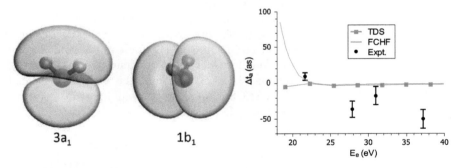

Fig. 7. Time delay difference between the $3a_1$ and $1b_1$ orbitals in the H_2O molecule (visualized on the left panel) as a function of the photon energy. The filled (red) circles—TDCS calculation, the (green) solid line—the FCHF calculation [17,52], the filled squares with error bars—experiment [17]

3   Tunneling Time and Attoclock

Measuring an offset angle of the peak photoelectron momentum distribution in the polarization plane of a close-to-circularly polarized laser field has been used to determine the tunneling time which the photoelectron spends under the barrier [3,53–55]. Once the photoelectron enters the tunnel at the peak value of the electric field, the most probable detection direction will be aligned with the vector potential at the instant of tunneling. This direction is tilted by 90° relative to the electric field at this instant. A measurable angular offset relative to this axis can be converted to the time the photoelectron spends under the barrier with the conversion rate of 7.4 attoseconds (1 as = 10^{-18} s) per 1° at 800 nm. Such a measurement is termed colloquially the attoclock. A similar reading can be obtained from an attoclock driven by a very short circularly polarized laser pulse [9,56–59].

The attoclock measurements were aimed to resolve the controversy of a finite tunneling time which has many decades of history [60]. This controversy is yet to be resolved with many conflicting reports of a finite time [6,61] as opposite to zero tunneling time [8,9,56,59]. A promising pathway to resolving this controversy is to examine an often neglected aspect of the attoclock measurement, the offset angle induced by the Coulomb field of the ion remainder. In theoretical calculations based on a numerical solution of the time-dependent Schrödinger equation (TDSE) [62–64], the Coulombic and tunneling components of the offset angle are inseparable. The Coulomb field contribution can be switched off however by replacing the atomic potential with the short-range Yukawa potential of the same binding strength. This procedure effectively eliminates the offset angle thus suggesting zero tunneling time [8,9]. In various classical simulations (the TIPIS model [65], backpropagation [56,57] or semiclassical simulations (the ARM model [9]), the Coulomb field is separable and its effect can be unambiguously determined.

3.1   Atomic Attoclock Simulations

Computational Technique   We solve numerically the TDSE

$$i\partial\Psi(\boldsymbol{r})/\partial t = \left[\hat{H}_{\mathrm{atom}} + \hat{H}_{\mathrm{int}}(t)\right]\Psi(\boldsymbol{r}), \tag{19}$$

where $\hat{H}_{\mathrm{atom}}$ describes the atomic target in the absence of the applied field and the interaction Hamiltonian is written in the velocity gage

$$\hat{H}_{\mathrm{int}}(t) = \boldsymbol{A}(t)\cdot\hat{\boldsymbol{p}}, \quad \boldsymbol{E}(t) = -\partial\boldsymbol{A}/\partial t . \tag{20}$$

Here the vector potential of the driving pulse is

$$\boldsymbol{A}(t) = -A_0 f(t)[\cos(\omega t)\hat{e}_x + \sin(\omega t)\hat{e}_y] \tag{21}$$

with the envelope function $f(t) = \cos^4(\omega t/4)$ for $-2\pi/\omega < t < 2\pi/\omega$ and zero elsewhere. The (peak) field intensity is given by $I = 2(\omega A_0)^2$ and the frequency

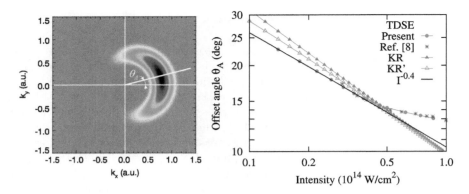

Fig. 8. Left: Photoelectron momentum distribution in the polarization plane at the driving field intensity $I = 8.6 \times 10^{13}$ W/cm^2 on hydrogen. The offset angle θ relative to the vector potential direction at the entrance of tunnel is marked. The coloration ranges from zero (red) to the maximum amplitude (black) linearly. Right: The attoclock offset angle θ_A as a function of the field intensity I from the present TDSE calculation (red filled circles), the H$_2$ set of [9] (blue asterisks), and the KR and KR$'$ models (filled and empty triangles). The present TDSE results are fitted with the I^{-n} dependence with $n = 0.41$

ω is taken to correspond to 800 nm radiation. At the tunneling entry $t = 0$, the electric field E_0 reaches its maximum in the $\hat{e}_y$ direction whereas the vector potential A_0 is largest in the $-\hat{e}_x$ direction. The rotating electric field of the driving pulse causes the photoelectron to make a single turn by 90° before it arrives to the detector in the $\hat{e}_x$ direction with momentum A_0. Photoelectron scattering in the Coulomb field adds an offset angle θ_A relative to this direction (left panel of Fig. 8).

Solution of the TDSE (19) is found using the iSURF method implemented in [29]. A typical calculation takes around 140 CPU hours on a high-performance, distributed-memory cluster. The solution of the TDSE is projected on the scattering states of the target atom thus forming the photoelectron momentum distribution. The 2D momentum distribution in the polarization plane $k^2 P(k_x, k_y)$ is shown on the left panel of Fig. 8. This distribution is integrated radially to obtain the angular distribution $P(\theta) = \int dk \, k^2 P(k_x, k_y)$. It is then fitted with a Gaussian to determine the peak position and this value assigned to the attoclock offset angle θ_A. The symmetry of $P(\theta)$ relative to θ_A is carefully monitored and serves as a test of the quality of the TDSE calculation.

Keldysh–Rutherford Model In a simplified classical model [66], the photoelectron is tunnel ionized and then scatters elastically on the ion remainder. We examine cases where the applied field is relatively weak and the pulse is short such that the actual trajectory of the field-driven photoelectron is similar to that of an elastically scattered particle. This allows us to take an estimate of the offset angle from the classical scattering formula [67]

$$\theta = 2 \int_{r_0}^{\infty} \frac{(\rho/r^2)\, dr}{[1 - (\rho/r)^2 - (2V/mv_\infty^2)]^{1/2}} - \pi. \tag{22}$$

Here ρ is the impact parameter, v_∞ is the velocity of the projectile at the source and the detector, and the point of the closest approach r_0 is the largest positive root of the denominator. In the case of the attractive Coulomb potential $V(r) = -Z/r$, (22) takes the form of the Rutherford formula [67]:

$$\tan \frac{\theta}{2} = \frac{1}{v_\infty^2} \frac{Z}{\rho}. \tag{23}$$

In a more general case of a screened Coulomb potential $V(r) = -Z/r \exp(-r/\lambda)$, the offset angle is given by a modified expression [68] which is, up to a typically small numerical correction (see 13 of [68]),

$$\tan \frac{\theta}{2} = \frac{1}{v_\infty^2} \frac{Z}{\rho} \exp(-1/z_0)\,. \tag{24}$$

Here z_0 is the root of $y(z) = 1 - (\rho/\lambda)^2 z^2 - (d/\lambda)z \exp(-1/z)$ and $d = 2Z/v_\infty$ is the so-called collision diameter. If the last term in the right-hand side of $y(z)$ can be neglected, (24) acquires a simple screening exponent $\exp(-\rho/\lambda)$ (Fig. 9).

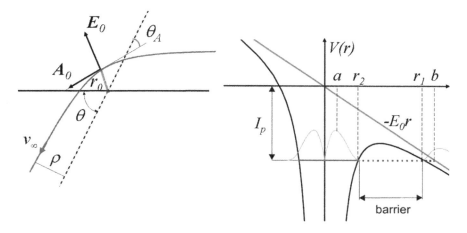

Fig. 9. Left: classical scattering trajectory of a particle in a central attractive potential. The scattering angle θ is defined by the impact parameter ρ and the asymptotic velocity v_∞. The tunnel ionized electron enters this trajectory at the point of the closest approach r_0 driven by the peak electric field $\boldsymbol{E_0}$ and arriving to the detector at the angle θ_A relative to the vector potential $\boldsymbol{A_0}$. Right: The Coulomb potential is tipped by the light field. A finite width potential barrier is created, through which the electron wave packet leaks out. I_p refers to the binding energy of the electron in an unperturbed atomic system. See text for further symbol definitions

Equations (23) and (24) can be readily applied to the case of tunneling ion-ization. First, we note that photoionization can be considered as half-scattering and the offset angle should be taken as one-half of the Rutherford or Yukawa scattering angles. Second, the point of the closest approach r_0 should be taken as the tunnel exit position which, for the Coulomb potential, is the largest root of the equation

$$Z/r + E_0 r = I_p \ , \quad r_{1,2} = b/2 \pm \sqrt{b^2/4 - ab} \ . \tag{25}$$

Here I_p is the ionization potential, E_0 is the peak value of the electric field, $b = I_p/E_0$, and $a = Z/I_p$. In the weak field limit, $r_1 = b \gg r_2 = a \simeq 1$, where a is the characteristic span of the atomic orbital. The onset of the over-the-barrier ionization (OBI) corresponds to $r_1 = r_2$ and $E_0 = I_p^2/(4Z)$ (right panel of Fig. 1). We note that the b parameter is used to evaluate the Keldysh tunneling time $\tau = b/v_{at}$ with $v_{at} = \sqrt{2I_p}$ which, in turn, defines the adiabaticity parameter $\gamma = \omega\tau$ [69]. With these assumptions, the attoclock offset angle in the case of the pure Coulomb potential takes the form

$$\theta_A = \frac{1}{2}\theta \simeq \frac{\omega^2}{E_0^2}\frac{Z}{\rho} = \frac{\omega^2}{E_0}\frac{Z}{I_p} \ . \tag{26}$$

The signature of this Keldysh–Rutherford (KR) formula is the field intensity dependence of the offset angle $\theta_A \propto E_0^{-1} \propto I^{-1/2}$. This dependence can be understood as a result of competition of the two terms in (23): the kinetic energy term $v_\infty^2/2$ and the potential energy term Z/ρ. As the field intensity grows, the kinetic energy grows linearly with I. Hence, this term alone would result in the I^{-1} dependence of the offset angle. This is partially compensated by the potential energy term as the width of the barrier decreases as $I^{-1/2}$. Hence, the resulting offset angle also decreases as $I^{-1/2}$. The recent attoclock measurement on the hydrogen atom [8] confirmed the $I^{-1/2}$ dependence experimentally.

Resulting values of the offset angle θ_A for hydrogen at various field intensi-ties are plotted on the right panel of Fig. 8. We find the present set of TDSE calculations to be hardly distinguishable from the set labeled H_2 of TDSE calcu-lations reported in [9]. This is contrasted with the two KR and KR' estimates. For the former, we simply plot (26), whereas in the latter we do not make the small angle approximation for the tangent function. Accordingly, both estimates converge together with increasing field intensity. The KR scales at all intensi-ties as $I^{-0.5}$ by construction. Fitting the KR' in the low-intensity range yields $I^{-0.44}$. The TDSE results display a similar dependency of $I^{-0.41}$ scaling for the same region but then flattens and deviates from both the KR and KR'. This is understandable as the KR model is expected to work for weak fields only when the field-driven trajectory is close to that involved in field-free scattering.

4 Conclusions and Further Directions

We demonstrated how an accurate numerical solution of the field-driven time-dependent Schrödinger equation can be used to gain valuable information on the attosecond time-resolved dynamics of atomic and molecular photoionization. This information is complementary to traditional photoemission studies with continuous synchrotron fields and brings one step closer to practical realization of the concept of a complete photoionization experiment. We focused our attention on the two key observables of attosecond chronoscopy: the Wigner time delay and the tunneling time. The Wigner time delay is unambiguously established in atomic [1] and molecular [17] photoionization and serves as a useful tool to study both one-electron potentials and collective many-electron effects. The tunneling time is still a subject of a considerable debate and controversy [9]. Our numerical simulation supported by the classical Keldysh–Rutherford model point to zero tunneling time.

In the future, the present single active electron model will be extended to include many-electron correlations and relativistic effects.

Acknowledgments. The author wishes to acknowledge a very fruitful collaboration with his long-term partners on this project, Dr. Igor Ivanov of Institute for Basic Science in Gwangj, Korea [8,9,13,18,25,64] and Dr. Vladislav Serov of Saratov State University, Russia [45,46]. Several important recent results are obtained with Mr. Alexander W. Bray [8,21,31,66], a Ph.D. student at the Australian National University. Experimental work on atomic hydrogen attoclock [8] was conducted in collaboration with Prof. Robert Sang, A/Prof. Igor Litvinyuk, and a Ph.D. student Mr. Satya Sainadh. Resources of the National Computational Infrastructure were employed.

References

1. Schultze, M., et al.: Delay in photoemission. Science **328**(5986), 1658–1662 (2010)
2. Klünder, K., et al.: Probing single-photon ionization on the attosecond time scale. Phys. Rev. Lett. **106**(14), 143002 (2011)
3. Eckle, P., et al.: Attosecond ionization and tunneling delay time measurements in helium. Science **322**(5907), 1525–1529 (2008)
4. Pazourek, R., Nagele, S., Burgdörfer, J.: Attosecond chronoscopy of photoemission. Rev. Mod. Phys. **87**, 765 (2015)
5. Dahlström, J.M., Guénot, D., Klünder, K., Gisselbrecht, M., Mauritsson, J., Huillier, A.L., Maquet, A., Taïeb, R.: Theory of attosecond delays in laser-assisted photoionization. Chem. Phys. **414**, 53–64 (2012)
6. Landsman, A.S., Keller, Ursula: Attosecond science and the tunnelling time problem. Phys. Rep. **547**, 1–24 (2015)
7. Jordan, I., Huppert, M., Pabst, S., Kheifets, A.S., Baykusheva, D., Wörner, H.J.: Spin-orbit delays in photoemission. Phys. Rev. A **95**, 013404 (2017)
8. Sainadh, U.S., Xu, H., Wang, X., Atia-Tul-Noor, Wallace, W.C., Douguet, N., Bray, A.W., Ivanov, I., Bartschat, K., Kheifets, A., Sang, R.T., Litvinyuk, I.V.: Attosecond angular streaking and tunnelling time in atomic hydrogen. Nature **568**, 75–77 (2019)

9. Torlina, L., Morales, F., Kaushal, J., Ivanov, I., Kheifets, A., Zielinski, A., Scrinzi, A., Muller, H.G., Sukiasyan, S., Ivanov, M., Smirnova, O.: Interpreting attoclock measurements of tunnelling times. Nat. Phys. **11**, 503–508 (2015)
10. Swoboda, M., Fordell, T., Klünder, K., Dahlström, J.M., Miranda, M., Buth, C., Schafer, K.J., Mauritsson, J., L'Huillier, A., Gisselbrecht, M.: Phase measurement of resonant two-photon ionization in helium. Phys. Rev. Lett. **104**, 103003 (2010)
11. Guénot, D., Klünder, K., Arnold, C.L., Kroon, D., Dahlström, J.M., Miranda, M., Fordell, T., Gisselbrecht, M., Johnsson, P., Mauritsson, J., Lindroth, E., Maquet, A., Taïeb, R., L'Huillier, A., Kheifets, A.S.: Photoemission-time-delay measurements and calculations close to the 3s-ionization-cross-section minimum in Ar. Phys. Rev. A **85**, 053424 (2012)
12. Guénot, D., Kroon, D., Balogh, E., Larsen, E.W., Kotur, M., Miranda, M., Fordell, T., Johnsson, P., Mauritsson, J., Gisselbrecht, M., Varjù, K., Arnold, C.L., Carette, T., Kheifets, A.S., Lindroth, E., L'Huillier, A., Dahlström, J.M.: Measurements of relative photoemission time delays in noble gas atoms. J. Phys. B **47**(24), 245602 (2014)
13. Palatchi, C., Dahlström, J.M., Kheifets, A.S., Ivanov, I.A., Canaday, D.M., Agostini, P., DiMauro, L.F.: Atomic delay in helium, neon, argon and krypton. J. Phys. B **47**(24), 245003 (2014)
14. Isinger, M., Squibb, R.J., Busto, D., Zhong, S., Harth, A., Kroon, D., Nandi, S., Arnold, C.L., Miranda, M., Dahlström, J.M., Lindroth, E., Feifel, R., Gisselbrecht, M., L'Huillier, A.: Photoionization in the time and frequency domain. Science **358**, 893 (2017)
15. Haessler, S., Fabre, B., Higuet, J., Caillat, J., Ruchon, T., Breger, P., Carré, B., Constant, E., Maquet, A., Mével, E., Salières, P., Taïeb, R., Mairesse, Y.: Phase-resolved attosecond near-threshold photoionization of molecular nitrogen. Phys. Rev. A **80**, 011404 (2009)
16. Caillat, Jérémie, Maquet, Alfred, Haessler, Stefan, Fabre, Baptiste, Ruchon, Thierry, Salières, Pascal, Mairesse, Yann, Taïeb, Richard: Attosecond resolved electron release in two-color near-threshold photoionization of n$_2$. Phys. Rev. Lett. **106**, 093002 (2011)
17. Huppert, M., Jordan, I., Baykusheva, D., von Conta, A., Wörner, H.J.: Attosecond delays in molecular photoionization. Phys. Rev. Lett. **117**, 093001 (2016)
18. Heuser, S., Galán, Á.J., Cirelli, C., Marante, C., Sabbar, M., Boge, R., Lucchini, M., Gallmann, L., Ivanov, I., Kheifets, A.S., Dahlström, J.M., Lindroth, E., Argenti, L., Martín, F., Keller, U.: Angular dependence of photoemission time delay in helium. Phys. Rev. A **94**, 063409 (2016)
19. Vos, J., Cattaneo, L., Patchkovskii, S., Zimmermann, T., Cirelli, C., Lucchini, M., Kheifets, A., Landsman, A.S., Keller, U.: Orientation-dependent stereo Wigner time delay in a small molecule. Science **360**(6395), 1326–1330 (2018)
20. Cirelli, C., Marante, C., Heuser, S., Petersson, C.L.M., Galán, A.J., Argenti, L., Zhong, S., Busto, D., Isinger, M., Nandi, S., Maclot, S., Rading, L., Johnsson, P., Gisselbrecht, M., Lucchini, M., Gallmann, L., Dahlström, J.: Anisotropic photoemission time delays close to a Fano resonance. Nature Commun. **9**, 955 (2018)
21. Jain, A., Gaumnitz, T., Bray, A.W., Kheifets, A.S., Wörner, H.J.: Photoionization delays in xenon using single-shot referencing in the collinear back-focusing geometry. Optics Lett. (2018)
22. Jain, A., Gaumnitz, T., Wörner, H.J.: Using a passively stable attosecond beamline for relative photoemission time delays at high XUV photon energies. Optics Expr. (2018)

23. Muller, H.G.: Reconstruction of attosecond harmonic beating by interference of two-photon transitions. Appl. Phys. B **74**, s17–s21 (2002)
24. Toma, E.S., Muller, H.G.: Calculation of matrix elements for mixed extreme-ultravioletinfrared two-photon above-threshold ionization of argon. J. Phys. B **35**(16), 3435 (2002)
25. Ivanov, I.A., Kheifets, A.S.: Angle-dependent time delay in two-color XUV+IR photoemission of He and Ne. Phys. Rev. A **96**, 013408 (2017)
26. Sarsa, A., Gálvez, F.J., Buendia, E.: Parameterized optimized effective potential for the ground state of the atoms He through Xe. At. Data Nucl. Data Tables **88**(1), 163–202 (2004)
27. Wendin, G., Starace, A.F.: Perturbation theory in a strong-interaction regime with application to 4d-subshell spectra of Ba and La. J. Phys. B **11**(24), 4119 (1978)
28. Deshmukh, P.C., Mandal, A., Saha, S., Kheifets, A.S., Dolmatov, V.K., Manson, S.T.: Attosecond time delay in the photoionization of endohedral atoms A@C_{60}: a probe of confinement resonances. Phys. Rev. A **89**, 053424 (2014)
29. Morales, F., Bredtmann, T., Patchkovskii, S.: iSURF: a family of infinite-time surface flux methods. J. Phys. B **49**(24), 245001 (2016)
30. Maquet, A., Taïeb, Richard: Two-colour IR+XUV spectroscopies: the soft-photon approximation. J. Mod. Opt. **54**(13–15), 1847–1857 (2007)
31. Bray, A.W., Naseem, F., Kheifets, Anatoli S.: Simulation of angular-resolved RABBITT measurements in noble-gas atoms. Phys. Rev. A **97**, 063404 (2018)
32. Dahlström, J.M., et al.: Theory of attosecond delays in laser-assisted photoionization. Chem. Phys. **414**, 53–64 (2013)
33. Zhang, C.-H., Thumm, U.: Streaking and Wigner time delays in photoemission from atoms and surfaces. Phys. Rev. A **84**, 033401 (2011)
34. Pazourek, R., Nagele, S., Burgdorfer, Joachim: Time-resolved photoemission on the attosecond scale: opportunities and challenges. Faraday Discuss. **163**, 353–376 (2013)
35. Amusia, M.Ya.: Atomic photoeffect. Plenum Press, New York (1990)
36. Codling, K., Houlgate, R.G., West, J.B., Woodruff, P.R.: Angular distribution and photoionization measurements on the 2p and 2s electrons in neon. J. Phys. B **9**(5), L83 (1976)
37. Houlgate, R.G., Codling, K., Marr, G.V., West, J.B.: Angular distribution and photoionization cross section measurements on the 3p and 3s subshells of argon. J. Phys. B **7**(17), L470 (1974)
38. Toffoli, D., Stener, M., Decleva, P.: Application of the relativistic time-dependent density functional theory to the photoionization of xenon. J. Phys. B **35**(5), 1275 (2002)
39. Mandal, A., Deshmukh, P.C., Kheifets, A.S., Dolmatov, V.K., Manson, S.T.: Angle-resolved Wigner time delay in atomic photoionization: the 4d subshell of free and confined Xe. Phys. Rev. A **96**, 053407 (2017)
40. Kilcoyne, A.L.D., Aguilar, A., Müller, A., Schippers, S., Cisneros, C., Alna'Washi, G., Aryal, N.B., Baral, K.K., Esteves, D.A., Thomas, C.M., Phaneuf, R.A.: Confinement resonances in photoionization of Xe@C_{60}^{+}. Phys. Rev. Lett. **105**, 213001 (2010)
41. Phaneuf, R.A., Kilcoyne, A.L.D., Aryal, N.B., Baral, K.K., Esteves-Macaluso, D.A., Thomas, C.M., Hellhund, J., Lomsadze, R., Gorczyca, T.W., Ballance, C.P., Manson, S.T., Hasoglu, M.F., Schippers, S., Müller, A.: Probing confinement resonances by photoionizing Xe inside a C_{60}^{+} molecular cage. Phys. Rev. A **88**, 053402 (2013)

42. Fano, U.: Propensity rules: an analytical approach. Phys. Rev. A **32**, 617–618 (1985)
43. Dahlström, J.M., Lindroth, E.: Study of attosecond delays using perturbation diagrams and exterior complex scaling. J. Phys. B **47**(12), 124012 (2014)
44. Kheifets, A., Mandal, A., Deshmukh, P.C., Dolmatov, V.K., Keating, D.A., Manson, Steven T.: Relativistic calculations of angle-dependent photoemission time delay. Phys. Rev. A **94**, 013423 (2016)
45. Serov, V.V., Kheifets, A.S.: Angular anisotropy of time delay in XUV+IR photoionization of H_2^+. Phys. Rev. A **93**, 063417 (2016)
46. Serov, V.V., Kheifets, Anatoli S.: Time delay in XUV/IR photoionization of H_2O. J. Chem. Phys. **147**(20), 204303 (2017)
47. Serov, V.V., Derbov, V.L., Joulakian, B.B., Vinitsky, S.I.: Wave-packet-evolution approach for single and double ionization of two-electron systems by fast electrons. Phys. Rev. A **75**, 012715 (2007)
48. Perdew, J.P., Zunger, A.: Self-interaction correction to density-functional approximations for many-electron systems. Phys. Rev. B **23**, 5048–5079 (1981)
49. Gunnarsson, O., Lundqvist, B.I.: Exchange and correlation in atoms, molecules, and solids by the spin-density-functional formalism. Phys. Rev. B **13**, 4274–4298 (1976)
50. Serov, V.V., Derbov, V.L., Sergeeva, T.A.: Interpretation of time delay in the ionization of two-center systems. Phys. Rev. A **87**, 063414 (2013)
51. Serov, V.V., Derbov, V.L., Sergeeva, T.A.: Interpretation of the time delay in the ionization of Coulomb systems by attosecond laser pulses. In: Advanced Lasers. Springer Series in Optical Sciences, vol. 193, pp. 213–230. Springer, Berlin (2015)
52. Baykusheva, D., Wörner, H.J.: Theory of attosecond delays in molecular photoionization. J. Chem. Phys. **146**(12), 124306 (2017)
53. Eckle, P., et al.: Attosecond angular streaking. Nat. Phys. **4**, 565–570 (2008)
54. Pfeiffer, A.N., Cirelli, C., Smolarski, M., Dimitrovski, D., Abu-samha, M., Madsen, L.B., Keller, U.: Attoclock reveals natural coordinates of the laser-induced tunneling current flow in atoms. Nat. Phys. **8**, 76–80 (2012)
55. Landsman, A.S., Weger, M., Maurer, J., Boge, R., Ludwig, A., Heuser, S., Cirelli, C., Gallmann, L., Keller, Ursula: Ultrafast resolution of tunneling delay time. Optica **1**(5), 343–349 (2014)
56. Ni, H., Saalmann, U., Rost, Jan-Michael: Tunneling ionization time resolved by backpropagation. Phys. Rev. Lett. **117**, 023002 (2016)
57. Ni, H., Saalmann, U., Rost, J.-M.: Tunneling exit characteristics from classical backpropagation of an ionized electron wave packet. Phys. Rev. A **97**, 013426 (2018)
58. Liu, J., Fu, Y., Chen, W., Lu, Z., Zhao, J., Yuan, J., Zhao, Z.: Offset angles of photocurrents generated in few-cycle circularly polarized laser fields. J. Phys. B **50**(5), 055602 (2017)
59. Eicke, N., Lein, Manfred: Trajectory-free ionization times in strong-field ionization. Phys. Rev. A **97**, 031402 (2018)
60. Landauer, R., Martin, Th.: Barrier interaction time in tunneling. Rev. Mod. Phys. **66**, 217–228 (1994)
61. Camus, N., Yakaboylu, E., Fechner, L., Klaiber, M., Laux, M., Mi, Y., Hatsagortsyan, K.Z., Pfeifer, T., Keitel, C.H., Moshammer, Robert: Experimental evidence for quantum tunneling time. Phys. Rev. Lett. **119**, 023201 (2017)
62. Martiny, C.P.J., Abu-samha, M., Madsen, L.B.: Counterintuitive angular shifts in the photoelectron momentum distribution for atoms in strong few-cycle circularly polarized laser pulses. J. Phys. B **42**(16), 161001 (2009)

63. Shvetsov-Shilovski, N.I., Dimitrovski, D., Madsen, L.B.: Ionization in elliptically polarized pulses: multielectron polarization effects and asymmetry of photoelectron momentum distributions. Phys. Rev. A **85**, 023428 (2012)

64. Ivanov, I.A., Kheifets, A.S.: Strong-field ionization of He by elliptically polarized light in attoclock configuration. Phys. Rev. A **89**, 021402 (2014)

65. Boge, R., Cirelli, C., Landsman, A.S., Heuser, S., Ludwig, A., Maurer, J., Weger, M., Gallmann, L., Keller, U.: Probing nonadiabatic effects in strong-field tunnel ionization. Phys. Rev. Lett. **111**, 103003 (2013)

66. Bray, A.W., Eckart, S., Kheifets, Anatoli S.: Keldysh-Rutherford model for the attoclock. Phys. Rev. Lett. **121**, 123201 (2018)

67. Landau, L.D., Lifshitz, E.M.: Mechanics. Theorecial Physics, vol. 1. Elsevier Science (1982)

68. Everhart, E., Stone, G., Carbone, R.J.: Classical calculation of differential cross section for scattering from a Coulomb potential with exponential screening. Phys. Rev. **99**, 1287–1290 (1955)

69. Popruzhenko, S.V.: Keldysh theory of strong field ionization: history, applications, difficulties and perspectives. J. Phys. B **47**(20), 204001 (2014)

Electron–Molecule Resonances: Current Developments

E. Krishnakumar[1]($\boxtimes$) and Vaibhav S. Prabhudesai[2]

[1] Raman Research Institute, C. V. Raman Avenue, Bangalore 560080, India
krishnakumar@rri.res.in, ekkumar@tifr.res.in
[2] Tata Institute of Fundamental Research, Homi Bhabha Road, Mumbai 400005,
India
vaibhav@tifr.res.in

Abstract. Electron–molecule resonances, which are short-lived excited states of molecular negative ions, have attracted increasing attention in recent times due to their complex dynamics as well as their role in wide variety of practical applications. Formation and decay of the resonance is the most efficient way of converting kinetic energy into chemical energy in a medium through the creation of vibrationally or electronically excited states, radicals and negative ions—all of which are chemically very active. It has been found that the energy specificity of this process allows chemical control by bond selective fragmentation of organic molecules. Though diverse experimental techniques have been used to study the resonances over the last few decades, recent advances have provided several new insights into the dynamics of these species. This review would provide a short overview of the role of these resonances in various areas of science and technology followed by some of the significant findings in recent times.

1 Introduction

1.1 Process, Dynamics, and Features

"Electron–molecule resonances" or negative ion resonances (NIR) in molecules are excited states of molecular negative ions. All the atoms and molecules are characterized with respect to their electron affinity which can be positive or negative. Those with positive electron affinity could form stable negative ions. These species stay as such until and unless interacted upon by radiation or other particles. It may be noted that even without positive electron affinity, atoms and molecules may have negative ion resonant states. The negative ions and their excited states are important species in any environment where free electrons are available, as in atmospheric or industrial plasmas, interstellar medium and liquid or solid medium subjected to high-energy radiation. Due to their importance, these species have been investigated for more than half a century [1–3]. Because of its importance in several areas of science and technology, the interest in this area is still thriving. Recent developments in new instrumentation and a number of important results have provided a new impetus to the study of NIRs in molecules.

P. C. Deshmukh et al. (eds.), *Quantum Collisions and Confinement of Atomic and Molecular Species, and Photons*, Springer Proceedings in Physics 230,
https://doi.org/10.1007/978-981-13-9969-5_2

A very recent review gives a reasonable account of this area in the last two decades [4]. In this article, we try to bring out some of the recent exciting developments in this area.

Depending on its energy, a free electron on its interaction with a molecule could impart its kinetic energy to the molecule and at the same time get captured forming the NIR. This is a discrete state lying in the continuum of the free electron + molecule system, and thus have finite lifetime against auto-detachment, which is loss of extra electron. This lifetime may vary from femtosecond to microseconds depending on the molecule and the given resonance. If the lifetime against auto-detachment is sufficiently large (of the order of vibrational motion time or larger), this electron–molecule collision complex could also dissociate to give a stable negative ion and one or more neutral fragments, depending on the nature of the molecule and its potential energy surface. The process starting from electron attachment and ending in the formation of a fragment negative ion is called dissociative electron attachment (DEA). The schematic of these processes is shown in Fig. 1. Here an electron of energy E_0 interacts with a molecule AB and gets captured to form the NIR. This state decays continuously releasing the extra electron through the process of auto-detachment. The electron thus released need not have the full energy of the free electron that was captured, as the neutral molecule may be left behind in excited rotational, vibrational, and/or electronic states. The ejected electron can be observed in a conventional scattered electron spectroscopy and could be used to obtain information on the resonance that was created as well as the state in which the neutral molecule was left behind.

The decay of the resonance leading to dissociation giving neutral and negative ion fragments (in this case, A and B^-) could be studied by negative ion mass spectrometry. A third channel for the decay of the resonance has been recently identified in polyatomic molecules in which the resonance decays producing non-radical neutral fragments by breaking more than one bonds (not a double bond) and a free electron. This has been called bond breaking by catalytic electron (BBCE) [5, 6]. The final products being neutral the process is difficult to be identified in general except by optical

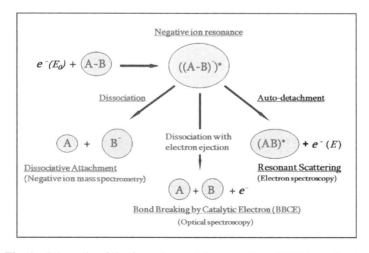

Fig. 1. Schematic of the formation and decay modes of NIR in molecules

spectroscopy or using other secondary probes. In this article, we will be mostly discussing the DEA channel and how it is used to study the electron–molecule resonances, but include a discussion on the recent identification of BBCE in an experiment.

The schematic of the decay of the resonance in terms of molecular potential energy curves for a diatomic system is shown in Fig. 2. Here the neutral molecule AB in its ground state and a generic repulsive state of the resonance AB^{-*} are shown. The width of the negative ion curve shows its finite lifetime against auto-detachment. Note that the electron affinity of B makes the energy of the NIR lower than that of the neutral state for inter-nuclear distance greater than R_c. The vertical dashed lines bounding the v = 0 state of the neutral (where most of the molecules would be energetically at room temperature) depict that the electron attachment process is Franck–Condon-type transition. The thick arrows marked **a** and **d** indicate auto-detachment and dissociation, respectively. The thin vertical arrows starting from AB^{-*} and ending on AB show how the molecule may be formed in vibrationally excited state as the electron is released. This auto-detachment decay can happen only till R_c, beyond which energetically it is disallowed.

The absolute cross sections for DEA, σ_{DEA} can be written as [7] a product of the electron attachment probability, σ_c and the survival probability p as

$$\sigma_{DEA}(\varepsilon) = \sigma_c(\varepsilon) \times p(\varepsilon)$$

and $p(\varepsilon)$ is given as

$$p(\varepsilon) = \exp\left(-\int_{E_\varepsilon}^{R_c} \frac{\Gamma_a(R)}{\hbar v(R)} dR\right) \cong \exp\left(-\frac{\tau_d}{\tau_a}\right)$$

where Γ_a the auto-detachment width, $v(R)$ the velocity of separation of A and B^-, τ_a the auto-detachment lifetime, and τ_d the dissociation time. These equations show that the

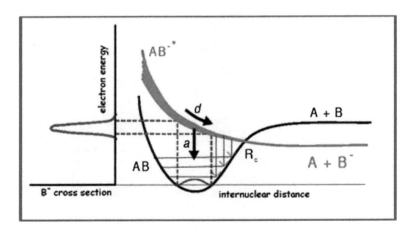

Fig. 2. Schematic of the DEA process

DEA cross sections are very sensitive to R_c, the inter-nuclear distance at which the electron attachment took place. Thus, initial vibrational excitation of the neutral molecule may lead to exponential change in the DEA cross section [8, 9]. In general, the cross section increases exponentially with decrease in dissociation time and increase in auto-detachment time. Note that depending on the isotopic composition of the molecule, there will be change in the dissociation time, thus leading to considerable isotope effect in the DEA cross section [10].

1.2 Applications

As described above, the electron–molecule resonance is a very efficient doorway in converting the electron kinetic energy into chemical energy by creating highly reactive species like electronically and vibrationally excited molecules, radicals, and negative ions. In a plasma environment, the free electrons are always present and they contribute to the plasma chemistry dominantly through the resonant attachment. Interaction of energetic photons or other particles in a condensed medium produces free electrons in an exponential manner through an ionization cascade till the entire energy is dissipated into the medium. While the ionization process creates positive ions and radicals, the cascade ionization process eventually creates low-energy electrons far more in numbers which then participate in the formation of NIRs and consequent production of reactive species. In many of the reactions induced through electron–molecule resonances, the role of electrons is not only in providing the necessary energy to overcome the barrier for a reaction but also in creating a pathway which may not have been possible otherwise. An excellent example of this is the possible formation of stable N_4^-, which has been recently identified in theoretical calculations [11].

There are several gaseous and condensed media in which the electron-induced chemistry plays a major role. One such medium is the industrial plasmas used for lighting, semiconductor etching, plasma-assisted chemical vapor deposition, and gas lasers [12]. The fluorocarbon plasmas used in the semiconductor industry have been in limelight due to the global warming potential of these molecules and consequent efforts to replace them with other molecules [4]. These and the new techniques being developed for nanolithography using focused electron beam and ion beam have brought the focus on electron interaction on several new molecules. The finding that low-energy electrons play a significant role in DNA damage [13] has resulted in increased interest in using DEA properties of molecules in radiation therapy. Single-molecule engineering using scanning tunneling microscopes (STM) is another area in which NIRs are found to play a major role [14]. It is also realized that the electron-induced processes are the inevitable links in the creation of molecules, including biological molecules in interstellar medium [15].

2 Theory

The resonances observed in electron–molecule collisions have been modeled using scattering theory with the attached electron getting captured in a state that can be described by a complex energy value—the real part describing the energy level and the imaginary part describing the lifetime of the resonance. Several approaches like the Schwinger multichannel method [16, 17], discrete momentum representation method [18], and R-matrix method [19] have been employed to determine and describe these resonance states. While these methods predict the symmetry, energy, and lifetime of the resonance they fall short in describing the DEA process as it involves treating the nuclear degrees of freedom and its coupling with the electronic one. To describe the DEA process, one needs to obtain the potential energy and the lifetime as a function of the nuclear coordinates of the excited negative ion states in addition to their variation as a function of geometry. The additional complication of the description of the extra electron, which would be done using defused orbitals, makes these calculations extremely challenging. Very low-energy electron attachment resulting in vibrational Feshbach resonance and subsequent dissociation has been described using R-matrix theories [20]. For higher energies, for diatomic molecules, the theoretical description has been achieved with a great degree of success [21, 22]. However, the same cannot be said about the polyatomic molecules. The complications of several nuclear degrees of freedom and their coupling with the electronic degrees of freedom have imposed severe constraints on the success of these calculations. There have been several attempts to address this issue by incorporating various approximations.

One such approximation is in describing the dissociation of the molecule as being confined to nuclear motion at the bond undergoing dissociation, while all other bonds are kept fixed. Such fixed nuclei approximation although reduces the computational efforts, it does not capture the very essence of the nuclear dynamics under the influence of the extra electron. In an advancement of this approach, Haxton et al. [23] have provided the possibility of incorporating the motion of all atoms in the molecule as exemplified in the detailed calculations of the DEA process in water molecule. For this purpose, they carried out calculation of the potential energy surface for the anion state using large-scale configuration interaction which gives the energy of the anion state. They also used the complex Kohn variational calculations to determine the width of the resonances. The idea behind these calculations being that if the resonance width is narrow in the range of nuclear geometries of the molecule, then the standard CI calculations can be carried out for the real part of the potential energy surfaces. For the corresponding imaginary part, the fixed nuclei complex Kohn variational method is used for limited geometries of the molecule, particularly in the Frank–Condon region with the neutral target potential energy surface. Knowing the potential energy surfaces of the anion and neutral ground state, one can describe the full dynamics of the DEA process using the local complex potential approximation. The DEA cross sections obtained using this method showed limited success [24]. The absolute cross sections for the H^- channel were consistent with the experimentally observed values but those for the O^- channel were found to be an order of magnitude lower than the measured values. One of the limitations of this approach is the local nature of the theory where the width of the resonance depends only on the geometry of the molecule. The electron energy dependence of the width cannot be described in this model.

In another set of efforts toward understanding the structure of these anion resonance states, the method of analytic continuation is used to obtain the resonance positions and widths [25]. One of the ways of achieving this analytic continuation is using the method of complex scaling [26]. A second approach is the addition of an optimized absorbing potential which separates the inner bound region of the anion from the outer part where wave function of the resonance does not vanish due to continuum states. The complex potential is chosen such that it does not affect the resonant state in the inner region but damps the asymptotically increasing continuum wave function in the outer region. This makes it bound in the inner region and insensitive to the artificially added complex potential. By suitably choosing a real and imaginary absorbing potential called continuum remover complex absorbing potential (CR-CAP), the problem of obtaining the resonance energies becomes the problem of solving the bound states in the finite inner region [27, 28]. After implementing such CR-CAP, the problem is handled with well-established quantum chemistry methods. In this way, not only the resonance energies can be derived for polyatomic molecules with fairly good accuracy, one can also calculate the molecular orbitals that are possibly occupied by the captured electron. This also helps in determining the possible dissociation paths for the given resonance as one can also calculate the relevant potential energy surfaces. The understanding of the nature of the resonance state in acetone and acetaldehyde that result in the formation of O^- from the carbonyl group (shown in Fig. 3) is an example of the usefulness of this technique [29].

The essence of molecular dynamics of the excited anion resonances is captured in the absolute cross sections and kinetic energy and angular distributions of the fragment anions produced in the DEA process. The potential energy surfaces obtained from the ab initio calculations for various resonances can give a fair amount of information on both these aspects. In the case of H_2O, apart from the absolute cross sections for the DEA channels, Haxton et al. [30] determined the expected angular distribution of the fragment anions. They calculated the relative differential cross section for the incoming electron to be captured for various molecular orientations with respect to the electron–momentum vector [31]. Under axial recoil approximation, these relative differential cross sections, also known as entrance channel amplitudes (shown in Fig. 4 for water), could be converted to the angular distribution of the fragment anion with respect to the electron beam axis after suitable molecular frame to laboratory frame transformation.

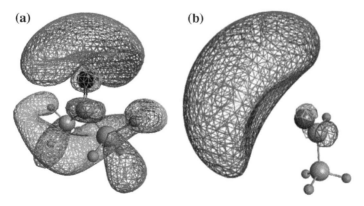

(a)   **(b)**

Fig. 3. Occupied molecular orbitals on electron attachment to **a** acetone at 8.2 eV and **b** acetaldehyde at about 9.5 eV. The resulting resonance leads to the formation of the O^- ion in each case. Here acetone undergoes a many body breakup, whereas acetaldehyde shows two body breakup (reprinted with permission from [29])

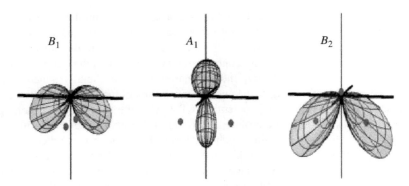

Fig. 4. Electron attachment probability calculated at the equilibrium geometry of the neutral water molecule for the three resonances as a function of polar angles about the center of mass. Dots denote positions of nuclei (Reprinted with permission from [31])

Usually, the experimentally obtained angular distributions were fitted with the functions obtained using the formalism presented by O'Malley and Taylor [32] for diatomic molecules and further adopted for the polyatomic molecules by Azria et al. [33]. Here the axial recoil approximation is understood to be essential to fit the angular distribution. The angular distribution function obtained this way essentially gives the same distribution that is deduced from the entrance channel amplitude method. For example, the angular distribution for the H^- production obtained [34] for the maximum kinetic energy ions from the A_1 resonance in water can be seen matching with the expected angular distribution using either method as shown in Figs. 4 and 5a, b. However, for the ions with kinetic energies lower than that expected based on the thermodynamics threshold have signature of the molecular dynamics on the NIR potential energy surface as in Fig. 5c. This has been captured by carrying out the classical trajectory calculations on the potential energy surface calculated earlier [31].

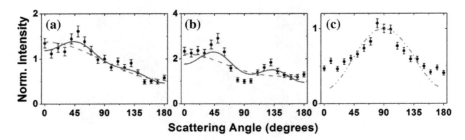

Fig. 5. Angular distribution obtained for H^- ions obtained from water molecules by DEA at **a** 8.5 eV and **b** 9 eV electron energy. The distributions shown are for the ions with kinetic energy release of 4 eV. The dashed line shows fit with s + p wave contribution to the $A_1 \rightarrow A_1$ transition and solid line shows the fit with s + p + d waves under axial recoil approximation. **c** Angular distribution for DEA at 9.5 eV electron energy with 2 eV of kinetic energy release. The dot-dashed line expected angular distribution obtained using classical trajectory calculations on the potential energy surface calculated for anion state [31] (Reprinted with permission from [34])

3 Experiments

As mentioned earlier, the very essence of these NIRs can be captured in the absolute cross sections and kinetic energy and angular distributions of the fragment anions from the dissociative electron attachment process. The absolute cross section provides a glimpse into the intricacies of the coupling of transient negative ion states with the auto-detachment continuum. The kinetic energy and angular distributions of the fragment anions carry footprints of the dynamics that these states undergo. Conventionally, mass spectrometers have been used to obtain the partial absolute cross sections for various fragment anions produced. Particularly, quadrupole mass spectrometer based setups have been used for obtaining relative cross sections with high mass resolution. This is essential when the molecules undergo fragmentation that has multiple channels available [35]. However, apart from the problem in dealing with H^- ions and ions with large kinetic energies, this method has not been able to provide absolute cross sections. It has been found that even obtaining relative cross sections with this technique runs into difficulties while dealing with ions produced with large kinetic energy. While the quadrupole mass spectrometer allows good mass resolution, the collection and transmission of all ions irrespective of their initial kinetic energies and angular distribution is a major issue with them. Time of flight technique with a segmented flight tube employing a pulsed electron beam and pulsed ion extraction was found to eliminate these problems successfully [36]. One improvement upon this technique is the use of momentum imaging of the ions ensuring complete collection and detection of the ions [10]. In both these methods, relative flow technique has been used to determine the absolute cross sections [37].

Earlier, the kinetic energy distributions of these ions were measured using ion energy analyzers coupled to a mass spectrometer, which were rotated about the interaction point in the crossed electron beam, molecular beam setup [38, 39]. Such measurements carried out in the plane containing the electron beam would provide the angle differential signal from the process. These methods were time-consuming and also restricted to limited range of angles, in addition to having systematic errors. The introduction of velocity map imaging (VMI) technique for measuring the momentum distribution of the fragment anions in low-energy electron collision experiment [40] has revolutionized the understanding of the DEA process. That this technique could cover entire 360° turned out to be very significant advantage as it provides data at 0° and 180°, which are very sensitive from symmetry considerations and enable identifying new resonances [41], distinguishing different overlapping resonances [42], and identifying new symmetry-based dynamics of the resonances [43, 44].

The basic idea behind the VMI technique [45] is to map the 3D Newton sphere in momentum space of the fragment ions produced on a 2D position-sensitive detector in such a way that the ions produced with a specific velocity are directed to fall at a given point on the detector. In order to obtain the original 3D velocity distribution, one makes the back transformation of the distribution obtained on the detector. The underlying requirement for this is to have an axis of symmetry in the original 3D distribution to be parallel to the detector plane. A variant of this technique, namely, velocity slice imaging (VSI) where a slice of the Newton sphere containing the symmetry axis parallel to the detector plane is mapped on the 2D position-sensitive detector [46]. This being a 2D-to-2D transformation, inversion of the measured data is not needed. While these techniques were successfully used for photodissociation experiments with pulsed lasers, employing them for low-energy electron collision needed several issues to be resolved—like how does one ensure that the electric field in the interaction region used for imaging does not affect the electron beam and how the huge secondary electron background is eliminated while detecting the negative ions. Apart from pulsing the electron beam and pulsing the imaging field in the interaction region, the bigger bottleneck of the secondary electron problem was solved by the clever use of a magnetic field [40]. A schematic of the velocity slice imaging is shown in Fig. 6. The success of this new experiment led to the development of several such or similar machines across the world [47–52] which have resulted in a spurt of important publications on the dynamics of the DEA process. This new development in experimentation led by TIFR has opened up a new window for studying the DEA process and is playing a major role in reviving this area of physics. Several of the new results obtained using this technique have been reviewed very recently [4].

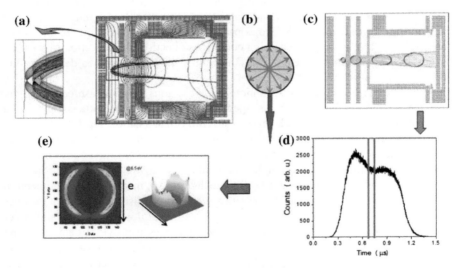

Fig. 6. Schematic of the velocity slice imaging technique for low-energy electron collisions. **a** SIMION modeling results for velocity focusing. The electrodes, equipotential surfaces, and the ion trajectories are shown. **b** Electron beam direction (blue arrow) and the central section of the Newton sphere. **c** Schematic of the trajectory of the Newton sphere. **d** Typical time of flight spectrum. The region bounded by the two lines in the center corresponds to the central slice of the Newton sphere. **e** Typical velocity slice image

4 Some Significant Results

4.1 Functional Group Dependence and Chemical Control Using Bond Selectivity

Chemical control has been the "holy grail" toward which a large number of different approaches have been made as it allows one to channel a given reaction in a specific way in order to increase the yield of a specific product and minimize the formation of unwanted by-products. The obvious and numerous advantages this will bring to the society have made it the ultimate goal in chemistry. A wide variety of approaches have been made toward this, the most prominent being use of lasers due to their monochromatic nature, high intensity in a very narrow spectral band, and, most importantly, the high level of coherence. Among several ideas that have been put forth and demonstrated, the most important one has been "coherent control" using femtosecond lasers [53–55]. These and various other techniques demonstrated till now involve a great deal of instrumentation and sophistication apart from the important question of scalability to make an impact for real-life applications. This is the context in which controlling chemical reactions using electrons becomes important.

Short reviews of chemical control using electrons have been given earlier [56, 57]. Electrons are ubiquitous and easily produced and their energy can be changed instantaneously and almost trivially by changing the potential on the accelerating electrode. As is well known, the DEA process can break a molecule even when the electron energy is less than the dissociation energy of the bond that is being broken. The extra energy that the electron needed comes from the electron affinity of the fragment. This, along with the different bond dissociation energies, leads to different threshold energies for different DEA channels. This provides a natural avenue for bond selection which is solely based on the dissociation threshold. One example of this is DEA to CS_2 [36] where S^- alone is produced below 4 eV, while all the possible fragment ions are produced above 4 eV. This property is easily manifested in chlorofluoro compounds in which the C–F bond dissociation energy is larger that of C–Cl bond dissociation and the electron affinity of Cl is slightly larger than F. This allows Cl^- to be formed at near-zero electron energies, while F^- is formed at electron energies above 2 eV [58, 59]. Such threshold energy dependence has been highlighted by an innovative experiment for complete chemical transformation of a molecular film of $1,2\text{-}C_2F_4Cl_2$ [60]. Another example of this is the site-selective dissociation of DNA bases by slow electrons (of energy less than 3 eV) in the H abstraction channel of the DEA process [61].

Around this time systematic measurements at TIFR, Mumbai [62, 63] on DEA to a number of organic molecules and their isotopomers showed that C–H, O–H, and N–H bonds in these molecules could be selectively broken, by controlling the electron energy. An example of the C–H and O–H bond selectivity is the data from acetic acid shown in Fig. 7. The H^- formation from acetic acid (CH_3COOH) shows three peaks at 6.6 eV, 8.5 eV, and 10 eV, respectively. In contrast, the H^- from CH_3COOD shows only the 10 eV peak. At the same time, D^- from CH_3COOD shows peaks at 6.5 and 8.5 eV and considerably reduced intensity at 10 eV. This clearly shows that the DEA process is being confined to either the C-site or the O-site depending on the electron energy. The measurements on other carboxylic acids and alcohols showed similar

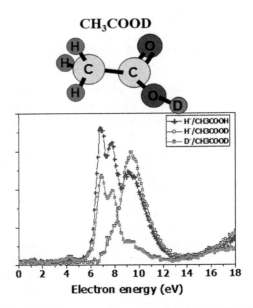

Fig. 7. Relative intensities of H⁻ and D⁻ from CH₃COOH and CH₃COOD as a function of electron energy. While H⁻ from CH₃COOH shows three peaks that from CH₃COOD shows only the peak at 10 eV and the first two peaks appear only in the D⁻ signal (Adapted with permission from [62])

behavior. On extending these measurements to amines, it was seen that such selectivity exists for the N-site as well, clearly distinguishing it from that at the C-site [62, 63]. It may be noted that in all these cases, the threshold energy needed to break any of the C–H, O–H, and N–H bonds is below 4 eV and what is seen here is not a threshold energy effect discussed above, where the selectivity is due to bond dissociation energy and the electron affinity of the fragment anion. The TIFR measurements showed that one can control the fragmentation in the energy range of 5 eV and above, which is well above the energy needed to break the C–H, N–H, and O–H bonds. This was a qualitative jump from the electron-based chemical control known till then. An analysis of the fundamental basis for this control led to the discovery of the new phenomenon of functional group dependence in the DEA process.

A comparison of the DEA data from the organic molecules in which the C–H, O–H, and N–H bonds were selectively broken with that from the precursor molecules of these bonds like CH₄, H₂O, and NH₃ showed that the resonant electron attachment property of these precursor molecules are retained in the respective sites in the organic molecules (Fig. 8). That is, the electron attachment properties of methyl, hydroxyl, and amine sites in organic molecules follow the pattern seen in methane, water, and ammonia, respectively, and are universal in nature. Further analysis showed that these observed properties arise from the formation of Feshbach resonances in which an electron is removed from a nonbonding orbital and two electrons (including the captured one) are placed in an antibonding orbital. This leads to a localization and

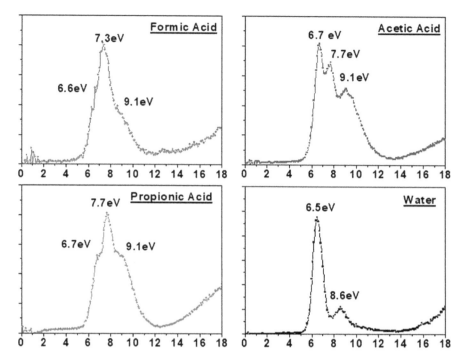

Fig. 8. Relative yield of H$^-$ from carboxylic acids and water. The vertical scale is relative intensity and the horizontal scale is electron energy in eV (Adapted with permission from [63])

speeding up the dissociation process so that there is minimum intramolecular energy redistribution (IVR), which has been historically a bane of chemical control using light.

This discovery of the functional group dependence in DEA provided a deeper and more robust basis for electron-based chemical control than known till then. That the bonds in organic molecules could be selectively cleaved based on the functional groups also opened up the possibility of developing new analytical techniques as well as greatly enhanced the potential of electron-based chemical control for industrial applications and in radiation therapy. This work has been one of the major milestones in the recent renaissance of low-energy electron–molecule collisions.

4.2   DEA to Halo Compounds

DEA to halogen compounds is particularly important since these molecules are used in dry etching of semiconductor chips. At the same time, these molecules are of concern in the chemistry of upper atmosphere leading to ozone destruction and as greenhouse gases. Consequently, DEA measurements on these molecules have attracted considerable interest. Here we summarize the highlights of measurements on two of these molecules using velocity slice imaging. The first one on CF$_3$I comes out as an excellent example of the contrast between photodissociation versus DEA and reveals how the dissociation product intensities are inverted by the presence of the extra electron [64]. The second one on CF_4 shows the interplay of symmetry and dynamics in the DEA process [43].

CF_3I: Due to its short lifetime in the atmosphere and hence reduced global warming potential as compared to tetrafluoromethane (CF_4), trifluoroiodomethane (CF_3I) is being considered as a replacement for the former as a plasma etchant in semiconductor industry. The relatively low lifetime of the molecule in the atmosphere is due to its efficient destruction by photodissociation leading to CF_3 radicals and iodine atoms in $^2P_{3/2}$ and $^2P_{1/2}$ states. It has been found that 84% of iodine atoms are produced in the excited $^2P_{1/2}$ state which lies at 0.94 eV above the ground state of $^2P_{3/2}$ [65]. This branching ratio is a direct result of the excitation probability of the molecule into the $^3Q_0$ and $^1Q_1$ states of the neutral on photoabsorption from where they dissociate unhampered to the respective dissociation limits (Fig. 9). Thus, the photodissociation of CF_3I could be shown as

$$CF_3I + h\nu \rightarrow CF_3I\left(^3Q_0\right) \rightarrow CF_3 + I\left(^2P_{1/2}\right) (\text{dominant channel})$$
$$\rightarrow CF_3I\left(^1Q_1\right) \rightarrow CF_3 + I\left(^2P_{3/2}\right) (\text{weak channel})$$

DEA measurements on CF_3I have shown the formation of I^- (at near-zero energy) and CF_3^- (at 3.8 eV) [66]. The resonance leading to CF_3^- at 3.8 eV is a core-excited resonance that is associated with electronic excitation to the antibonding σ^*_{CF3-I} LUMO [66], which could result in a $^2A_1$ or ^2E state identical to the $\sigma \rightarrow \sigma^*$ and $n \rightarrow \sigma^*$ excitation of the A-band transition in the neutral. The transition of this type in the neutral molecule on photoabsorption leads to the $^3Q_0$ and $^3Q_1$ states shown in Fig. 9. In analogy with the photoexcitation, the negative ion resonant states, $^2A_1$ and ^2E leading to the formation of CF_3^- could be qualitatively represented in the figure by dashed lines.

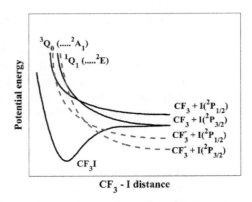

Fig. 9. Schematic of the potential energy curves as a function of C–I distance of CF_3I and CF_3I^- which are involved in photodissociation and DEA, respectively. In addition to the neutral ground state, the $^3Q_0$ and $^1Q_1$ excited states of neutral molecule (solid lines) and the corresponding $^2A_1$ and ^2E core-excited negative ion states (dotted lines) are shown (Reproduced with permission from [64])

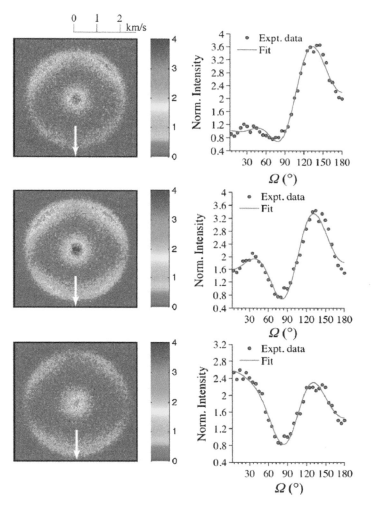

Fig. 10. Velocity slice images (left) and the respective angular distributions (right) of CF_3^- at electron energies of 3.3 eV (top), 3.8 eV (middle) and 4.3 eV (bottom). The white arrows in the images show the electron beam direction and the solid lines in the angular distribution figures are fit for A_1 symmetry (Reproduced with permission from [64])

The momentum images of CF_3^- obtained with electrons of energy around 3.8 eV (Fig. 10) showed an inner blob and an outer ring [64]. The kinetic energy analysis of the images proved that the outer ring corresponds to the production of the neutral atomic iodine in the ground state ($I\ ^2P_{3/2}$) and the inner blob corresponds to the excited $I\ (^2P_{1/2})$. By time-integrating VSI data corresponding to the inner and outer structures, the total contributions of each of the two (inner and outer) Newton spheres could be obtained which showed that 96–98% of the total CF_3^- signal corresponds to the outer ring—that is, $I\ (^2P_{3/2})$ channel. On fitting the angular distribution of the ions corresponding to the outer ring, it was seen that the corresponding resonant state is definitely of A_1 symmetry and not of E symmetry. This allowed the DEA process leading to CF_3^- formation to be described as

$$CF_3I + e^- \rightarrow CF_3I^- * \left(^2A\right) \rightarrow CF_3 + I\left(^2P_{1/2}\right) (\text{weak channel})$$
$$\rightarrow CF_3I^- * \left(^2A\right) \rightarrow CF_3I^{-*}\left(^2E\right) \rightarrow CF_3 + I\left(^2P_{3/2}\right) (\text{dominant channel})$$

Thus, in contrast to photodissociation, DEA leads to the production of I atoms mostly in the ground state. This is a clear example of how the electron-induced chemistry could be different from photochemistry. It is interesting to note that the contrasting behavior of electron vis a vis the photon is due to the electron spin, which allows both the negative ion resonant states to have the same spin multiplicity. The $^3Q_0$ state which has the dominant oscillator strength cannot transfer its excitation to the $^1Q_1$ state since they have different spin multiplicity. However, the $^2A_1$ resonant state created with similar core excitation as that of $^3Q_0$ state (and hence dominant transition probability for formation) has a conical intersection with the ^2E resonant state (which has low probability for formation by electron attachment, similar to its neutral counterpart $^1Q_1$). This conical intersection between A_1 and E states allows dissociation to the $I(^2P_{3/2})$ limit, as shown in Fig. 9 [64].

CF$_4$: Being an effective plasma etchant gas, CF_4 has been investigated in several low-energy electron scattering studies, including DEA [67, 68]. DEA to CF_4 is known to produce F^- and CF_3^- with fairly high cross section and F_2^- with small cross section between 4 and 10 eV [69]. While it is a very important molecule in practical applications, the high-level tetrahedral symmetry it possesses makes it an interesting one from basic physics considerations.

One of the symmetry-based effects in molecules which have been of considerable interest is the Jahn–Teller effect. This has been observed in optical spectroscopy and photochemistry of molecules [70]. The question was if this effect can be seen in a scattering process and DEA studies on CF_4 using momentum imaging proved that it could be observed in the electron attachment and subsequent dissociation process. Signatures of the Jahn–Teller effect were suspected to be present in DEA to CH_4 [71] which has identical symmetry properties like CF_4, but could not be clearly proved.

The momentum images of F^- and CF_3^- at different electron energies across the respective DEA peaks were analyzed both for the kinetic energy distribution of the fragments and their angular distributions [43, 72]. F^- momentum distribution shows the presence of two distinct but overlapping NIRs in the range of 4–10 eV—one going through two-body dissociation and the other one decaying through three-body fragmentation. CF_3^-, on the other hand, appears to be formed from a single resonance between 5 and 9 eV. The mutual forward–backward mirror symmetry of the momentum images of the two fragments (Fig. 11) and that both these are produced in same electron energy range showed that they are the products of the same NIR. These angular distributions of F^- and particularly that of CF_3^- could not be fitted well for the CF_4^- resonance state in the T_d symmetry. A good fit could be obtained only by allowing for a superposition of T_d and C_{3v} symmetries (Fig. 12). The quantum mechanical superposition of the two symmetries in the angular distribution can manifest only if during the attachment process the electron encounters both these symmetries in the molecule. What it means is that degenerate states of higher symmetry are

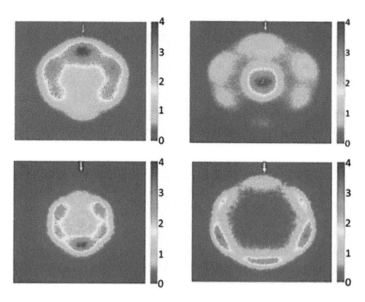

Fig. 11. Velocity slice images of F$^-$ (upper) at 5.5 eV (left) and 7.5 eV (right) and for CF$_3^-$ (lower) at 5.5 eV (left) and 8.5 eV (right). The arrows at the top show the direction of electron beam (Adapted with permission from [43])

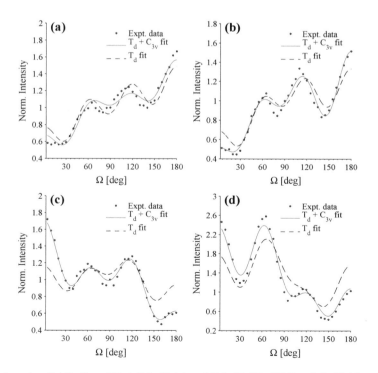

Fig. 12. Angular distributions F$^-$ at 5.5 eV (**a**) and 7.5 eV (**b**); CF$_3^-$ at 5.5 eV (**c**) and 7.5 eV (**d**) (Reprinted with permission from [43])

split allowing for the capture of an electron with a different angular momentum which was not possible in the pure T_d symmetry. The lifting of degeneracy due to structural change is typical of Jahn–Teller effect. This is possible since CF_4^- has C_{3v} symmetry with one of the C–F bond being elongated. Because of the dynamics and the corresponding Jahn–Teller distortion, the angular distribution clearly showed the superposition of the T_d and C_{3v} symmetries, the first such instance seen in electron scattering.

The presence of the Jahn–Teller effect could also be determined from the kinetic energy distributions of the two fragments as a function of electron energy across the respective DEA peaks. The kinetic energy release from the two-body dissociation channel of F^- showed a direct proportionality to the incident electron energy across the resonance, though a well-defined fraction of the excess energy is seen to be shared by the internal excitation of CF_3 neutral fragment. In the case of CF_3^- channel also, a part of the excess energy is shared between the internal excitation of CF_3^- and the kinetic energy of the fragments. However, this sharing seemed to have a strong dependence on the incident electron energy (Fig. 13). Up to certain electron energy, decided by the stability and geometric structure of the CF_3^- relatively small energy appears as kinetic energy of the fragments. Above the threshold where CF_3^- could be formed in the geometry similar to the CF_3 part of CF_4, most of the excess energy appears as kinetic energy of the fragments. This dynamics of the change in geometry of the CF_3 part as the CF_4^- undergoes dissociation is also manifested in the angular distribution of the fragments as Jahn–Teller effect [43, 72].

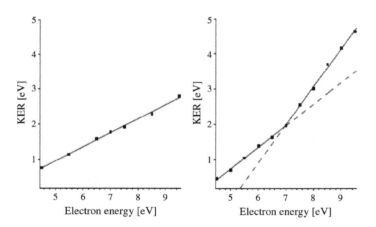

Fig. 13. Kinetic energy release determined from the velocity slice images in the F^- channel (left) and CF_3^- channel (right). While the F^- channel shows monotonic increase with electron energy, the CF_3^- channel shows different behavior above and below 7 eV of electron energy (Adapted with permission from [43])

4.3   Formation of Molecular Oxygen by DEA to CO_2

DEA to CO_2 has been reported to produce O^- as the predominant product at energies of 4, 8, and 13 eV [73] and at higher energies O_2^- and C^- are also reported [74]. The molecular dynamics leading to the O^- channel has been studied recently in great detail in terms of angular distribution and kinetic energy distribution using the VSI technique [51, 75, 76]. From the earlier measurements [74], the C^- channel appearing at 16, 17, and 18.7 eV, respectively, was assigned to three-body dissociation leading to $C^- + O + O$ based on its occurrence at the relatively high electron energy.

In the very recent velocity slice imaging studies of C^- ions, Wang et al. [77] observed two distinct structures, as shown in Fig. 14. The data from two electron energies are shown in the figure. The more intense unresolved structure in the images in the left panel is due to three-body fragmentation ($C^- + O + O$) while the structures outside it are due to two-body fragmentation ($C^- + O_2$). The two-body fragmentation with O_2 $(X^3\Sigma_g^-)$ has the thermodynamic threshold of 10.18 eV. The finer structures in the two-body channel are due to vibrational excitation of O_2 $(X^3\Sigma_g^-)$ as marked in the kinetic energy distributions in the right panels. The data showed that near 16 eV the O_2

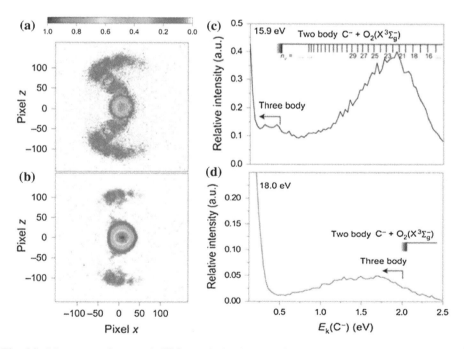

Fig. 14. Momentum image of C^- ions obtained at **a** 15.9 eV and **b** 18 eV. Corresponding kinetic energy distributions are shown in **c** and **d,** respectively. The more intense unresolved structure in the images is due to three-body fragmentation ($C^- + O + O$) while the structures outside it are due to two-body fragmentation ($C^- + O_2$). The finer structures in the two-body channel are due to vibrational excitation of O_2. It may be noted that at lower electron energy the O_2 production is more efficient (Adapted with permission from [77])

production channel has cross section almost equal to that for the three-body fragmentation channel, which has the thermodynamic threshold at 15.28 eV. As the electron energy is increased, the relative contribution from the O_2 production channel was found to be decreasing. The detailed analysis of the observed angular distribution of the energetic C^- ions showed that up to 17.5 eV, the two-body dissociation appeared to start with little bend in the O–C–O, whereas for higher electron energies, the parent anion undergoes substantial bending before dissociation.

The observation that O_2 is being produced by DEA to CO_2 is of interest in the context of the early evolution of the planet Earth. It was assumed that in early Earth molecular oxygen was produced primarily by recombination of two O atoms in the presence of a third body [78]. Recently, photodissociation of CO_2 has been proposed to be another process responsible for producing O_2 on early Earth [79]. The observation of O_2 production by DEA from CO_2 shows the presence of yet another process in the early evolution of Earth's atmosphere [77].

4.4   Quantum Coherence in Single-Electron Attachment

Dissociative electron attachment to H_2 has been extensively studied in the past. The H^- channel shows three peaks in the ion yield curve at 4, 10, and 14 eV of electron energy. Of particular interest is the 14 eV peak which results from the dissociation of H_2^- anion into H^- and H atom in excited (n = 2) state. So far, all reports in electron scattering measurements assign this resonance to the $^2\Sigma_g^+$ state [80]. Based on the formalism for fragment anion angular distribution under axial recoil approximation [32], the angular distribution expected from this resonance ($^1\Sigma_g^+ \rightarrow {}^2\Sigma_g^+$ transition) is isotropic if we consider the lowest partial wave (s-wave) transfer in resonance formation. However, the observed angular distribution (Fig. 15) at the 14 eV peak shows the ion intensity peaking at 0° and 180°. But more importantly the distribution shows forward–backward asymmetry, which is a feature not expected from a homo-nuclear

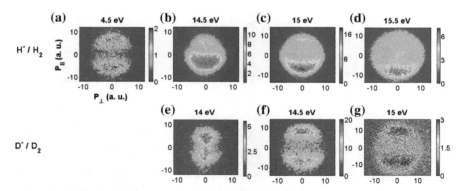

Fig. 15. Velocity slice images (converted to momentum images) of H^- from H_2 and D^- from D_2 at different electron energies. The electron beam direction is from bottom to top of the figure. The data at 4.5 eV for H_2 is shown for comparison, which shows the expected symmetry in the forward–backward lobes [44]

diatomic molecule due to its inversion symmetry. This asymmetry was found to be mostly absent or even reversed based on attaching electron energy in the case of D_2.

These results, which were first of its kind and surprising to begin with, were explained due to the very quantum behavior of the attaching electron and subsequent interference between two quantum paths of DEA. A single electron on attachment to a molecule creates a coherent superposition of two NIRs ($^2\Sigma_g^+$ and $^2\Sigma_u^+$) of opposite parity that dissociate to the same limit. The two NIRs that participate in this process necessarily overlap in the Frank–Condon region. The even ($^2\Sigma_g^+$) and odd ($^2\Sigma_u^+$) parity states are determined by the quantum of angular momentum (in this case, 0 and 1, respectively) transferred to the molecule by the attaching electron. The single coherent state created by the superposition of the two NIRs propagates through two different paths decided by the potential energy curves of the individual NIRs and results in a phase difference at the dissociation limit. This phase difference leads to the interference and the observed forward–backward asymmetry in the angular distributions. The phase difference is dependent on how fast the wave packet of the resonance proceeds along the two paths, which in turn depends on the total energy available based on the energy of the attaching electron, the path traversed, and the mass of the two nuclei. Thus, one expects to see a difference in the interference pattern and the angular distribution as a function of the electron energy as well as the isotopic composition.

This in a way explains the difference between H_2 and D_2. Yet another effect leading to the difference in H_2 and D_2 data is the fact that the two NIRs are continuously decaying through auto-detachment. In fact, the probability for formation of D^- from D_2 is only about one-third of H^- from H_2 [10], implying more loss through auto-detachment in the case of D_2 as compared to that of H_2. The cause of this is again the mass difference between the two, with the wave packet of the D_2 resonant system taking longer time to reach the dissociation limit as compared to that of H_2. If the lifetime of any of the two resonances is comparable or shorter than the dissociation time scale, the different dissociation time will substantially change the relative amplitude of each path. The net effect of this is a difference in the contrast of the asymmetry between H_2 and D_2 as observed in the experiment. A qualitative model of this using empirical potential energy curves for the resonances is shown in Fig. 16. It is also clear that the shorter lived $^2\Sigma_g^+$ resonance contributes substantially in the auto-detachment channel leaving its signature on the angular distribution of the ejected electron [80], while the longer lived $^2\Sigma_u^+$ makes dominant contribution to the DEA channel as observed in its angular distribution [44].

4.5 Bond Breaking by Catalytic Electrons

As discussed in Sect. 1, a new mode of decay of the NIR was proposed in which the final products are non-radical neutrals and a free electron. The initial proposal of this came in a reaction in which the free electron attachment was followed by breaking two bonds in an organic molecule producing two neutral fragments and a free electron [5]. This model was further extended to another system in a concerted barrierless four-bond-breaking mechanism giving rise to three neutral products and a free electron [6]. It appears that in these cases, the electron is captured into a multicentre antibonding orbital which weakens the corresponding bonds and they consequently relax.

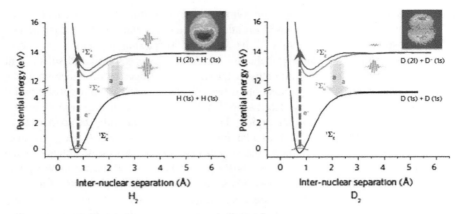

Fig. 16. Schematic of the DEA process in H_2 and D_2 at 14 eV. Two quantum paths arising from the electron attachment lead to the same dissociation limit causing their interference. The amplitude of each path is dependent on the dissociation time and the individual lifetime of the participating NIRs against auto-detachment. The change in amplitude is evident for the D_2 case when compared with H_2. The resulting forward–backward asymmetry in the angular distribution can be seen in the VSI images shown in the inset [44]

At the same time, there is a transfer of electron density from the bonding orbitals to antibonding orbitals during the course of this and the system evolves with the anti-bonding orbitals becoming lower in energy compared to the bonding orbitals, thus leading to dissociation, with concomitant release of the extra electron. It appears that the whole process is triggered by the attachment of the free electron. The electron thus acts as a catalyst in this chemical reaction.

The first experimental evidence for such catalytic action of free electron was observed in the low-energy electron interaction with the condensed formic acid [81]. In this experiment, cold formic acid films were irradiated with electrons of a given energy with simultaneous monitoring of the film by Fourier transform infra-red (FTIR) spectroscopy under UHV conditions. CO_2 was found to be the main product. The CO_2 production was found to peak at 6 and 11 eV (Fig. 17). At least the one at 6 eV could not account for by the DEA process since no DEA channel is known to be active either in the gas phase or in the condensed phase at that energy. The only other resonant process available to explain the experimental data was the BBCE and the theoretical calculations supported this further (Fig. 18). It appears that here the BBCE is leading the formic acid to form CO_2 and H_2 with the excess electron leaving the system.

The catalytic action of low-energy electron can be a doorway to many enzymatic activities in the biological systems. The resonant nature of this process gives an additional edge making it an interesting candidate for control of electron-induced chemistry.

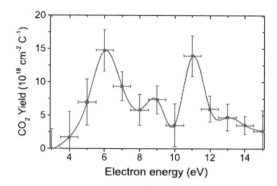

Fig. 17. The CO_2 yield plotted as a function of energy of the electron that irradiates hundred monolayers thick formic acid film deposited at 50 K on gold (111) substrate. For point in the plot, several freshly prepared films are irradiated and the procedure is repeated for all electron energies (Reprinted with permission from [81])

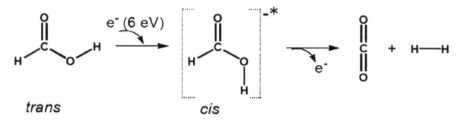

Fig. 18. Schematic of the theoretical model in the conversion of formic acid to CO_2 and H_2 by the catalytic action of the free electron (Reprinted with permission from [81])

5   Future Outlook

As seen so far, the invention of new experimental techniques has changed our understanding of electron–molecule resonances, particularly their dynamics. New frontiers of electron–molecule resonant interaction have also come forth in the form of catalytic action of low-energy electrons and bond selectivity in the fragmentation observed in the DEA process. Although a lot of new information has been unraveled, there are several aspects of electron–molecule resonances and particularly their dynamics that are not yet understood. Following are the possible directions to be pursued in the future to uncover rich details of unknowns of NIR state.

5.1   Improving the Resolution of Momentum Imaging

Most of the details of the DEA process are presented in the kinetic energy and angular distributions of the fragment anions. The momentum imaging technique unravels these details. However, these details are limited by the resolution of the imaging process which is primarily due to limited resolution of the electron beam energy. The first and the foremost step to improve our understanding of the DEA process would be to improve the energy resolution of the electron beam. One aspect this would definitely bring out is the details of the internal energy of the molecular fragments with vibrational state resolution. This could lead to qualitative improvement in the understanding of the dissociation dynamics of the NIRs in polyatomic molecules.

In many polyatomic molecules, the dissociation results in ions with low kinetic energies. Along with improvement in electron energy resolution, improving the momentum imaging apparatus in its capability of magnifying low-energy ion images may provide further details of the dissociation process. Marrying these two advancements together would provide unprecedented details of the dynamics. Moreover, imaging spectrometers with larger mass resolutions will also help in looking at varieties of fragments that are generally formed in the DEA to larger molecules.

5.2   Detecting Neutral Fragments

So far, the DEA dynamics has been deduced by obtaining the cross sections as well as kinetic energy and angular distributions of the fragment anions. However, when the DEA leads to multiple channels, detecting neutral products are essential to fully understand the kinematics and dynamics of the process. Li et al. [82] have devised a method of obtaining the details of the neutral fragment using electron impact ionization following DEA. They have demonstrated this technique for DEA to CCl_4 for near-zero energy electron attachment where the cross section is very large. Other possibility of detecting the neutrals would be resonant-enhanced multiphoton ionization (REMPI). Implementation of such scheme along with the momentum imaging can provide great details of the DEA process. Apart from that, this method can also provide information about electron impact neutral dissociation of molecules.

5.3   Electron Collision with Excited Molecules and Radicals

Most of the studies in electron collisions on molecules have been carried out from their ground states, though it is well known that information on the electron attachment to excited molecules in the higher vibrational and electronic states is needed for many practical applications involving plasmas. From pure physics point of view, electron attachment to excited molecules allows accessing selectively (depending on the excited state) a wider Franck–Condon region to probe the DEA process as a whole and the NIR states in particular for their dynamics. As discussed earlier, the site selectivity of dissociation observed in DEA provides a possible way to control chemical reactions. With the initial excitation of vibrational or electronic degree of freedom, one can enhance this control and even realize a mode selective chemistry which was envisaged after the discovery of lasers.

Limited DEA studies from vibrationally excited molecules have been carried out using heated gases [2] and in one case on vibrational state-selected molecules [83]. The DEA experiments to excited molecules were reviewed about 20 years back [84] and unfortunately after the only consistent efforts including the first measurement of absolute cross sections from an excited state [85–88] very little work has been reported on the DEA to excited molecules. The only exceptions are a couple of reports on effects of vibrational excitations by measuring the temperature dependences of the DEA processes [89, 90]. As for DEA to radicals and unstable molecules, there has been limited success [91–93].

In most of these experiments, the main bottleneck has been in preparing the molecules in selected excited states with sufficient target densities. Optical pumping is the only way to obtain high level of state selectivity. Techniques based on coherent control or adiabatic passage [94] are a way forward in this respect. The excited state DEA studies coupled with momentum imaging technique hold forth great possibilities in revolutionizing the understanding of electron–molecule resonances and for enhancing the potential of electron-controlled chemistry.

Acknowledgements. E.K. acknowledges the Raja Ramanna Fellowship. V. S. P. acknowledges funding from Dept. of Atomic Energy, Govt. of India.

References

1. Schulz, G.J.: Resonances in electron impact on diatomic molecules. Rev. Mod. Phys. **45**(3), 423–486 (1973)
2. Christophorou, L. (ed.): Electron-molecule Interactions and Their Applications, vol. 1. Academic Press, New York (1984)
3. Illenberger, E., Momigny, J.: Gaseous Molecular Ions: An Introduction to Elementary Processes Induced by Ionization. Springer (1992)
4. Fabrikant, I.I., Eden, S., Mason, N.J., Fedor, J.: Recent progress in dissociative electron attachment: from diatomics to biolmolecules. Adv. At. Mol. Opt. Phys. **66**, 545–657 (2017)
5. Davis, D., Vysotskiy, V.P., Sajeev, Y., Cederbaum, L.S.: Electron impact catalytic dissociation: two-bond breaking by a low-energy catalytic electron. Angew. Chem. Int. Ed. **50**(18), 4119–4122 (2011)
6. Davis, D., Vysotskiy, V.P., Sajeev, Y., Cederbaum, L.S.: A one-step four-bond-breaking reaction catalysed by an electron. Angew. Chem. Int. Ed. **51**(32), 8003–8007 (2012)
7. O'Malley, T.F.: Theory of dissociative attachment. Phys. Rev. **150**(1), 14–29 (1966)
8. Allan, M., Wong, S.F.: Effect of vibrational and rotational excitation on dissociative attachment in hydrogen. Phys. Rev. Lett. **41**, 1791–1794 (1978)
9. Hall, R.I., Čadež, I., Landau, M., Pichou, F., Schermann, C.: Vibrational excitation of hydrogen via recombinative desorption of atomic hydrogen gas on a metal surface. Phys. Rev. Lett. **60**(4), 337–340 (1988)
10. Krishnakumar, E., Denifl, S., Čadež, I., Markelj, S., Mason, N.J.: Dissociative electron attachment cross section for H_2 and D_2. Phys. Rev. Lett. **106**(24), 243201 (4 p) (2011)
11. Sajeev, Y.: Cycloaddition of molecular dinitrogens: formation of tetrazete anion (N_4^-; D_{2h}) through associative electron attachment. Chem. Phys. **117**(15–16), 2162–2166 (2019)
12. Christophorou, L. (ed.): Electron-molecule interactions and their applications, vol. 2. Academic Press, New York (1984)
13. Boudaiffa, B., Cloutier, P., Hunting, D., Huels, M.A., Sanche, L.: Resonant formation of DNA strand breaks by low-energy (3 to 20 eV) electrons. Science **287**(5458), 1658–1660 (2000)
14. Martel, R., Avouris, P., Lyo, I.-W.: Molecularly adsorbed oxygen species on Si(111)-(7×7): STM-induced dissociative attachment studies. Science **272**(5260), 385–388 (1996)

15. Boyer, M.C., Rivas, N., Tran, A.A., Verish, C.A., Arumainayagam, C.R.: The role of low-energy (≤ 20 eV) electrons in astrochemistry. Surf. Sci. **652**, 26–32 (2016)

16. Takatsuka, K., McKoy, V.: Extension of the Schwinger variational principle beyond the static-exchange approximation. Phys. Rev. A **24**(5), 2473–2480 (1981)

17. da Costa, R.F., Varella, M.T. do N., Bettega, M.H.F., Lima, M.P.A.: Recent advances in the application of the Schwinger multichannel method with pseudopotentials to electron-molecule collisions. Eur. Phys. J. D **69**, 159 (2015)

18. Lane, N.F.: The theory of electron-molecule collisions. Rev. Mod. Phys. **52**(1), 29–119 (1980)

19. Tennyson, J.: Electron-molecule collision calculations using the R-matrix method. Phys. Rep. **491**(2–3), 29–76 (2010)

20. Fabrikant, I.I.: Recent progress in theory of dissociative attachment: from diatomics to biomolecules. J. Phys. Conf. Ser. **204**, 012004 (2010)

21. Domcke, W.: Theory of resonance and threshold effects in electron-molecule collisions: the projection-operator approach. Phys. Rep. **208**(2), 97–188 (1991)

22. Fabrikant, I.: Quasiclassical R-matrix theory of inelastic processes in collisions of electrons with HCl molecules. Phys. Rev. A **43**(7), 3478–3486 (1991)

23. Haxton, D.J., Zhang, Z., McCurdy, C.W., Rescigno, T.N.: Complex potential surface for the $^2B_1$ metastable state of the water anion. Phys. Rev. A **69**(6), 062713 (11 p) (2004)

24. Haxton, D.J., Zhang, Z., Meyer, H.-D., Rescigno, T.N., McCurdy, C.W.: Dynamics of dissociative attachment of electron to water through the $^2B_1$ metastable state of the anion. Phys. Rev. A **69**(6), 062714 (16 p) (2004)

25. Santra, R., Cederbaum, L.: Non-Hermitian electronic theory and applications to clusters. Phys. Rep. **368**(1), 1–117 (2002)

26. Aguilar, J., Combes, J.M.: A class of analytic perturbations for one-body Schrodinger Hamiltonians. Commun. Math. Phys. **22**(4), 269–279 (1971)

27. Sajeev, Y., Vysotskiy, V., Cederbaum, L.S., Moiseyev, N.: Continuum remover-complex absorbing potential: efficient removal of the nonphysical stabilization points. J. Chem. Phys. **131**(2), 211102 (4 p) (2009)

28. Sajeev, Y.: Real-valued continuum remover potential: an improved L2-stabilization method for the chemistry of electronic resonance states. Chem. Phys. Lett. **587**, 105–112 (2013)

29. Prabhudesai, V.S., Tadsare, V., Ghosh, S., Gope, K., Davis, D., Krishnakumar, E.: Dissociative electron attachment studies on acetone. J. Chem. Phys. **141**(16), 164320 (7 p) (2014)

30. Haxton, D.J., McCurdy, C.W., Rescigno, T.N.: Angular dependence of dissociative electron attachment to polyatomic molecules: application to the $^2B_1$ metastable state of the H_2O and H_2S anions. Phys. Rev. A **73**(6), 062724 (15 p) (2006)

31. Adaniya, H., Rudek, B., Osipov, T., Haxton, D.J., Weber, T., Rescigno, T.N., McCurdy, C. W., Belkacem, A.: Imaging the molecular dynamics of dissociative electron attachment to water. Phys. Rev. Lett. **103**(23), 233201 (4 p) (2009)

32. O'Malley, T.F., Taylor, H.S.: Angular dependence of scattering products in electron-molecule resonant excitation and in dissociative attachment. Phys. Rev. **176**(1), 207–221 (1968)

33. Azria, R., Coat, Y.L., Lefevre, G., Simon, D.: Dissociative electron attachment on H_2S: energy and angular distributions of H^- ions. J. Phys. B At. Mol. Phys. **12**, 679–687 (1979)

34. Ram, N.B., Prabhudesai, V.S.., Krishnakumar, E.: Comment on "Imaging the molecular dynamics of dissociative electron attachment to water." Phys. Rev. Lett. **106**(4), 049301 (1 p) (2011)

35. Hotop, H., Ruf, M.-W., Allan, M., Fabrikant, I.I.: Resonance and threshold phenomena in low-energy electron collisions with molecules and clusters. Adv. At. Mol. Opt. Phys. **49**, 86–216 (2003)

36. Krishnakumar, E., Nagesha, K.: Dissociative attachment of electrons to CS_2. J. Phys. B At. Mol. Opt. Phys. **25**(7), 1645–1660 (1992)

37. Srivastava, S.K., Chutjian, A., Trajmar, S.: Absolute elastic differential electron scattering cross sections in the intermediate energy region. I. H_2. J. Chem. Phys. **63**(6), 2659–2665 (1975)

38. Van Brunt, R.J., Kieffer, L.J.: Angular distribution of O^- from dissociative electron attachment to O_2. Phys. Rev. A **2**(5), 1899–1905 (1970)

39. Cadez, I., Tronc, M., Hall, R.I.: Dissociative electron attachment in CO: angular distribution of the O^- ions. J. Phys. B **8**(5), L73–L76 (1975)

40. Nandi, D., Prabhudesai, V.S., Krishnakumar, E., Chatterjee, A.: Velocity slice imaging for dissociative electron attachment. Rev. Sci. Instrum. **76**(5), 053107 (8 p) (2005)

41. Prabhudesai, V.S., Nandi, D., Krishnakumar, E.: On the presence of the $^4\Sigma_u^-$ resonance in dissociative electron attachment to O_2. J. Phys. B At. Mol. Opt. Phys. **39**(14), L277–L283 (2006)

42. Gope, K., Prabhudesai, V.S., Mason, N.J., Krishnakumar, E.: Probing the resonant states of Cl_2 using velocity slice imaging. J. Phys. B At. Mol. Opt. Phys. **49**(1), 015201 (9 p) (2016)

43. Ómarsson, F.H., Szymanska, E., Mason, N.J., Krishnakumar, E., Ingólfsson, O.: Quantum superposition of target and product states in reactive electron scattering from CF_4 revealed through velocity slice imaging. Phys. Rev. Lett. **111**(6), 063201 (4 p) (2013)

44. Krishnakumar, E., Prabhudesai, V.S., Mason, N.J.: Symmetry breaking by quantum coherence in single electron attachment. Nat. Phys. **14**, 149–153 (2018)

45. Eppink, A.T.J.B., Parker, D.H.: Velocity map imaging of ions and electrons using electrostatic lenses: application to photoelectron and photofragment ion imaging of molecular oxygen. Rev. Sci. Instrum. **68**(9), 3477–3484 (1997)

46. Gebhardt, C.R., Rakitzis, T.P., Samartzis, P.C., Ladopoulos, V., Kitsopoulos, T.N.: Slice imaging: a new approach to ion imaging and velocity mapping. Rev. Sci. Instrum. **72**(10), 3848–3453 (2001)

47. Adaniya, H., Slaughter, D.S., Osipov, T., Weber, T., Belkacem, A.: A momentum imaging microscope for dissociative electron attachment. Rev. Sci. Instrum. **83**(2), 023106 (2012)

48. Wu, B., Xia, L., Li, H.-K., Zeng, X.-J., Tian, S.X.: Positive/negative ion velocity mapping apparatus for electron molecule reactions. Rev. Sci. Instrum. **83**(1), 013108 (2012)

49. Moradmand, A., Williams, A., Landers, A.L., Fogle, M.: Momentum-imaging apparatus for the study of dissociative electron attachment dynamics. Rev. Sci. Instrum. **84**(3), 033104 (2013)

50. Szyman´ska. E., Prabhudesai, V.S., Mason, N.J., Krishnakumar, E.: Dissociative electron attachment to acetaldehyde, CH_3CHO: a laboratory study using the velocity map imaging technique. Phys. Chem. Chem. Phys. **15**(3), 998–1005 (2013)

51. Nag, P., Nandi, D.: Dissociation dynamics in the dissociative electron attachment to carbon dioxide. Phys. Rev. A **91**(5), 052705 (7 p) (2015)

52. Rescigno, T.N., Trevisan, C.S., Orel, A.E., Slaughter, D.S., Adaniya, H., Belkacem, A., Weyland, M., Dorn, A., McCurdy, C.W.: Dynamics of dissociative electron attachment to ammonia. Phys. Rev. A **93**(5), 052704 (10 p) (2016)

53. Tanner, D.J., Rice, S.A.: Coherent control of selectivity of chemical reaction via control of wave packet evolution. J. Chem. Phys. **83**(10), 5013–5018 (1985)

54. Potter, E.D., Herek, J.L., Pedersen, S., Liu, Q., Zewail, A.H.: Femtosecond laser control of a chemical reaction. Nature **355**, 66–68 (1992)

55. Rabitz, H., de Vivie-Riedle, R., Motzkus, M., Kompa, K.: Whither the future of controlling quantum phenomena? Science **288**(5467), 824–828 (2000)
56. Krishnakumar, E.: J. Phys. Conf. Ser. **185**, 012022 (8 p) (2009)
57. Mason, N.J.: Electron-induced chemistry: a forward look. Int. J. Mass Spectrom. **277**(1–3), 31–34 (2008)
58. Illenberger, E.: Electron-attachment reactions in molecular clusters. Chem. Rev. **92**(7), 1589–1609 (1992)
59. Aflatooni, K., Burrow, P.D.: Dissociative electron attachment in chlorofluoromethane and the correlation with vertical attachment energies. Int. J. Mass Spectrom. **205**(1–3), 149–161 (2001)
60. Balog, R., Illenberger, E.: Complete chemical transformation of a molecular film by subexcitation electrons (<3 eV). Phys. Rev. Lett. **91**(21), 213201 (4 p) (2003)
61. Abdul–Carime, H., Gohlke, S., Illenberger, E.: Site-specific dissociation of DNA bases by slow electrons at early stages of irradiation. Phys. Rev. Lett. **92**(16), 168103 (4 p) (2004)
62. Prabhudesai, V.S., Kelkar, A.H., Nandi, D., Krishnakumar, E.: Functional group dependent site specific fragmentation of molecules by low energy electrons. Phys. Rev. Lett. **95**(14), 143202 (4 p) (2005)
63. Prabhudesai, V.S., Kelkar, A.H., Nandi, D., Krishnakumar, E.: Functional group dependent dissociative electron attachment to simple organic molecules. J. Chem. Phys. **128**(15), 154309 (7 p) (2008)
64. Ómarsson, F.H., Mason, N.J., Krishnakumar, E., Ingólfsson, O.: State selectivity and dynamics in dissociative electron attachment to CF_3I revealed through velocity slice imaging. Angew. Chem. Int. Ed. **53**(45), 12051–12054 (2014)
65. Gedanken, A.: The magnetic circular dichroism of the A band in CF_3I, C_2H_5I and t-BuI. Chem. Phys. Lett. **137**(5), 462–466 (1987)
66. Oster, T., Ingolfsson, O., Meinke, M., Jaffke, T., Illenberger, E.: Anion formation from gaseous and condensed CF_3I on low energy electron impact. J. Chem. Phys. **99**(7), 5141–5150 (1993) (and references therein)
67. Christophorou, L.G., Olthoff, J.K., Rao, M.V.V.S.: Electron interactions with CF_4. J. Phys. Chem. Ref. Data **25**(5), 1341–1388 (1996)
68. Christophorou, L.G., Olthoff, J.K.: Electron interactions with plasma processing gases: an update for CF_4, CHF_3, C_2F_6, and C_3F_8. J. Phys. Chem. Ref. Data **28**(4), 967–982 (1999)
69. Iga, I., Rao, M., Srivastava, S.K., Nogueira, J.C.: Formation of negative ions by electron impact on SiF_4 and CF_4. Z. Phys. D **24**(2), 111–115 (1992)
70. Mckinlay, R.G., Z̈ure, J.M., Paterson, M.J.: Vibronic coupling in inorganic systems. Photochemistry, conical intersections, and the Jahn-Teller and pseudo-Jahn-Teller effects. Adv. Inorg. Chem. **62**, 351–390 (2010)
71. Ram, N.B., Krishnakumar, E.: Dissociative electron attachment to methane probed using velocity slice imaging. Chem. Phys. Lett. **511**(1–3), 22–27 (2011)
72. Ómarsson, F.H., Szymanska, E., Mason, N.J., Krishnakumar, E., Ingólfsson, O.: Velocity slice imaging study on dissociative electron attachment to CF_4. Eur. Phys. J. D **68**(8), 101 (2014)
73. Chantry, P.J.: Dissociative attachment in carbon dioxide. J. Chem. Phys. **57**(8), 3180–3186 (1972)
74. Spence, D., Schulz, G.J.: Cross section for production of O_2^- and C^- by dissociative electron attachment in CO_2: an observation the Renner-Teller effect. J. Chem. Phys. **60**(1), 216–220 (1974)
75. Wu, B., Xia, L., Wang, Y.-F., Li, H.-K., Zeng, X.-J., Tian, S.X.: Renner-Teller effect on dissociative electron attachment to carbon dioxide. Phys. Rev. A **85**(5), 052709 (2012)

76. Moradmand, A., Slaughter, D.S., Haxton, D.J., Rescigno, T.N., McCurdy, C.W., Weber, Th, Matsika, S., Landers, A.L., Belkacem, A., Fogle, M.: Dissociative electron attachment to carbon dioxide via the $^2\Pi_g$ shape resonance. Phys. Rev. A **88**(3), 032703 (2013)

77. Wang, X.-D., Gao, X.-F., Xuan, C.-J., Tian, S.X.: Dissociative electron attachment to CO_2 produces molecular oxygen. Nat. Chem. **8**, 258–263 (2016)

78. Holland, H.D.: The oxygenation of the atmosphere and oceans. Phil. Trans. R. Soc. B **361**, 903–915 (2006)

79. Lu, Z., Chang, Y.C., Yin, Q.-Z., Ng, C.Y., Jackson, W.M.: Evidence for direct molecular oxygen production in CO_2 photodissociation. Science **346**(6205), 61–64 (2014)

80. Weingartshofer, A., Ehrhardt, H., Hermann, V., Linder, F.: Measurements of absolute cross sections for (e, H_2) collision processes: formation and decay of H_2^- resonances. Phys. Rev. A **2**(2), 204–304 (1970)

81. Davis, D., Kundu, S., Prabhudesai, V.S., Sajeev, Y. Krishnakumar, E.: Formation of CO_2 from formic acid through catalytic electron channel. J Chem. Phys. **149**(6), 064308 (8 p) (2018)

82. Li, Z., Milosavljević, A.R., Carmichael, I., Ptasinska, S.: Characterization of neutral radicals from a dissociative electron attachment process. Phys. Rev. Lett. **119**(5), 053402 (5 p) (2017)

83. Külz, M., Keil, M., Kortyna, A., Schellhaa, B., Hauck, J., Bergmann, K., Meyer, W., Weyh, D.: Dissociative attachment of low-energy electrons to state-selected diatomic molecules. Phys. Rev. A **53**(5), 3324 (1996)

84. Christophorou, L.G., Olthoff, J.K.: Electron interactions with excited atoms and molecules. Adv. At. Mol. Opt. Phy. **44**, 155–293 (2000)

85. Krishnakumar, E., Kumar, S.V.K., Rangwala, S.A., Mitra, S.K.: Cross sections for dissociative attachment of excited and ground electronic states of SO_2. Phys. Rev. A **56**(3), 1945–1953 (1997)

86. Krishnakumar, E., Kumar, S.V.K., Rangwala, S.A., Mitra, S.K.: Excited state dissociative attachment and couplings of electronic states of SO_2. J. Phys. B At. Mol. Opt. Phys. **29**(17), L657–L665 (1996)

87. Rangwala, S.A., Kumar S.V.K., Krishnakumar, E.: Dissociative electron attachment to electronically excited CS_2. Phys. Rev. A **64**(1), 012707 (5 p) (2001)

88. Kumar, S.V.K., Ashoka, V.S., Krishnakumar, E.: Dissociative attachment of electrons to vibronically excited SO_2. Phys. Rev. A **70**(5), 052715 (7 p) (2004)

89. Bald, I., Langer, J., Tegeder, P., Ingolfsson, O.: From isolated molecules through clusters and condensates to the building blocks of life. Int. J. Mass Spectrum. **277**(1–3), 4–25 (2008)

90. Rosa, A., Barszczewska, W., Nandi, D., Ashoka, V.S., Kumar, S.V.K., Krishnakumar, E., Bruning, F., Illenberger, E.: Unusual temperature dependence in dissociative electron attachment to 1,4-chlorobromobenzene. Chem. Phys. Lett. **342**(5–6), 536–544 (2001)

91. Haughey, S.A., Field, T.A., Langer, J., Shuman, N.S., Miller, T.M., Friedman, J.F., Viggiano, A.A.: Dissociative electron attachment to C_2F_5 radicals. J. Chem. Phys. **137**(5), 054310 (8 p) (2012)

92. Varambhia, H.N., Faure, A., Graupner, K., Field, T., Tennyson, J.: Experimental observation of dissociative electron attachment to S_2O and S_2O_2 with a new spectrometer for unstable molecules. Mon. Not. R. Astron. Soc. **403**(3), 1409–1412 (2010)

93. Field, T., Slattery, A.E., Adams, D.J., Morrison, D.D.: Experimental observation of dissociative electron attachment to S_2O and S_2O_2 with a new spectrometer for unstable molecules. J. Phys. B At. Mol. Opt. Phys. **38**(3), 255–264 (2005)

94. Vitanov, N.V., Halfmann, T., Shore, B.W., Bergmann, K.: Laser-induced population transfer by adiabatic passage techniques. Annu. Rev. Phys. Chem. **52**, 763–809 (2001)

Electron Collisions with CO Molecule: An R-Matrix Study Using a Large Basis Set

Amar Dora[1] and Jonathan Tennyson[2(✉)]

[1] Department of Chemistry, North Orissa University, Baripada 757003, Odisha, India
amardora@gmail.com
[2] Department of Physics and Astronomy, University College London, Gower St.,
London WC1E 6BT, UK
j.tennyson@ucl.ac.uk

Abstract. Fixed-nuclei **R**-matrix calculations are performed at the equilibrium geometry of carbon monoxide using the very large cc-pV6Z Gaussian basis set. Results from a close-coupling model involving 27 low-lying target states indicate the presence of three $^2\Sigma^+$ resonances at 10.1 eV (width 0.1 eV), 10.38 eV (0.0005 eV), and 11.15 eV (0.005 eV), a $^2\Delta$ resonance at 13.3 eV (0.1 eV) and two $^2\Pi$ resonances at 1.9 eV (1.3 eV) and 12.8 eV (0.1 eV). These new results are in very good agreement with many experimental studies but in contrast to a previous calculation using a smaller cc-pVTZ basis set where we found only one $^2\Sigma^+$ resonances at 12.9 eV. This is the first time that any theoretical study has reported these high lying $^2\Sigma^+$ resonances in agreement to experiment and reported detection of a $^2\Delta$ resonance. Total, elastic and electronic excitation cross sections of CO by electron impact are also presented.

1 Introduction

The study of various processes involving collision of electrons with molecules is of fundamental interest to many areas of science and technology. The cross sections of these processes get significantly enhanced in certain narrow collision energy range due to the formation of metastable anionic states, called resonances, which are formed due to temporary capture of the colliding electron by the molecule. Therefore, the resonances play an important role in processes such as dissociative electron attachment (DEA) and their detection and characterization is a major part of any electron–molecule collision study.

Resonances in low-energy electron collision with the CO molecule have been studied by several groups in the past. A classic review of resonances in CO, along with other diatomic molecules, was given by Schulz [1]. In addition, a compilation of available cross sections of different processes in electron–CO collisions is given by Itikawa [2]. Most of the works on CO have been directed toward the lowest lying and well-known $^2\Pi$ resonance at 1.6 eV and on the elastic and vibrational excitation processes associated with it. In the low and intermediate energy range,

© Springer Nature Singapore Pte Ltd. 2019
P. C. Deshmukh et al. (eds.), *Quantum Collisions and Confinement of Atomic and Molecular Species, and Photons*, Springer Proceedings in Physics 230,
https://doi.org/10.1007/978-981-13-9969-5_3

there have been quite a few experimental [3–6] and theoretical [7–11] studies on electron impact electronic excitation of CO. However, there is very little consensus among them on the presence and characteristics of the higher lying resonances.

Electron collision with CO can produce anionic fragments through the DEA process as both C and O can form stable anions. Formation of these atomic anions as result of electron collisions has indeed been observed for a long time and their production cross sections have been measured by several groups [12–17]. These cross sections show a broad peak in the 9–12 eV energy range. The presence of a $^2\Sigma^+$ resonance at 10.04 eV has been established by Sanche and Schulz [18] and is thought to contribute to this DEA peak. Furthermore, shape resonances at 10.4 and 10.7 eV and Feshbach resonances at 11.3 and 12.2 eV were reported from experiments [1,18,19]; these may also contribute to DEA.

In the DEA process, O^- ion production is favored and the cross section of C^- ion is very weak. Therefore, recently there have been attempts by experimentalists [17,20,21] to establish the nature of resonances through the measurement of O^- angular distributions. However, these groups arrived at different conclusions from the analysis of their results while using essentially the same experimental technique of velocity time-sliced imaging method. Newer and more precise measurements made by Gope et al. [22] have helped to settle the above controversy [23,24], but the problem calls for accurate theoretical calculations.

In our recent attempt [25] to study the higher lying resonances in CO, our best model, close-coupling method with 50 target states represented using a cc-pVTZ basis set, found only one $^2\Sigma^+$ resonance at 12.9 eV in addition to the low-lying $^2\Pi$ at 1.7 eV. In this study, we report on a similar fixed-nuclei calculation but with a much larger cc-pV6Z basis set. Here, we find clear signatures of a number of narrow resonances lying in the 10–13 eV range that are in great agreement with previous experimental findings.

2   Theory

In this section, we describe the theory briefly. Details of the **R**-matrix method and its implementation in the UK molecular R-matrix codes [26,27] can be found in the review article by Tennyson [28].

The **R**-matrix method, which is employed here, divides the space around the electron+target system into an inner region, inside which the exchange and correlation effects among all $N+1$ electrons is explicitly considered, and an outer region, where the scattering electron is considered to be interacting with the multipolar potential of the N-electron target. In the inner region, the scattering wave function, ψ_k^{N+1}, is represented using a close-coupling (CC) expansion:

$$\psi_k^{N+1} = \mathcal{A}\sum_{ij} a_{ijk}\Phi_i^N(\mathbf{x}_1...\mathbf{x}_N)u_{ij}(\mathbf{x}_{N+1}) + \sum_i b_{ik}\chi_i^{N+1}(\mathbf{x}_1...\mathbf{x}_{N+1}), \quad (1)$$

where $\mathbf{x}_i$ are the space-spin coordinates of the electrons. The Φ_i^N in (1) represents the wave function of the i-th target state, the u_{ij} are the continuum orbitals representing the scattering electron and $\mathcal{A}$ is the anti-symmetrization operator. The χ_i^{N+1}, in the second term, are called L^2 configurations. These configurations are constructed by occupation of the target molecular orbitals (MOs) by all $N+1$ electrons. The a_{ijk} and b_{ik} are variational parameters which are obtained from the diagonalization of the scattering Hamiltonian [29].

At the boundary of the inner region sphere, the R-matrix is calculated from the boundary amplitude of the inner region wave functions and the R-matrix poles. The R-matrix is then propagated outward and matched to analytical asymptotic scattering functions. From this matching, the K-matrix is calculated as a function of scattering energy. The K-matrix is a key quantity and other scattering observables can be obtained from this.

In the following section, we use two quantities, the eigenphase sum and the time delay, to detect and fit the resonances. The fitting of eigenphase sum to the Breit–Wigner form yields the resonance position and width, which is done automatically by the module RESON [30] in the UKRmol codes. On the other hand, the time delay is fitted to a Lorentzian function through the module TIMEDEL [31,32] to get the same resonance parameters.

3   Calculation and Results

As in the previous paper [25] (hereafter referred as paper-I), here we report on **R**-matrix calculation at the equilibrium geometry of CO, $R_{eq} = 2.1323$ a$_0$. As mentioned above, we use the UK molecular R-matrix codes (also called UKRmol codes) [26,27,33] for the scattering calculations. The necessary target molecular orbitals are obtained from MOLPRO [34] using the largest supported cc-pV6Z basis set which has no augmented diffused functions. The calculations were performed in the C$_{2v}$ point group since neither MOLPRO nor the Gaussian version of the UKRmol codes can use the full C$_{\infty v}$ symmetry of CO molecule. However, the target and resonant states in the C$_{2v}$ symmetry can be clearly correlated with their C$_{\infty v}$ counterparts. Therefore, these states are reported below in C$_{\infty v}$ symmetry.

3.1   Target Results

The target molecular orbitals necessary in **R**-matrix calculations are obtained from MOLPRO using the state-averaged complete active space (CAS) self-consistent field (SCF) method. We used the LQUANT option available in MOLPRO CASSCF program to specify the L_z quantum number and hence obtained the target states in $C_{\infty v}$ symmetry. A total of 27 low-lying target states in $C_{\infty v}$ symmetry were computed. These are 4 $^1\Sigma^+$, 2 $^1\Sigma^-$, 5 $^1\Pi$, 2 $^1\Delta$, 4 $^3\Sigma^+$, 3 $^3\Sigma^-$, 5 $^3\Pi$, and 2 $^3\Delta$. In terms of C$_{2v}$ assignments, this amounts to 41 target states once the degenerate states of Π and Δ symmetry states are accounted for. The CASSCF active space is the same as in paper-I, where 10 valence electrons are distributed freely over 10 valence orbitals. The active space configuration is defined as $(1a_1 - 2a_1)^4 (3a_1 - 6a_1, 1b_1 - 3b_1, 1b_2 - 3b_2)^{10}$.

Table 1. The CASSCF ground state energy (in E_h), the lowest 11 vertical excitation energies (in eV), and ground state dipole moments (μ in D) of CO calculated using cc-pV6Z basis set. The target results using cc-pVTZ basis set are also included for comparison. The experimental data are from Nielsen et al. [35]

State	cc-pVTZ	cc-pV6Z	Expt
$X\,^1\Sigma^+$	-112.8565508	-112.8599842	
$1\,^3\Pi$	6.31	6.43	6.32 $(a\,^3\Pi)$
$1\,^3\Sigma^+$	8.39	8.36	8.51 $(a'\,^3\Sigma^+)$
$1\,^1\Pi$	8.83	8.97	8.51 $(A\,^1\Pi)$
$1\,^3\Delta$	9.23	9.22	9.36 $(d\,^3\Delta)$
$1\,^3\Sigma^-$	9.60	9.60	9.88 $(e\,^3\Sigma^-)$
$1\,^1\Sigma^-$	9.97	9.95	9.88 $(I\,^1\Sigma^-)$
$1\,^1\Delta$	10.00	10.00	10.23 $(D\,^1\Delta)$
$2\,^3\Sigma^+$	12.90	10.39	10.40 $(b\,^3\Sigma^+)$
$2\,^1\Sigma^+$	13.76	11.16	10.78 $(B\,^1\Sigma^+)$
$2\,^3\Pi$	12.29	11.34	
$2\,^1\Pi$	13.72	11.84	
μ	0.291	0.238	0.122

Table 1 presents the target ground state energy (in E_h), vertical excitation energies to lowest 11 states (in eV), and the ground state dipole moment (in Debye) for the cc-pV6Z basis set. It also includes the target results from paper-I obtained with cc-pVTZ basis set along with the experimental values as reported by Nielsen et al. [35] for the purpose of comparison. As can be seen in the table, the first seven excitation energies in both cc-pVTZ and cc-pV6Z basis sets are in excellent agreement with each other and also to the experimental values. However, if we compare the $2\,^3\Sigma^+$ state, which is designated as $b\,^3\Sigma^+$ in spectroscopic notation, the CASSCF value of 10.39 eV with cc-pV6Z basis set is very close to the experimentally determined value of 10.4 eV, while that in case of cc-pVTZ basis set is 12.9 eV. Similarly, the energy for $2\,^1\Sigma^+$ state from the cc-pV6Z basis set is in much better agreement with experiment than that from the cc-pVTZ basis set. Notwithstanding the above argument, it may be pointed out that the vertical excitation energies of the first four target states obtained using the much lower cc-pVTZ basis set are slightly closer to the experimentally estimated values than those from the larger cc-pV6Z basis set. This might happen given that the excitation energies are obtained by finding the difference between the absolute energies of the ground state and the excited states. Of course, the absolute energies of all the states calculated with cc-pV6Z basis set are smaller than those calculated with the cc-pVTZ basis set, as to be expected from a variational method. Moreover, the theoretically calculated ground state dipole moment of CO is twice that of the experimental value despite use of the large cc-pV6Z basis set. We assign this difference to our use of the CASSCF method. In order to obtain accurate dipole moments close to the experiments, one would need to calculate highly correlated wave functions as obtained in multi-reference configuration interaction methods. However, we did not pursue this aspect as it was not the purpose of this work.

In the **R**-matrix calculations, the relative energy positions of different excited states with respect to the ground state are very important. Since resonances are often closely associated with a parent target state [36], getting correct target positions is of tremendous importance to get resonance positions correctly. In the paper-I, we found a very narrow Feshbach resonance lying extremely close to the $2\,^3\Sigma^+$ target state at 12.9 eV, and therefore we assigned the later to be its parent. In the present calculation, we get this target state very close to the experimental value of 10.4 eV, and therefore we expect this resonance to be near to 10.4 eV. It is known from experiments that the $b\,^3\Sigma^+$ $(2\,^3\Sigma^+)$ state of CO has Rydberg character. In the present calculation with cc-pV6Z basis set, which includes higher angular momentum basis functions up to i-functions $(l = 6)$, the Rydberg nature of the state is represented correctly.

3.2   Scattering Results

In the following, we present the results obtained from the close-coupling (CC) scattering calculation using the abovementioned 27 target states. Here, we do not explore on calculations with different scattering models such as the static exchange and static exchange with polarization models, as these are unsuitable for describing the higher lying resonances that are associated with excited target states. Such studies have been already performed previously in paper-I where extensive testing with respect to basis sets, molecular orbitals, **R**-matrix radius, etc. has been performed.

In the present CC calculation, an **R**-matrix sphere of radius $a = 12\ a_0$ is used and the continuum functions, appropriate for this radius [37], using partial waves up to $\ell \leq 4$ are included. The necessary occupied and virtual target molecular orbitals are obtained from MOLPRO by doing the state-averaged CASSCF calculation for the 27 target states which carry equal weights.

The positions and widths of the resonances that are found in this study are listed in Table 2. We have used both the eigenphase sum and the time delay to find the resonance parameters. Table 2 contains the fitted resonance parameters from both RESON and TIMEDEL programs. As can be seen in the table, both the methods give almost identical values in most of the cases. To show the resonances obtained in the calculation, we provide the eigenphase sum and the time-delay plots as a function of scattering energy.

Figure 1 shows the eigenphase sum and the time delay for the $^2B_1$ symmetry. $^2B_1$ and $^2B_2$ symmetry results are identical as these constitute the two degenerate parts of the $^2\Pi$ symmetry. The lowest lying $^2\Pi$ resonance is well known from many experiments and theoretical calculations. In our calculation, the position of this resonance is found by RESON to be at 1.87 eV with a width of 1.29 eV, while TIMEDEL missed this resonance. Most of the experiments report the position of this resonance to be at 1.6 eV. It is worth pointing here that in our previous work using the cc-pVTZ basis set we obtained this resonance at

Table 2. Positions (and widths) of the resonances as detected by the RESON module in fitting of eigenphase sums and by the TIMEDEL module in fitting of the time delays. All quantities are in eV

Symmetry	RESON	TIMEDEL
$^2A_1$	10.1019 (0.1126)	10.0924 (0.1277)
	10.3850 (0.00049)	–
	11.1579 (0.0048)	11.1576 (0.0047)
	13.2969 (0.0977)	13.2916 (0.1310)
$^2A_2$	13.3135 (0.1641)	13.2823 (0.1456)
$^2B_1$	1.8744 (1.2916)	–
	12.8306 (0.0889)	12.8302 (0.0891)

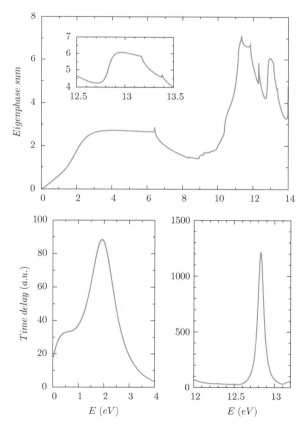

Fig. 1. $^2B_1$ symmetry: the upper panel shows the eigenphase sum and the lower shows the time delay as a function of scattering energy. The inset in the upper panel clearly shows the second $^2\Pi$ resonance at 12.83 eV

1.73 eV, which is lower than the present value and closer to the experimentally reported data. We can explain this on the basis that the **R**-matrix method relies on the difference between the target and scattering states energies. Although each part is variational and gives lower absolute energies for both the target and scattering states, the difference between them can lead to resonances moving to higher energy as the basis set is improved. This behavior has also been observed in other studies [38]. In addition to the lowest one, we find a higher energy $^2\Pi$ resonance lying at 12.83 eV with a narrow width of 0.1 eV. This can be seen in both the eigenphase sum and the time-delay plots.

The total and elastic cross sections from the $^2\Pi$ ($^2B_1+^2B_2$) symmetry is presented in Fig. 2 and the dominant electron impact electronic excitation cross sections in Fig. 3. These figures show the effect of the $^2\Pi$ resonances on different

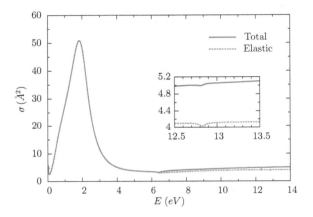

Fig. 2. Total and elastic cross sections for $^2B_1+^2B_2$ ($^2\Pi$) symmetry. The inset shows the effect of the second $^2\Pi$ resonance on the cross sections clearly

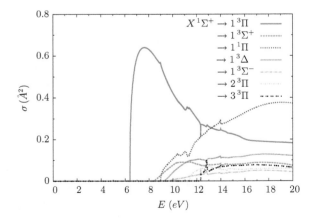

Fig. 3. The dominant electron impact excitation cross sections for $^2B_1+^2B_2$ ($^2\Pi$) symmetry

cross sections. The lowest $^2\Pi$ resonance causes a huge increase in the total and elastic cross section while the higher $^2\Pi$ resonance has very little effect on these cross sections.

The eigenphase sum and the time-delay plots for the $^2A_1$ and $^2A_2$ symmetries are shown together in Fig. 4. For the $^2A_1$ symmetry, we find four resonances at 10.1 eV (width 0.1 eV), 10.385 eV (width 0.00048 eV), 11.158 eV (width 0.0048 eV), and 13.29 eV (0.1 eV), while for the $^2A_2$ symmetry we find a single resonance at around 13.3 eV (width 0.1 eV). Both the $^2A_1$ and $^2A_2$ scattering calculations find a common resonance at around 13.3 eV with a width of approximately 0.1 eV. Therefore, we assign them to be the two components of a $^2\Delta$ resonance. The first three $^2A_1$ resonances at 10.1, 10.385, and 11.158 eV are assigned to be of $^2\Sigma^+$ symmetry in the natural $C_{\infty v}$ symmetry of CO molecule. The second and third $^2\Sigma^+$ resonances are very narrow and lie extremely close to the $2\,^3\Sigma^+$

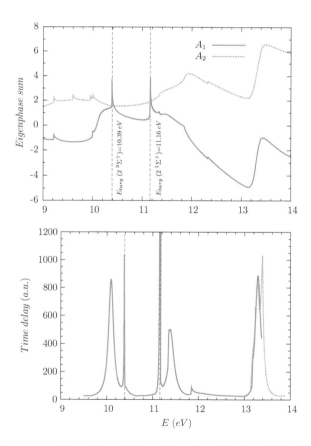

Fig. 4. Eigenphase sum (upper) and time-delay (lower) plots for both $^2A_1$ and $^2A_2$ symmetries. The vertical dashed lines indicate the excitation thresholds for the $2\,^3\Sigma^+$ and $2\,^1\Sigma^+$ target states

and $2\,^1\Sigma^+$ target states, respectively. We assign these target states to be their respective parents. With this the binding energy of the second $^2\Sigma^+$ resonance with respect to its parent $2\,^3\Sigma^+$ state would be 0.001 eV, while that of the third $^2\Sigma^+$ resonance with respect to its parent $2\,^1\Sigma^+$ state would be 0.006 eV. All these resonances are clearly seen in Fig. 4 where we have also marked the parent target states.

The effect of the $^2\Sigma^+$ and $^2\Delta$ resonances can be seen in the total and elastic cross sections shown for $^2A_1$ symmetry in Fig. 5. The dominant electron impact excitation cross sections for $^2A_1$ symmetry is shown in Fig. 6. Similarly, Fig. 7 shows the total and elastic cross sections, and Fig. 8 shows the dominant electron impact excitation cross sections of CO in the $^2A_2$ symmetry.

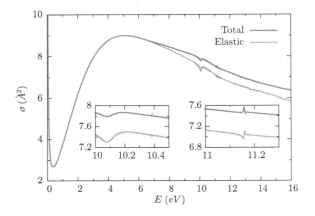

Fig. 5. Total and elastic cross section for $^2A_1$ symmetry. The insets show the effect of the resonances on the cross sections

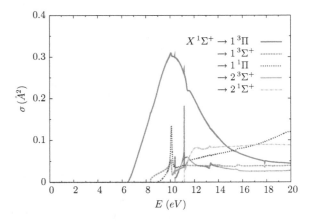

Fig. 6. The dominant electron impact excitation cross sections for $^2A_1$ symmetry

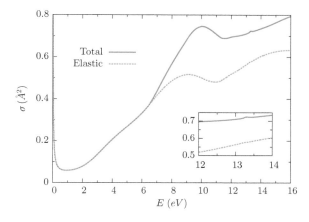

Fig. 7. Total and elastic cross sections of CO in the $^2A_2$ symmetry

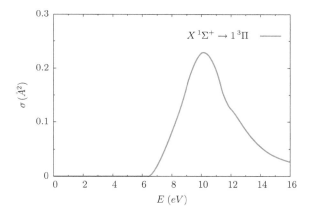

Fig. 8. The dominant electron impact excitation cross sections of CO in the $^2A_2$ symmetry

4   Conclusion

We have undertaken an R-matrix study using a very large basis set of cc-pV6Z to identify higher energy resonances in electron scattering with carbon monoxide molecule. Our fixed-nuclei calculation at the equilibrium geometry of CO indicates the presence of three $^2\Sigma^+$ resonances, one $^2\Delta$ resonance, and one $^2\Pi$ resonance in the energy range of 10–14 eV, in addition to the low-lying $^2\Pi$ resonance near 2 eV. The presence of a number of resonances above 10 eV has been reported by many experiments; however, this is the first theoretical study to identify them clearly. The large cc-pV6Z basis set employed here effectively represents the Rydberg like target states of CO, and therefore helps in getting the Rydberg resonances in CO correctly. All these high-energy resonances have very narrow widths, and therefore can have longer lifetimes to result in dissociation giving anionic fragments. It is our purpose to study the DEA process from these resonances and we are already computing resonance parameters as a function of internuclear distance.

The work clearly shows the need to use extended basis sets to accurately represent both highly excited target states and the associated resonances. This presents something of a challenge to R-matrix calculations due to the requirement that the target wave function must be entirely enclosed by the R-matrix sphere. However, we note the development of a new UK R-matrix code based on the use of B-splines for the continuum basis [39] which facilitates the uses of greatly extended inner regions.

Acknowledgements. AD gratefully acknowledges UCL for the use of its computational resources.

References

1. Schulz, G.J.: Rev. Mod. Phys. **45**, 423 (1973). https://doi.org/10.1103/RevModPhys.45.423
2. Itikawa, Y.: J. Phys. Chem. Ref. Data **44**, 013105 (2015). https://doi.org/10.1063/1.4913926
3. Skubenich, V.V.: Opt. Spectrosc. **23**, 540 (1967)
4. Newman, D.S., Zubek, M., King, G.C.: J. Phys. B At. Mol. Opt. Phys. **16**, 2247 (1983). https://doi.org/10.1088/0022-3700/16/12/019
5. Polley, J.P., Bailey, T.L.: Phys. Rev. A **37**, 733 (1988). https://doi.org/10.1103/PhysRevA.37.733
6. Middleton, A.G., Brunger, M.J., Teubner, P.J.O.: J. Phys. B At. Mol. Opt. Phys. **26**, 1743 (1993). https://doi.org/10.1088/0953-4075/26/11/009
7. Weatherford, C.A., Huo, W.M.: Phys. Rev. A **41**, 186 (1990). https://doi.org/10.1103/PhysRevA.41.186
8. Morgan, L.A., Tennyson, J.: J. Phys. B At. Mol. Opt. Phys. **26**, 2429 (1993)
9. Lee, M.T., Iga, I., Brescansin, L.M., Machado, L.E., Machado, F.B.C.: J. Molec. Struct. (THEOCHEM) **585**, 181 (2002). https://doi.org/10.1016/S0166-1280(02)00045-3
10. Gracas, M., Martins, R., Maniero, A.M., Machado, L.E., Vianna, J.D.M.: Brazilian. J. Phys. **35**, 945 (2005)
11. Riahi, R., Teulet, P., Jaidane, N., Gleizes, A.: Eur. Phys. J. D **56**, 67 (2010). https://doi.org/10.1140/epjd/e2009-00267-5
12. Rapp, D., Briglia, D.D.: J. Chem. Phys. **43**, 1480 (1965). https://doi.org/10.1063/1.1696958
13. Chantry, P.J.: Phys. Rev. **172**, 125 (1968)
14. Stamatovic, A., Schulz, G.J.: J. Chem. Phys. **53**, 2663 (1970). https://doi.org/10.1063/1.1674387
15. Cadex, I., Tronc, M., Hall, R.I.: J. Phys. B At. Mol. Opt. Phys. **8**, L73 (1975). https://doi.org/10.1088/0022-3700/8/5/003
16. Hall, R.I., Cadex, I., Schermann, C., Tronc, M.: Phys. Rev. A **15**, 599 (1977). https://doi.org/10.1103/PhysRevA.15.599

17. Nag, P., Nandi, D.: Phys. Chem. Chem. Phys. **17**, 7130 (2015). https://doi.org/10.1039/c4cp05678g
18. Sanche, L., Schulz, G.J.: Phys. Rev. Lett. **26**, 943 (1971). https://doi.org/10.1103/PhysRevLett.26.943
19. Mazeau, J., Gresteau, F., Joyez, G., Reinhardt, J., Hall, R.I.: J. Phys. B At. Mol. Phys. **5**, 1890 (1972). https://doi.org/10.1088/0022-3700/5/10/018
20. Tian, S.X., Wu, B., Xia, L., Wang, Y.F., Li, H.K., Zeng, X.J., Luo, Y., Yang, J.: Phys. Rev. A **88**, 012708 (2013). https://doi.org/10.1103/PhysRevA.88.012708
21. Wang, X.D., Xuan, C.J., Luo, Y., Tian, S.X.: J. Chem. Phys. **143**, 066101 (2015). https://doi.org/10.1063/1.4928639
22. Gope, K., Tadsare, V., Prabhudesai, V.S., Mason, N.J., Krishnakumar, E.: Eur. Phys. J. D **70**, 134 (2016). https://doi.org/10.1140/epjd/e2016-70180-y
23. Nag, P., Nandi, D.: Phys. Rev. A **91**, 056701 (2015). https://doi.org/10.1103/PhysRevA.91.056701
24. Tian, S.X., Luo, Y.: Phys. Rev. A **91**, 056702 (2015). https://doi.org/10.1103/PhysRevA.91.056702
25. Dora, A., Tennyson, J., Chakrabarti, K.: Eur. Phys. J. D **70**, 197 (2016)
26. Morgan, L.A., Tennyson, J., Gillan, C.J.: Comput. Phys. Commun. **114**, 120 (1998)
27. Carr, J.M., Galiatsatos, P.G., Gorfinkiel, J.D., Harvey, A.G., Lysaght, M.A., Madden, D., Mašín, Z., Plummer, M., Tennyson, J.: Eur. Phys. J. D **66**, 58 (2012)
28. Tennyson, J.: Phys. Rep. **491**, 29 (2010)
29. Tennyson, J.: J. Phys. B At. Mol. Opt. Phys. **29**, 1817 (1996)
30. Tennyson, J., Noble, C.J.: Comput. Phys. Commun. **33**, 421 (1984)
31. Stibbe, D.T., Tennyson, J.: Comput. Phys. Commun. **114**, 236 (1998)
32. Little, D.A., Tennyson, J., Plummer, M., Sunderland, A.: Comput. Phys. Commun. **215**, 137 (2017). https://doi.org/10.1016/j.cpc.2017.01.005
33. Tennyson, J., Morgan, L.A.: Phil. Trans. A **357**, 1161 (1999)
34. Werner, H.J., Knowles, P.J., Manby, F.R., Schütz, M., et al.: Molpro, version 2006.1, a package of ab initio programs (2006). See http://www.molpro.net
35. Nielsen, E.S., Jørgensen, P., Oddershede, J.: J. Chem. Phys. **73**, 6238 (1980). https://doi.org/10.1063/1.440119
36. Stibbe, D.T., Tennyson, J.: J. Phys. B At. Mol. Opt. Phys. **30**, L301 (1997)
37. Faure, A., Gorfinkiel, J.D., Morgan, L.A., Tennyson, J.: Comput. Phys. Commun. **144**, 224 (2002)
38. Dora, A., Bryjko, L., van Mourik, T., Tennyson, J.: J. Chem. Phys. **130**, 164307 (2009)
39. Darby-Lewis, D., Masin, Z., Tennyson, J.: J. Phys. B At. Mol. Opt. Phys. **50**, 175201 (2017). https://doi.org/10.1088/1361-6455/aa8161

Dynamic Polarizabilities and Magic Wavelengths of Sr$^+$ for Focused Vortex Light

Anal Bhowmik$^{(\boxtimes)}$ and Sonjoy Majumder

Department of Physics, Indian Institute of Technology Kharagpur,
Kharagpur 721302, India
analbhowmik@phy.iitkgp.ernet.in, sonjoym@phy.iitkgp.ernet.in

Abstract. A theory of dynamic polarizability for trapping relevant states of Sr$^+$ is presented here when the ions interact with a focused optical vortex. The coupling between the orbital and spin angular momentum of the optical vortex varies with focusing angle of the beam and is studied in the calculation of the magic wavelengths for $5s_{1/2} \rightarrow 4d_{3/2,5/2}$ transitions of Sr$^+$. The initial state of our interest here is $5s_{1/2}$ with $m_J = -1/2$ of which is different possible trapping state compare to our recent work on Sr$^+$ [Phys. Rev. A **97**, 022511 (2018)]. We find a variation in magic wavelengths and the corresponding polarizabilities with different combinations of orbital and spin angular momentum of the vortex beam. The variation is very significant when the wavelengths of the beam are in the infrared region of the electromagnetic spectrum. The calculated magic wavelengths will help the experimentalists to trap the ion for performing the high-precision spectroscopic measurements.

1 Introduction

Optical trapping of atoms or ions has been extensively used in high- precision spectroscopic measurements [1,2]. But the mechanism of trapping using a laser light inevitably produces a shift in the energy levels of the atoms involved in absorption. The shift is called the stark shift. In general, the shift is different for these energy states of the atom. Thus, naturally it will influence the fidelity of the precision measurement experiments due to nonachievement of exact resonance. However, this drawback can be diminished if the atoms are trapped at magic wavelengths of the laser beam, for which the differential AC stark shift of an atomic transition effectively vanishes. Therefore, the magic wavelengths have significant applications in atomic clocks [3–5], atomic magnetometers [6], and atomic interferometers [7].

All the previous studies of magic wavelengths for trapping of different atoms or ions are obtained for the Gaussian modes of a laser [8–11]. In this work, we

© Springer Nature Singapore Pte Ltd. 2019
P. C. Deshmukh et al. (eds.), *Quantum Collisions and Confinement of Atomic and Molecular Species, and Photons*, Springer Proceedings in Physics 230,
https://doi.org/10.1007/978-981-13-9969-5_4

determine the magic wavelengths of the transitions $5s_{1/2,-1/2} \rightarrow 4d_{3/2,m_J}$ and $5s_{1/2,-1/2} \rightarrow 4d_{5/2,m_J}$ of Sr$^+$ ion, assuming the external light field is a circularly polarized focused optical vortex such as Laguerre-Gaussian (LG) beam [12]. Since the stark shifts will be different for the states $5s_{1/2,-1/2}$ and $5s_{1/2,+1/2}$, different laser frequencies (magic frequency) should be applied to minimize the systematic errors in the experiments involved the state $5s_{1/2,-1/2}$ compare to $5s_{1/2,+1/2}$ state. Therefore, it is important to quantify the magic wavelengths of the transitions $5s_{1/2,-1/2} \rightarrow 4d_{3/2,m_J}$ and $5s_{1/2,-1/2} \rightarrow 4d_{5/2,m_J}$ of Sr$^+$, as we have already reported the magic wavelengths related to $5s_{1/2,+1/2}$ state [13]. However, the special property of optical vortex is that, apart from the polarization (i.e., spin angular momentum (SAM)), the optical vortex carries orbital angular momentum (OAM) due to its helical phase front [14]. Now, it is well known that during the interaction of a paraxial LG beam with atoms or ions (which are below its recoil limit), the quadrupole transition is the lowest order transition, where the OAM of the LG beam affects the electronic motion [14,15]. Therefore, the OAM of a paraxial LG beam does not influence the dipole polarizability of an atomic state. Hence, in case of paraxial LG beam, the dipole polarizability and the magic wavelengths solely depend on the SAM of the beam. But unlike the paraxial LG beam, the OAM and SAM of the optical vortex get coupled when the beam is focused [12]. This leads to the transfer of OAM to the electronic motion of the atoms in the dipole transition level and creates an impact on the polarizability of an atomic state [12]. Further, the coupling of angular momenta increases with the focusing angle. However, in this work, we quantify all these effects of OAM and SAM on the polarizability of an atomic state regarding magic wavelengths.

2 Theory

If an atom or ion is placed in an external oscillating electric field $E(\omega)$, then the second-order shift in a particular energy level of the atom or ion is proportional to the square of the electric field, $E^2(\omega)$. The proportional coefficient is called the dynamic polarizability $\alpha(\omega)$ of the atomic or ionic energy state at frequency ω of the external electric field and it can be written as [16]

$$\alpha(\omega) = \alpha_c(\omega) + \alpha_{vc}(\omega) + \alpha_v(\omega), \tag{1}$$

where $\alpha_c(\omega)$ and $\alpha_v(\omega)$ are dynamic core polarizability of the ionic core and dynamic valence polarizability of the single valence system, respectively. This ionic core is obtained by removing the valence electron from the system. $\alpha_{vc}(\omega)$ is the correction [17] in core polarizability in the presence of the valence electron. As the core electrons are tightly bound to the nucleus, the presence of a valence electron is expected not to change the core polarizability significantly. Thus, we consider α_{vc} in the present method of calculations without variation of ω. $\alpha_v(\omega)$ is calculated using the external electric field of focused LG beam [12]. In case of focused LG beam, OAM, and SAM are no longer separately a good quantum number as they get coupled to each other. Therefore, the effect of total angular momentum (OAM+SAM) can be seen on $\alpha_v(\omega)$, which can be expressed as [13]

$$\alpha_v(\omega) = 2A_0\alpha_v^0(\omega) + 2 \times \left(\frac{m_J}{2J_v}\right) A_1\alpha_v^1(\omega) + 2 \times \left(\frac{3m_J^2 - J_v(J_v+1)}{2J_v(2J_v-1)}\right) A_2\alpha_v^2(\omega), \quad (2)$$

where J_v is the total angular momentum of the state ψ_v and m_J is its magnetic component. The coefficients A_is are $A_0 = \left[\{I_0^{(l)}\}^2 + \{I_{\pm2}^{(l)}\}^2 + 2\{I_{\pm1}^{(l)}\}^2\right]$, $A_1 = \left[\pm\{I_0^{(l)}\}^2 \mp \{I_{\pm2}^{(l)}\}^2\right]$ and $A_2 = \left[\{I_0^{(l)}\}^2 + \{I_{\pm2}^{(l)}\}^2 - 2\{I_{\pm1}^{(l)}\}^2\right]$. The parameter $I_m^{(l)}$, where m takes the values 0, ±1, and ±2, depends on focusing angle (θ_{max}) by Bhowmik et al. [12] and Zhao et al. [18]

$$I_m^{(l)}(r'_\perp, z') = \int_0^{\theta_{max}} d\theta \left(\frac{\sqrt{2}r'_\perp}{w_0 \sin\theta}\right)^{|l|} (\sin\theta)^{|l|+1}\sqrt{\cos\theta}\, g_{|m|}(\theta) J_{l+m}(kr'_\perp \sin\theta)e^{ikz'\cos\theta}.$$

$$(3)$$

Here $r'_\perp$ is the projection of $\mathbf{r}'$ on the xy plane, w_0 is the waist of the paraxial circularly polarized LG beam which is focused by a high numerical aperture. The angular functions are $g_0(\theta) = 1 + \cos\theta$, $g_1(\theta) = \sin\theta$ and $g_2(\theta) = 1 - \cos\theta$. $\alpha_v^0(\omega)$, $\alpha_v^1(\omega)$ and $\alpha_v^2(\omega)$ introduced in (2) are the scalar, vector and tensor parts, respectively, of the valence polarizability and are expressed as [16,19]

$$\alpha_v^0(\omega) = \frac{2}{3(2J_v+1)} \sum_n \frac{|\langle\psi_v||d||\psi_n\rangle|^2 \times (\epsilon_n - \epsilon_v)}{(\epsilon_n - \epsilon_v)^2 - \omega^2}, \quad (4)$$

$$\alpha_v^1(\omega) = -\sqrt{\frac{6J_v}{(J_v+1)(2J_v+1)}} \sum_n (-1)^{J_n+J_v} \left\{\begin{matrix} J_v & 1 & J_v \\ 1 & J_n & 1 \end{matrix}\right\} \frac{|\langle\psi_v||d||\psi_n\rangle|^2 \times 2\omega}{(\epsilon_n - \epsilon_v)^2 - \omega^2}, \quad (5)$$

and

$$\alpha_v^2(\omega) = 4\sqrt{\frac{5J_v(2J_v-1)}{6(J_v+1)(2J_v+1)(2J_v+3)}} \sum_n (-1)^{J_n+J_v}$$

$$\left\{\begin{matrix} J_v & 1 & J_n \\ 1 & J_v & 2 \end{matrix}\right\} \frac{|\langle\psi_v||d||\psi_n\rangle|^2 \times (\epsilon_n - \epsilon_v)}{(\epsilon_n - \epsilon_v)^2 - \omega^2}. \quad (6)$$

Henceforth, whenever we mention about SAM or OAM in the following text, it is considered to be the angular momentum of the paraxial LG beam before passing through the focusing lens.

3   Numerical Results and Discussions

The aim of this work is to calculate the dynamic polarizabilities of the $5s_{1/2}$, $4d_{3/2}$, and $4d_{5/2}$ states for different magnetic sublevels of Sr^+. The scalar, vector, and tensor parts of the valence polarizabilities are calculated using (4), (5), and (6). The precise estimations of these three parts of the valence polarizability depend on the accuracy of the unperturbed energy levels and the dipole matrices among them. In order to evaluate these properties, we use correlation exhaustive relativistic coupled cluster (RCC) theory [20–24] with wave operators associated with single and double and partial triple excitations in linear and nonlinear forms. The wavefunctions calculated by the RCC method can produce highly precise $E1$ transition amplitudes as discussed in our recent work [13]. Calculation in this reference yields that the static core polarizability $(\alpha_c(0))$ of the ion is 6.103 a.u., and the static core-valence parts of the polarizabilities $(\alpha_{vc}(0))$ for the states $5s_{\frac{1}{2}}$, $4d_{\frac{3}{2}}$ and $4d_{\frac{5}{2}}$ are -0.25 a.u., -0.38 a.u., and -0.42 a.u., respectively.

In order to determine the precise values of dynamic valence polarizabilities, we require calculating a large number of dipole matrix elements. Another way to say, the running index n in (4)–(6) is turning out to be around 25 for Sr^+ to obtain accurate valence polarizability. Since the RCC method is computationally very expensive, we break our total calculations of valence polarizability into three parts depending on their significance in the sums of (4)–(6). The first part includes the most important contributing terms to the valence polarizabilities, which involve the $E1$ matrix elements associated with the intermediate states from 5^2P to 8^2P and 4^2F to 6^2F. Therefore, these matrix elements are calculated using the correlation exhaustive RCC method. The second part consists of the comparatively less significant terms associated with $E1$ matrix elements in the polarizability expressions arising from intermediate states from 9^2P to 12^2P and 7^2F to 12^2F. Thus we calculate the second part using second-order relativistic many-body perturbation theory [25]. The last part, whose contributions are comparatively further small to the valence polarizability, includes the intermediate states from $n = 13$ to 25, are computed using the Dirac Fock wavefunctions.

In Figs. 1 and 2, we present the variations of total polarizabilities of $5s_{1/2,-1/2}$, $4d_{3/2,m_J}$, and $4d_{5/2,m_J}$ (for different magnetic quantum numbers, m_J, of the states) states with the frequency of the external field of the focused LG beam. The focusing angle of the LG beam is considered $50°$ in both the figures. The combinations of angular momenta of the paraxial LG beam have been chosen as $(OAM, SAM) = (+1, +1)$ and $(+1, -1)$ in Fig. 1 and Fig. 2, respectively. The resonances occur in the plots due to the $5s_{1/2} \rightarrow 5p_{1/2,3/2}$ transitions for $5s_{1/2}$ state, $4d_{3/2} \rightarrow 5p_{1/2,3/2}$ transitions for $4d_{3/2}$ state and $4d_{5/2} \rightarrow 5p_{3/2}$ transitions for $4d_{5/2}$ state. The plots show a number of intersections between the polarizabilities of $5s_{\frac{1}{2}}$ at $m_J = -1/2$ and different multiplets of $4d_{\frac{3}{2},\frac{5}{2}}$ states. These intersections indicate magic wavelengths, at which the difference in the stark shifts of the two related states vanishes. Figures show that magic wavelengths which fall in the infrared region of the electromagnetic spectrum have large polarizabilities compared to the magic wavelengths of the visible or ultraviolet region. These magic wavelengths with high polarizabilities will be more effective to trap the ion, and thus they are highly recommended for trapping. These two figures are given as an example. Similar plots are studied for different focusing angles, say $60°$ and $70°$, and corresponding magic wavelengths are discussed later in this paper.

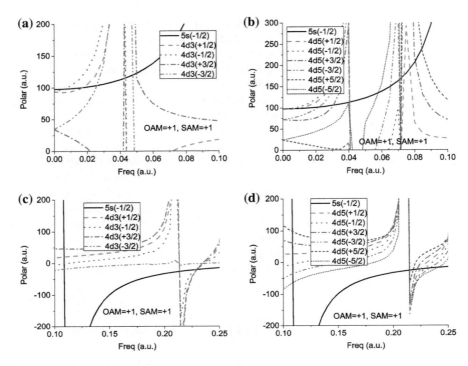

Fig. 1. Variation of polarizabilities (Polar) of $5s_{\frac{1}{2}}$ and $4d_{\frac{3}{2},\frac{5}{2}}$ states with frequency (Freq) are plotted when the focusing angle of LG beam is $50°$ with OAM = +1 and SAM = +1. The brackets indicate the magnitudes of different magnetic components. Figure **a** and **c** are for the $5s_{\frac{1}{2}}$ and $4d_{\frac{3}{2}}$ states, and Figure **b** and **d** are for the $5s_{\frac{1}{2}}$ and $4d_{\frac{5}{2}}$ states

In Tables 1, 2 and 3, we have listed a large number of magic wavelengths along with their corresponding polarizabilities, when the focusing angles of LG beam are $50°$, $60°$ and $70°$. The Table 1 is for the transition $5s_{1/2} \rightarrow 4d_{3/2}$, and the combinations of OAM and SAM are $(+1, +1)$, $(+1, -1)$, $(+2, +1)$ and $(+2, -1)$. In Tables 2 and 3, the transition is $5s_{1/2} \rightarrow 4d_{5/2}$ but the combinations of OAM and SAM are $((+1, +1), (+1, -1))$ and $((+2, +1), (+2, -1))$, respectively. The m_J value of $5s_{1/2}$ is considered $-1/2$ throughout this paper and the tables show totally distinct set of magic wavelengths compared to the results published [13] considering $m_J = 1/2$. There are five sets of magic wavelengths obtained for each of the multiplets of $4d_{3/2}$ for all combinations of angular momenta and focusing angles of the LG beam in the given frequency range. Whereas, in the same range of wavelength spectrum for $4d_{5/2}$ state, our calculations show seven sets of magic wavelengths (see Tables 2 and 3) for most of the multiplets. Since the resonance transition of $5s_{\frac{1}{2}} \rightarrow 4d_{\frac{3}{2}}, 4d_{\frac{5}{2}}$ are 687 nm and 674 nm, respectively, thus the ion is attracted and trapped to the high- intensity (low intensity) region of the

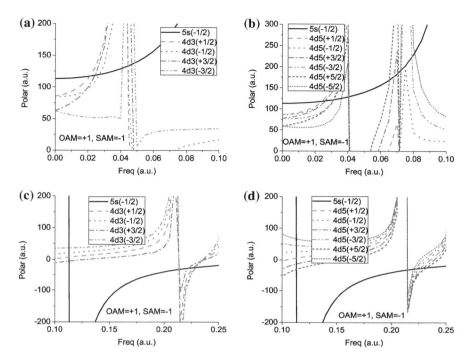

Fig. 2. Variation of polarizabilities (Polar) of $5s_{\frac{1}{2}}$ and $4d_{\frac{3}{2},\frac{5}{2}}$ states with frequency (Freq) are plotted when the focusing angle of LG beam is $50°$ with OAM $= +1$ and SAM $= -1$. The brackets indicate the magnitudes of different magnetic components. Figure **a** and **c** are for the $5s_{\frac{1}{2}}$ and $4d_{\frac{3}{2}}$ states, and Figure **b** and **d** are for the $5s_{\frac{1}{2}}$ and $4d_{\frac{5}{2}}$ states

LG beam when the magic wavelength is larger (smaller) than the resonance wavelength.

Since this work is about the finding of suitable magic wavelengths for trapping, we only give an estimation of the theoretical uncertainty in the calculated magic wavelengths. Here we collect the most important set of $E1$ matrix elements, which include $5s_{\frac{1}{2}} \rightarrow 5p_{\frac{1}{2},\frac{3}{2}}$ transitions for $5s_{\frac{1}{2}}$ state; $4d_{\frac{3}{2}} \rightarrow 5p_{\frac{1}{2},\frac{3}{2}}$ and $4d_{\frac{3}{2}} \rightarrow 4f_{\frac{5}{2}}$ transitions for $4d_{\frac{3}{2}}$ state; $4d_{\frac{5}{2}} \rightarrow 5p_{\frac{3}{2}}$ and $4d_{\frac{5}{2}} \rightarrow 4f_{\frac{5}{2},\frac{7}{2}}$ transitions for $4d_{\frac{5}{2}}$ state. We compare our RCC results with the SDpT values calculated by Safronova [26] and further apply those $E1$ matrix elements in place of our present RCC values to recalculate the magic wavelengths. This approach leads to the theoretical uncertainty in our calculated magic wavelength values is about $\pm 1\%$.

Table 1. Magic wavelengths (in nm) of Sr$^+$ for focusing angles 50°, 60°, and 70° of the LG beam for the transitions $5s_{1/2}(-1/2) \rightarrow 4d_{3/2}(m_J)$

Non-paraxial LG beam

State ($4d_{3/2}(m_J)$)	$\lambda_{magic}^{50°}$	α	$\lambda_{magic}^{60°}$	α	$\lambda_{magic}^{70°}$	α	State ($4d_{3/2}(m_J)$)	$\lambda_{magic}^{50°}$	α	$\lambda_{magic}^{60°}$	α	$\lambda_{magic}^{70°}$	α
	OAM $= +1$, SAM $= +1$							OAM $= +1$, SAM $= -1$					
(+1/2)	2680.20	100.36	2201.13	104.81	1963.94	109.41	(+1/2)	2169.68	117.01	2080.52	118.77	2016.08	120.32
	1069.56	114.82	1054.71	119.84	1047.43	124.19		1037.89	132.76	1023.90	134.76	1021.60	136.67
	422.27	20.89	421.10	20.50	417.25	20.13		404.65	7.19	404.65	7.69	404.65	8.47
	213.61	−26.26	214.31	−27.87	213.51	−28.82		213.21	−32.25	213.11	−32.87	214.01	−33.72
	198.27	−20.34	198.45	−21.32	198.71	−22.37		200.72	−26.41	200.72	−26.69	200.81	−27.23
(−1/2)	8136.31	98.17	4339.37	102.62	3120.78	106.42	(−1/2)	1693.80	119.80	1668.99	121.56	1650.85	122.76
	1069.56	114.98	1042.64	119.56	1040.26	124.45		1021.60	133.40	1019.31	135.28	1014.77	137.07
	421.88	−0.84	420.71	0.33	416.87	2.29		404.65	20.87	404.65	20.76	404.65	20.22
	213.11	−26.13	214.11	−27.80	213.21	−28.51		213.21	−32.25	213.11	−32.87	214.31	−33.80
	200.19	−20.96	200.28	−22.04	200.37	−23.13		199.31	−25.72	199.40	−26.33	199.49	−26.61
(+3/2)	1074.61	114.75	1079.70	117.94	1082.26	122.18	(+3/2)	1875.04	118.41	1875.04	119.83	1875.04	121.14
	911.27	122.62	916.77	126.69	920.47	130.40		971.50	136.00	977.75	137.55	977.75	138.96
	422.67	46.61	421.49	44.61	417.25	42.08		404.29	−5.33	404.29	−4.11	404.29	−2.46
	213.91	−26.39	214.31	−27.87	213.61	−28.82		212.91	−31.91	212.91	−32.53	213.31	−33.37
	198.71	−20.47	198.71	−21.40	198.71	−22.51		204.78	−28.38	204.23	−28.50	203.77	−28.53
(−3/2)	1759.20	103.64	1786.80	106.72	1800.92	110.43	(−3/2)	1114.02	129.69	1130.60	130.60	1199.04	130.05
	957.21	120.05	957.21	124.22	963.28	127.90		935.59	138.21	939.45	139.68	945.30	141.12
	421.49	−17.31	420.71	−15.76	416.87	−12.98		405.01	35.93	405.01	34.62	405.01	33.22
					212.52	−28.35		213.21	−32.25	213.11	−32.87	214.31	−33.80
					208.53	−26.52		198.97	−25.49	199.05	−25.99	199.14	−26.42

(continued)

Table 1. (continued)

Non-paraxial LG beam

State ($4d_{3/2}(m_J)$)	$\lambda_{magic}^{50°}$	α	$\lambda_{magic}^{60°}$	α	$\lambda_{magic}^{70°}$	α	State ($4d_{3/2}(m_J)$)	$\lambda_{magic}^{50°}$	α	$\lambda_{magic}^{60°}$	α	$\lambda_{magic}^{70°}$	α
	OAM = +1, SAM = +1							OAM = +1, SAM = −1					
	OAM = +2, SAM = +1							OAM = +2, SAM = −1					
(+1/2)	2462.88	101.83	2052.40	107.37	1822.53	113.66	(+1/2)	2149.21	117.80	2052.40	119.38	1955.51	121.23
	1057.15	116.84	1052.27	122.12	1037.89	127.22		1021.60	134.04	1023.90	136.07	1019.31	137.47
	419.55	20.60	418.01	20.67	415.72	21.18		404.65	7.20	404.65	7.77	404.65	8.96
	213.21	−26.96	213.21	−27.86	213.11	−29.72		213.81	−32.84	213.31	−33.02	213.81	−33.82
	198.36	−20.70	1986.20	−21.98	198.88	−23.52		200.81	−26.65	200.72	−26.97	200.81	−27.44
(−1/2)	6417.37	100.05	3704.34	104.54	2696.06	110.28	(−1/2)	1687.53	119.86	1662.90	121.92	1633.10	123.97
	1047.43	117.88	1040.26	122.15	1037.89	127.22		1021.60	134.04	1019.31	136.45	1008.04	138.16
	419.17	0.34	417.63	1.67	415.35	3.45		405.01	20.78	405.01	20.42	405.01	20.08
	213.21	−26.96	213.21	−27.86	213.11	−29.72		214.21	−33.10	213.71	−33.31	214.11	−34.02
	200.19	−21.41	200.37	−22.68	200.54	−24.18		199.31	−25.96	199.49	−26.46	199.58	−26.88
(+3/2)	1077.15	116.11	1082.26	119.92	1090.03	125.50	(+3/2)	1875.04	118.48	1875.04	120.23	1859.73	121.61
	913.09	123.69	918.62	129.07	924.21	133.98		975.66	136.02	977.75	138.62	979.86	139.45
	419.94	45.49	418.40	43.48	415.72	39.44		404.29	−4.96	404.65	−3.42	404.65	−1.53
	213.21	−26.96	213.21	−27.86	213.11	−29.72		213.01	−32.49	213.01	−32.97	213.21	−33.53
	198.71	−20.82	198.71	−22.04	198.79	−23.48		204.60	−28.39	203.95	−28.51	203.50	−28.66
(−3/2)	1772.89	105.38	1793.83	109.29	1822.53	113.66	(−3/2)	1119.49	130.00	1139.08	131.07	1238.13	130.00
	955.21	122.16	961.25	125.59	967.37	131.68		937.52	138.77	941.39	140.31	947.26	141.51
	419.17	−16.99	417.63	−14.64	415.35	−11.16		405.01	35.64	405.01	33.98	405.01	32.12
			212.22	−27.86	212.71	−29.20		214.31	−33.18	213.81	−33.42	214.21	−34.17
			209.78	−26.52	207.48	−26.86		199.05	−25.81	199.05	−26.17	199.14	−26.63

Table 2. Magic wavelengths (in nm) of Sr$^+$ for different focusing angles 50°, 60° and 70° of the LG beam for the transitions $5s_{1/2}(-1/2) \rightarrow 4d_{5/2}(m_J)$

Non-paraxial LG beam

State ($4d_{5/2}(m_J)$)	$\lambda^{50°}_{magic}$	α	$\lambda^{60°}_{magic}$	α	$\lambda^{70°}_{magic}$	α	State ($4d_{5/2}(m_J)$)	$\lambda^{50°}_{magic}$	α	$\lambda^{60°}_{magic}$	α	$\lambda^{70°}_{magic}$	α
	OAM = +1, SAM = +1							OAM = +1, SAM = −1					
(+1/2)	5841.46	98.42	2920.73	102.94	2255.61	108.35	(+1/2)	1875.04	118.28	1808.07	120.28	1766.02	21.76
	1077.15	114.82	1116.75	116.27	1105.91	121.27		1079.70	131.17	1082.26	132.32	1072.08	134.34
	617.39	175.83	616.55	182.48	614.06	189.51							
	589.44	192.13	592.50	194.22	596.38	199.86							
	419.55	27.71	418.78	27.40	417.25	26.66		404.65	6.03	404.65	6.82	404.29	7.99
	212.32	−25.66	212.12	−27.38	212.32	−28.26		212.32	−31.54	212.32	−32.58	212.32	−32.52
	202.59	−21.92	202.59	−22.97	202.68	−24.18		200.90	−26.58	200.99	−26.96	201.16	−27.34
(−1/2)	14697.86	97.94	4032.16	102.43	2744.78	107.01	(−1/2)	1766.02	119.13	1719.37	120.76	1687.53	122.43
	1077.15	114.82	1116.75	116.27	1105.91	121.27		1079.70	130.99	1082.26	132.32	1072.08	134.34
	419.55	−4.56	418.40	−2.99	416.87	−0.42		404.65	25.76	404.65	25.14	404.29	25.28
	212.32	−25.66	212.12	−27.38	212.12	−28.26		212.32	−31.54	212.32	32.58	212.32	−32.52
	199.05	−20.64	199.49	−21.48	199.75	−22.95		202.86	−27.35	202.86	−27.76	202.86	−28.18
(+3/2)	1489.00	106.01	1484.15	109.22	1479.33	113.58	(+3/2)	1739.06	119.54	1719.37	120.76	1697.59	122.35
	1095.27	113.95	1125.02	116.27	1116.75	121.17		1082.26	130.85	1087.43	132.17	1072.08	134.34
	631.07	170.76	631.07	176.44	630.20	182.83		668.08	177.61	664.19	180.87		
	566.00	208.97	570.25	212.34	570.97	217.89		648.13	184.20	649.05	186.09		
	420.33	61.77	419.17	59.39	417.25	55.85		404.65	−12.78	404.65	−11.43	404.29	−8.59
	212.32	−25.66	212.12	−27.38	212.12	−28.26		212.32	−31.54	212.32	−32.58	212.32	−32.52
	205.61	−23.12	205.43	−24.00	205.06	−25.42		198.88	−25.54	199.14	−26.08	199.40	−26.58

(continued)

Table 2. (continued)

Non-paraxial LG beam

OAM = +1, SAM = +1

State ($4d_{5/2}(m_J)$)	$\lambda^{50°}_{magic}$	α	$\lambda^{60°}_{magic}$	α	$\lambda^{70°}_{magic}$	α
(−3/2)	2109.41	101.90	2007.20	105.43	1890.60	110.08
	1084.84	114.38	1116.75	116.27	1114.02	121.23
	694.56	150.71	691.40	155.63	687.23	163.29
	643.55	165.68	644.46	169.06	646.29	175.62
	419.17	−35.11	418.01	−31.33	416.48	−26.87
	212.32	−25.66	212.12	−27.38	212.12	−28.26
	195.22	−19.20	195.89	−20.45	196.82	−21.76
(+5/2)	635.47	168.88	633.70	176.44	633.70	181.31
	537.94	237.89	542.42	241.88	547.64	243.35
	420.33	98.74	419.55	93.34	417.63	86.15
	212.32	−25.66	212.12	−27.38	212.12	−28.26
	208.91	−24.48	208.34	−25.49	207.86	−26.20
(−5/2)			1228.12	113.97	1279.87	116.64
			1130.60	116.13	1127.81	120.74
	786.93	134.41	760.66	140.86	740.87	151.66
	640.83	167.80	641.74	173.42	642.64	177.94
	419.17	−62.67	418.01	−58.24	416.10	−51.20
	212.32	−25.66	212.12	−27.38	212.12	−28.26
	189.06	−17.17	190.56	−18.38	192.49	−20.11

OAM = +1, SAM = −1

State ($4d_{5/2}(m_J)$)	$\lambda^{50°}_{magic}$	α	$\lambda^{60°}_{magic}$	α	$\lambda^{70°}_{magic}$	α
(−3/2)	1479.33	121.89	1474.54	123.15	1479.33	124.73
	1087.43	130.75	1092.65	132.01	1077.15	134.14
	628.46	192.56	626.73	196.34	625.01	198.77
	575.29	222.87	577.48	225.62	578.95	226.73
	405.01	45.89	404.65	43.96	404.29	41.92
	212.32	−31.54	212.32	−32.58	212.32	−32.52
	204.50	−28.12	204.32	−28.49	204.23	−28.82
(+5/2)	1455.70	122.30	1474.54	123.15	1493.88	124.25
	1097.91	130.25	1097.91	131.69	1082.26	133.88
	697.75	169.21	694.56	171.81	690.35	175.53
	643.55	186.01	644.46	187.88	644.46	190.28
	404.65	−32.05	404.65	−28.77	403.93	−24.62
	212.32	−31.54	212.32	−32.58	212.32	−32.52
	196.48	−24.61	196.99	−25.12	197.67	−25.82
(−5/2)	1262.14	125.78	1276.28	127.13	1287.10	127.99
	1111.30	129.79	1111.30	131.27	1095.27	133.42
	631.95	190.75	631.07	193.16	631.07	196.65
	563.21	233.46	569.54	231.24	570.25	232.41
	405.01	67.44	404.65	64.11	404.65	60.55
	212.32	−31.54	212.32	−32.58	212.32	−32.52
	206.82	−29.23	206.45	−29.52	205.98	−29.63

Table 3. Magic wavelengths (in nm) of Sr$^+$ for different focusing angles 50°, 60°, and 70° of the LG beam for the transitions $5s_{1/2}(-1/2)$ → $4d_{5/2}(m_J)$

Non-paraxial LG beam

State ($4d_{5/2}(m_J)$)	$\lambda^{50°}_{\text{magic}}$	α	$\lambda^{60°}_{\text{magic}}$	α	$\lambda^{70°}_{\text{magic}}$	α
	OAM = +2, SAM = +1					
(+1/2)	3927.88	100.17	2517.31	106.00	2052.40	112.18
	1077.15	115.89	1103.23	119.70	1069.56	126.17
	617.39	177.85	614.89	186.71	612.41	196.04
	590.96	193.35	594.05	198.04	600.31	202.71
	419.55	27.04	418.01	26.84	415.72	26.52
	212.22	−26.23	212.32	−27.71	212.22	−29.44
	202.59	−22.30	202.68	−23.71	202.77	−24.98
(−1/2)	6328.24	99.71	3120.78	105.14	2289.62	111.36
	1077.15	116.05	1103.23	119.70	1069.56	126.17
	419.17	−4.40	417.63	−2.19	415.72	0.68
	212.22	−26.23	212.32	−27.71	212.22	−29.44
	199.14	−20.73	199.66	−22.34	200.10	−23.84
(+3/2)	1484.15	107.39	1479.33	111.66	1479.33	116.62
	1095.27	115.33	1114.02	119.05	1082.26	125.77
	631.07	172.27	630.20	179.39	629.33	188.18
	567.41	209.40	570.25	214.61	572.40	222.72
	419.94	61.30	418.40	57.72	416.10	53.64
	212.22	−26.23	212.32	−27.71	212.22	−29.44
	205.52	−23.18	205.24	−24.35	204.87	−25.69

State ($4d_{5/2}(m_J)$)	$\lambda^{50°}_{\text{magic}}$	α	$\lambda^{60°}_{\text{magic}}$	α	$\lambda^{70°}_{\text{magic}}$	α
	OAM = +2, SAM = −1					
(+1/2)	1852.17	118.93	1793.83	120.89	1739.06	122.92
	1087.43	131.29	1074.61	133.39	1079.70	135.08
	404.29	5.89	404.29	6.90	404.29	8.39
	212.22	−32.29	212.52	−32.56	212.22	−33.14
	200.90	−26.66	201.07	−27.13	201.34	−27.69
(−1/2)	1739.06	119.79	1700.13	121.56	1650.85	123.61
	1090.03	131.10	1077.15	133.27	1079.70	135.05
	404.29	26.20	404.29	25.68	404.29	24.43
	212.22	−32.29	212.52	−32.56	212.22	−33.14
	202.86	−27.49	202.86	−27.92	202.86	−28.32
(+3/2)	1732.45	119.83	1706.49	121.50	1681.30	123.42
	1092.65	130.97	1077.15	133.19	1082.26	134.95
	667.11	179.06	660.34	183.18		
	648.13	184.71	649.05	186.68		
	404.29	−12.25	404.29	−9.90	404.29	−7.10
	212.22	−32.29	212.52	−32.56	212.22	−33.14
	198.97	−25.73	199.31	−26.30	199.66	−26.87

(continued)

Table 3. (continued)

Non-paraxial LG beam

OAM = +2, SAM = +1

State $(4d_{5/2}(m_J))$	$\lambda^{50°}_{magic}$	α	$\lambda^{60°}_{magic}$	α	$\lambda^{70°}_{magic}$	α
(−3/2)	2071.06	103.32	1947.15	107.98	1815.27	113.42
	1087.43	115.55	1111.30	119.44	1074.61	126.06
	694.56	152.80	689.31	159.51	681.07	169.57
	644.46	167.38	646.29	173.59	646.29	180.33
	418.78	−34.39	417.25	−29.38	415.35	−23.95
	212.22	−26.23	212.32	−27.71	212.22	−29.44
	195.30	−19.68	196.31	−21.02	197.50	−22.74
(+5/2)					1165.30	123.12
					1136.24	123.94
	635.47	171.08	633.70	178.77	632.82	186.03
	539.21	239.06	545.02	242.09	552.28	243.70
	420.33	97.01	418.78	90.44	416.48	80.76
	212.22	−26.23	212.32	−27.71	212.22	−29.44
	208.72	−24.75	208.15	−26.05	207.58	−27.21
(−5/2)	1168.29	113.24	1262.14	115.47	1328.38	119.09
	1136.24	113.69	1127.81	118.73	1103.23	124.89
	776.21	137.65	750.63	146.66	725.53	158.38
	640.83	168.57	641.74	174.84	642.64	182.21
	418.40	−61.56	417.25	−55.41	414.97	−46.65
	212.22	−26.23	212.32	−27.71	212.22	−29.44
	189.45	−17.39	191.44	−19.02	193.80	−21.77

OAM = +2, SAM = −1

State $(4d_{5/2}(m_J))$	$\lambda^{50°}_{magic}$	α	$\lambda^{60°}_{magic}$	α	$\lambda^{70°}_{magic}$	α
(−3/2)	1474.54	122.38	1479.33	123.94	1479.33	125.58
	1097.91	130.71	1082.26	132.99	1084.84	134.74
	627.59	193.70	626.73	196.76	625.01	200.22
	576.02	223.86	577.48	226.28	579.69	227.60
	404.29	45.43	404.65	43.94	404.65	41.00
	212.22	−32.29	212.52	−32.56	212.22	−33.14
	204.41	−28.21	204.32	−28.63	204.14	−29.02
(+5/2)	1460.36	122.73	1484.15	123.80	1498.79	125.28
	1105.91	130.59	1090.03	132.75	1090.03	134.53
	695.62	170.93	691.40	174.00	686.20	178.48
	643.55	187.15	644.46	189.10	645.37	191.67
	403.93	−30.40	403.93	−26.88	404.29	−22.59
	212.22	−32.29	212.52	−32.56	212.22	−33.14
	196.56	−24.70	197.33	−25.37	198.02	−26.09
(−5/2)	1265.65	126.00	1279.87	127.27	1294.41	128.65
	1116.75	130.07	1100.56	132.28	1100.56	134.26
	631.95	191.87	631.95	194.58	630.20	197.79
	565.30	233.57	569.54	231.96	570.97	234.22
	404.29	66.53	404.65	62.78	404.65	57.52
	212.22	−32.29	212.52	−32.56	212.22	−33.14
	206.73	−29.36	206.17	−29.50	205.70	−29.76

4   Conclusions

In conclusions, we find a wide list of magic wavelengths for the transitions $5s_{1/2}(m_J = -1/2) \rightarrow 4d_{3/2}(m_J)$ and $5s_{1/2}(-1/2) \rightarrow 4d_{5/2}(m_J)$ of the Sr$^+$ ion. We have found here a quite distinct set of values of magic wavelengths compared to the same with the initial state $5s_{1/2}(m_J = 1/2)$. These magic wavelengths fall from infrared to vacuum-ultraviolet range in the electromagnetic spectrum and will help experimentalists to trap ions in either at the high intensity or the low-intensity region of the LG beam. The variations in the magic wavelength are found tunable by varying the OAM, SAM, and focusing angles of the LG beam. An appreciable amount of deviations concerning focusing angles in the infrared magic wavelengths and corresponding polarizabilities are observed. As these infrared magic wavelengths have significant large values of polarizabilities, they are recommended as the best for trapping in the high precision experiments.

References

1. Champenois, C., Houssin, M., Lisowski, C., Knoop, M., Hagel, G., Vedel, M., Vedel, F.: Phys. Lett. A **331**, 298 (2004)
2. Chou, C.W., Hume, D.B., Koelemeij, J.C.J., Wineland, D.J., Rosenband, T.: Phys. Rev. Lett. **104**, 070802 (2010)
3. Rosenbusch, P., Ghezali, S., Dzuba, V.A., Flambaum, V.V., Beloy, K., Derevianko, A.: Phys. Rev. A **79**, 013404 (2009)
4. Margolis, H.S.: J. Phys. B **42**, 154017 (2009)
5. Nicholson, T., Campbell, S., Hutson, R., Marti, G., Bloom, B., McNally, R., Zhang, W., Barrett, M., Safronova, M., Strouse, G., et al.: Nat. Commun. **6**, 6896 (2015)
6. Dong, R.C., Wei, R., Du, Y.B., Zou, F., Lin, J.D., Wang, Y.Z.: Appl. Phys. Lett. **106**, 152402 (2015)
7. Biedermann, G.W., Wu, X., Deslauriers, L., Roy, S., Mahadeswaraswamy, C., Kasevich, M.A.: Phys. Rev. A **91**, 033629 (2015)
8. Arora, B., Safronova, M.S., Clark, C.W.: Phys. Rev. A **76**, 052509 (2007)
9. Ludlow, A.D., et al.: Science **319**, 1805 (2008)
10. Flambaum, V.V., Dzuba, V.A., Derevianko, A.: Phys. Rev. Lett. **101**, 220801 (2008)
11. Arora, B., Sahoo, B.K.: Phys. Rev. A **86**, 033416 (2012)
12. Bhowmik, A., Mondal, P.K., Majumder, S., Deb, B.: Phys. Rev. A **93**, 063852 (2016)
13. Bhowmik, A., Dutta, N.N., Majumder, S.: Phys. Rev. A **97**, 022511 (2018)
14. Mondal, P.K., Deb, B., Majumder, S.: Phys. Rev. A **89**, 063418 (2014)
15. Schmiegelow, C.T., Schulz, J., Kaufmann, H., Ruster, T., Poschinger, U.G., Schmidt-Kaler, F.: Nat. Commun. **7**, 12998 (2016)
16. Dutta, N.N., Roy, S., Deshmukh, P.C.: Phys. Rev. A **92**, 052510 (2015)
17. Safronova, M.S., Safronova, U.I.: Phys. Rev. A **83**, 012503 (2011)
18. Zhao, Y., Edgar, J.S., Jeffries, G.D.M., McGloin, D., Chiu, D.T.: Phys. Rev. Lett. **99**, 073901 (2007)
19. Mitroy, J., Safronova, M.S., Clark, C.W.: J. Phys. B **43**, 202001 (2010)
20. Dutta, N.N., Majumder, S.: Ind. J. Phys. **90**, 373 (2016)

21. Bhowmik, A., Roy, S., Dutta, N.N., Majumder, S.: J. Phys. B: At. Mol. Opt. Phys. **50**, 125005 (2017)
22. Bhowmik, A., Dutta, N.N., Roy, S.: Astrophys. J. **836**, 125 (2017)
23. Das, A., Bhowmik, A., Dutta, N.N., Majumder, S.: J. Phys. B: At. Mol. Opt. Phys. **51**(51), 025001 (2018)
24. Biswas, S., Das, A., Bhowmik, A., Majumder, S.: Mon. Not. R. Astron. Soc. **477**, 5605 (2018)
25. Johnson, W.R., Liu, Z.W., Sapirstein, J.: At. Data Nucl. Data Tables **64**, 279 (1996)
26. Safronova, U.I.: Phys. Rev. A **82**, 022504 (2010)

Spin–Orbit Interaction Features in Near-Threshold Photoionization Dynamics: An Energy- and Angle-Dependent Study

Ankur Mandal[1(✉)], Soumyajit Saha[2], and P. C. Deshmukh[1,3]

[1] Indian Institute of Science Education and Research Tirupati, Tirupati 517507, India
ankur@iisertirupati.ac.in,
pcd@iittp.ac.in
[2] Indian Institute of Technology Madras, Chennai 600036, India
soumyajit147@gmail.com
[3] Indian Institute of Technology Tirupati, Tirupati 517506, India
http://www.physics.iitm.ac.in/labs/amp/

Abstract. In this proceeding, we present a study of the spin–orbit interaction dominated features in time domain photoelectron dynamics. In particular, a case study of Wigner–Eisenbud–Smith (WES) time delay in the photoionization of Xe $4d$ subshell in the near-threshold region is presented and discussed. We show that the Spin–Orbit Interaction Activated Interchannel Coupling (SOIAIC) produces a prominent resonance-like structure in the WES time delay spectrum.

1 Introduction

Resonances are ubiquitous in nature. Sensitive energy-dependence of the oscillator strengths of atomic transitions cause resonance features in observables such as cross-section, photoelectron angular distribution asymmetry parameter, etc. These resonances could be due to the direct interference between different transition channels, or mediated by the correlated electron dynamics. When an electromagnetic field interacts with an atomic system, there are various possible pathways through which the system may evolve such as ionization of the atomic system, its excitation to a higher energy bound state, autoionization, multiple ionization, etc. Observation of real-time dynamics of these light–matter interactions in molecules and atoms has become possible very recently after exciting advances in ultrafast laser techniques especially in the past two decades.

Many experimental and theoretical studies have been carried out in the past decade to observe and understand the mechanism of photon–atomic-system interaction. The time an electron takes to be detached from the potential of the residual subsystem is a quantity of fundamental interest. We shall not review the

© Springer Nature Singapore Pte Ltd. 2019
P. C. Deshmukh et al. (eds.), *Quantum Collisions and Confinement of Atomic and Molecular Species, and Photons*, Springer Proceedings in Physics 230,
https://doi.org/10.1007/978-981-13-9969-5_5

original literature extensively. For more details of the attosecond pump-probe like measurement and different theoretical analyses, the reader is referred to [1–8].

In the pump-probe technique, the total observed time delay can be written as

$$\tau_{atomic} \cong \tau_{WES} + \tau_{measurement}, \tag{1}$$

where τ_{atomic} is the total time delay measured in the experiment, τ_{WES} is the intrinsic time delay associated to the one photon (pump) transition, and $\tau_{measurement}$ is the delay induced by the measurement process (probe).

For scattering by a spherically symmetric potential, the WES time delay for the lth partial wave is given by

$$t_{l,WES}(E) = 2\hbar \frac{\partial \delta_l(E)}{\partial E}. \tag{2}$$

Since photoionization is interpreted as half scattering [9], the time delay is given by the energy derivative of the phase of the complex matrix element of the coupling operator (here $\mathbf{d}$ representing the dipole operator);

$$t_{WES}(E, \hat{\Omega}) = \hbar \frac{d}{dE} arg \left[\left\langle \psi_f(E, \hat{\Omega}) | \mathbf{d} | \psi_i \right\rangle \right] \tag{3}$$

where ψ_i and ψ_f are the initial and the final state wave functions, respectively [7].

The WES time delay comes from the phase accumulated by the photoelectron due to its interaction with the residual ion. For photoionization of a neutral atom, the potential between the residual ion and the photoelectron is Coulombic even at a large distance, while it is dominated by inter-electronic interactions in the small r region. The asymptotic phase shift due to the scattering by a pure Coulomb-potential is

$$\delta_{l,Coulomb}(E, r) = \sigma_l^C(E) + \frac{Z}{k} \ln 2kr \tag{4}$$

where $\sigma_l^C(E) = arg \Gamma(1 + l - i\frac{Z}{k})$ is the "Coulomb phaseshift". The total time delay due the scattering in this case is

$$t_{l,Coulomb}(E, r) = \hbar \frac{\partial \delta_{l,Coulomb}(E, r)}{\partial E} = \hbar \left[\frac{\partial \sigma_l^C(E)}{\partial E} + \frac{Z}{(2E)^{3/2}} \left\{ 1 - \ln(2kr) \right\} \right]. \tag{5}$$

$\hbar \frac{\partial \sigma_l^C(E)}{\partial E}$ is the part of the time delay attributed to the "Coulomb phaseshift". In the case of a Coulomb potential, the term $\hbar \frac{Z}{(2E)^{3/2}} \left\{ 1 - \ln(2kr) \right\}$ also contributes to the time delay.

In streaking experiments, for a short-range potential the logarithmic Coulomb term does not contribute to the time delay, and $\tau_{Streaking}$ is, therefore, the same as τ_{WES} corresponding to the short-range potential [7]. When the Coulomb term is present

$$\tau_{Streaking} = \tau_{WES}^{SR+C} + \tau_{CLC}, \tag{6}$$

where $\tau_{WES}^{SR+C} = \hbar \frac{\partial \sigma_l^{SR+C}(E)}{\partial E}$ and τ_{CLC} is the time shift due to the motion of electron in residual Coulomb along with the Laser field (CLC: Coulomb Laser Coupling); $\sigma_l^{SR+C}(E) = \eta_{SR} + arg\Gamma\left[1 + l - iZ/k\right]$. η_{SR} is the phase shift due to the short- range part of the potential due to the correlated many-body interaction of multi-electronic system. For one-electron systems $\eta_{SR} = 0$. Together with the pump-probe experimental observation and an estimate of cc/CLC time delay one can get an understanding of the light–matter interaction (pump-system).

These attosecond pump-probe experiments have opened up a new domain in which information which was time integrated into earlier measurement schemes can now be captured [7, 10–15]. Theoretically, the atomic processes are described by the complex transition matrix elements, each of which has an absolute magnitude, and a phase. Accurate computation of these matrix elements requires the inclusion of important electron interactions/correlations to obtain satisfactory agreement with experiments.

The relativistic effects in the photoionization of atoms are especially important for high-Z atoms in the periodic table. One of the most important manifestations of the relativistic effects is the spin–orbit splitting of the nl subshell. Important consequences of this include the splitting of spectral lines, nonzero photoionization cross-section near Cooper minima, etc. [16–19]. An important leveraging of the spin–orbit interaction in the photoionization from an atomic spin–orbit doublet has been seen in experiments and is understood in terms of some electron correlation effects that were not known earlier [20–22]. The fine-structure splitting of the nd subshell and the delayed maximum in its photoionization cross-section involves a significant influence on the oscillator strengths in the usually strong $l+1/2$ channels by that from the $l-1/2$ photoionization channels. The $l-1/2$ cross-section is also affected, but somewhat less significantly. The leveraging of the spin–orbit interaction affects also the spin polarization and the angular distribution of photoelectrons [23, 24]. The correlation-induced leveraging of spin–orbit interactions is also observed in the photoionization of the $3d$ subshell of Xe; and it is predicted that as we go inner from outer subshells this effect should get to be more prominent, since spin–orbit splitting is more [20].

The immediate threshold energy region is highly influenced by the CLC/cc time delay, which is important for measurement. Among some other possibilities, the time delay relative to different subshells, or relative to different atomic species, at different angles can be studied. Such studies would provide valuable information about the one photon transition [25–28]. The spin–orbit split subshell properties/structures have been studied, and interchannel interaction effects have been observed in various observables of the experiment such as the photoionization cross-section, angular distribution asymmetry parameter, and the spin polarization parameters. Most of these observations/studies were carried out in the energy domain [20–24], now possible to assess in the time domain. The spin–orbit interaction affects the photoionization cross-section in the $l + 1/2$ state but not quite the photoelectron angular distribution, even if it depends on the relative phases in the different dipole channels. So the natural question comes: how is the energy variation of the Wigner–Eisenbud–Smith (WES) time delay in the near-threshold region affected by the Spin–Orbit Activated Inter-Channel Coupling (SOIAIC)? In a recent study, it was found that the three relativistic channels from the $4d$ subshell to ϵf final states show very different time delays at energies near the thresholds. Similar results have been found in the photoionization of [29,30]. These studies reveal the dynamical differences among the individual channels brought about by interchannel coupling. In a recent work [31], the spin–orbit interaction induced confinement resonances were investigated in the WES time delay in photoionization from high Z atoms.

In essence, it is found that WES time delay is highly sensitive to the SOIAIC effect in the case of photoionization from endohedral atoms. In general, the interference between different channels with the outgoing final state of different angular momentum causes the dependence of the WES time delay on the angle of photoelectron ejection with respect to the polarization of light. The present study is one of the very first attempts to understand the SOIAIC effect in photoionization time delay in specific dipole-allowed channels and investigate the energy and the angle dependence of the WES photoionization time delay.

This paper is organized in the following manner. In the next section, we present a brief theoretical formulation. Results for the angle and energy dependence of WES time delay for photoemission from inner $4d_{3/2}$ and $4d_{5/2}$ subshells of Xe at two different levels of truncation are presented and discussed in Sect. 3. Conclusions are presented in the Sect. 4.

2   Theoretical Method

2.1   Photoionization Amplitude

Detailed theoretical formalism that provides the foundation for the present work has been presented elsewhere [25,30]. Electric dipole transitions from an nd initial state leads to the following six ionization channels: $nd_{j=3/2} \rightarrow \epsilon p_{1/2}, \epsilon p_{3/2}, \epsilon f_{5/2}$ and $nd_{j=5/2} \rightarrow \epsilon p_{3/2}, \epsilon f_{5/2}, \epsilon f_{7/2}$. The following expressions for the $nd_{3/2}$ ionization amplitudes [30] are of importance for the present work. In (7), the superscript 1+ of T' represents electric multipole transition with spin-up final state (similarly 1− represents spin-down final state in other cases). The subscripts 10 represent the unit angular momentum photon with plane polarization. The subscript $nd_{3/2,1/2}$ of $\left[T_{10}'^{(1+)}\right]$ represent the initial state nd_{jm} from which the photoionization is taking place. $D_{nd_{3/2} \rightarrow \epsilon p_{3/2}}$ is used as a shorthand notation for the reduced dipole matrix element of the transition $nd_{3/2} \rightarrow \epsilon p_{3/2}$, which contains the partial wave phase shift of the outgoing photoelectron and the electron-correlation induced phase. In our calculations, these phases are the DHF (Dirac-Hartree-Fock) phase (Coulomb phaseshift and short range phase due to the exchange and non-Coulombic part of the potential for a multi-electronic atom) and the short range many-body-correlation correction to the DHF phase (in total similar to σ_l^{SR+C} as defined earlier). Y_{lm} are the spherical harmonics.

This scheme of notation has been followed for writing the expressions for all $T's$:

$$\left[T_{10}'^{(1+)}\right]_{nd_{3/2,1/2}} = -\frac{1}{3\sqrt{2}}D_{nd_{3/2}\to\epsilon p_{1/2}}Y_{10}(\hat{p}) + \frac{1}{3}\sqrt{\frac{1}{10}}D_{nd_{3/2}\to\epsilon p_{3/2}}Y_{10}(\hat{p})$$
$$+ \sqrt{\frac{3}{70}}D_{nd_{3/2}\to\epsilon f_{5/2}}Y_{30}(\hat{p}). \tag{7}$$

$$\left[T_{10}'^{(1-)}\right]_{nd_{3/2,1/2}} = \frac{1}{3}D_{nd_{3/2}\to\epsilon p_{1/2}}Y_{11}(\hat{p}) + \frac{1}{6\sqrt{5}}D_{nd_{3/2}\to\epsilon p_{3/2}}Y_{11}(\hat{p})$$
$$- \sqrt{\frac{2}{35}}D_{nd_{3/2}\to\epsilon f_{5/2}}Y_{31}(\hat{p}). \tag{8}$$

$$\left[T_{10}'^{(1+)}\right]_{nd_{3/2,3/2}} = \frac{1}{2}\sqrt{\frac{3}{5}}D_{nd_{3/2}\to\epsilon p_{3/2}}Y_{11}(\hat{p}) + \sqrt{\frac{2}{105}}D_{nd_{3/2}\to\epsilon f_{5/2}}Y_{31}(\hat{p}). \tag{9}$$

$$\left[T_{10}'^{(1-)}\right]_{nd_{3/2,3/2}} = -\sqrt{\frac{1}{21}}D_{nd_{3/2}\to\epsilon f_{5/2}}Y_{32}(\hat{p}). \tag{10}$$

The angle-resolved amplitudes for the $nd_{5/2}$ initial state take the following forms:

$$\left[T_{10}'^{(1+)}\right]_{nd_{5/2,1/2}} = \frac{1}{\sqrt{15}}D_{nd_{5/2}\to\epsilon p_{3/2}}Y_{10}(\hat{p}) - \frac{1}{7\sqrt{10}}D_{nd_{5/2}\to\epsilon f_{5/2}}Y_{30}(\hat{p})$$
$$- \frac{\sqrt{2}}{7}D_{nd_{5/2}\to\epsilon f_{7/2}}Y_{30}(\hat{p}). \tag{11}$$

$$\left[T_{10}'^{(1-)}\right]_{nd_{5/2,1/2}} = \frac{1}{\sqrt{30}}D_{nd_{5/2}\to\epsilon p_{3/2}}Y_{11}(\hat{p}) + \frac{1}{7}\sqrt{\frac{2}{15}}D_{nd_{5/2}\to\epsilon f_{5/2}}Y_{31}(\hat{p})$$
$$- \frac{1}{7}\sqrt{\frac{3}{2}}D_{nd_{5/2}\to\epsilon f_{7/2}}Y_{31}(\hat{p}). \tag{12}$$

$$\left[T_{10}'^{(1+)}\right]_{nd_{5/2,3/2}} = \frac{1}{\sqrt{15}}D_{nd_{5/2}\to\epsilon p_{3/2}}Y_{11}(\hat{p}) - \frac{1}{7}\sqrt{\frac{3}{5}}D_{nd_{5/2}\to\epsilon f_{5/2}}Y_{31}(\hat{p})$$
$$- \frac{5}{14\sqrt{3}}D_{nd_{5/2}\to\epsilon f_{7/2}}Y_{31}(\hat{p}). \tag{13}$$

$$\left[T_{10}'^{(1-)}\right]_{nd_{5/2,3/2}} = \frac{1}{7}\sqrt{\frac{3}{2}}D_{nd_{5/2}\to\epsilon f_{5/2}}Y_{32}(\hat{p}) - \frac{1}{14}\sqrt{\frac{10}{3}}D_{nd_{5/2}\to\epsilon f_{7/2}}Y_{32}(\hat{p}). \tag{14}$$

$$\left[T_{10}'^{(1+)}\right]_{nd_{5/2,5/2}} = -\frac{1}{7}\sqrt{\frac{5}{6}}D_{nd_{5/2}\to\epsilon f_{5/2}}Y_{32}(\hat{p}) - \frac{\sqrt{6}}{14}D_{nd_{5/2}\to\epsilon f_{7/2}}Y_{32}(\hat{p}). \tag{15}$$

$$\left[T_{10}'^{(1-)}\right]_{nd_{5/2,5/2}} = \frac{\sqrt{5}}{7}D_{nd_{5/2}\to\epsilon f_{5/2}}Y_{33}(\hat{p}) - \frac{1}{14}D_{nd_{5/2}\to\epsilon f_{7/2}}Y_{33}(\hat{p}). \tag{16}$$

The corresponding amplitudes with the possible $m = -1/2, -3/2, -5/2$ projection will have similar structures. The associated photoelectron group delay, which is the WES time delay [32–35] is defined as $\frac{d\eta}{dE}$, where $\eta = \tan^{-1}\left[\frac{\mathrm{Im}T_{10}^{1\pm}}{\mathrm{Re}T_{10}^{1\pm}}\right]$

The angle-dependent time delay for the case where neither the orientation of the residual ion nor the spin ν of the photoelectron is detected is given for the nd_j state as

$$\tau_{nd_j}(\theta) = \frac{\sum_{m,\nu} \tau_{nd_{j,m,\nu}}(\theta) \left| \left[T_{10}^{\prime(1\nu)} \right]_{nd_{j,m}} \right|^2}{\sum_{m,\nu} \left| \left[T_{10}^{\prime(1\nu)} \right]_{nd_{j,m}} \right|^2} \tag{17}$$

which is the weighted average over the initial m-states and final state spins of the photoelectron. We have presented the expressions only for positive $m = 1/2, 3/2, 5/2$. The corresponding time delay expressions with negative m are identical.

2.2   Level of Truncation

Since the photoionization of atoms involves correlated many-electron dynamics, the ab-initio Relativistic Random Phase Approximation (RRPA), which includes both the relativistic and the many-electron correlation effects on the same footing [36–38], is applied for calculating the complex matrix elements of the dipole transitions. For including the select final state correlations, calculations have been performed at the following two levels of truncation in RRPA:

Level 1 [10 channel coupled]

$$4d_{5/2} \rightarrow \epsilon f_{7/2}, \epsilon f_{5/2}, \epsilon p_{3/2}$$
$$5p_{3/2} \rightarrow \epsilon d_{5/2}, \epsilon d_{3/2}, \epsilon s_{1/2}$$
$$5p_{1/2} \rightarrow \epsilon d_{3/2}, \epsilon s_{1/2}$$
$$5s_{1/2} \rightarrow \epsilon p_{3/2}, \epsilon p_{1/2}$$

$$4d_{3/2} \rightarrow \epsilon f_{5/2}, \epsilon p_{3/2}, \epsilon p_{1/2}$$
$$5p_{3/2} \rightarrow \epsilon d_{5/2}, \epsilon d_{3/2}, \epsilon s_{1/2}$$
$$5p_{1/2} \rightarrow \epsilon d_{3/2}, \epsilon s_{1/2}$$
$$5s_{1/2} \rightarrow \epsilon p_{3/2}, \epsilon p_{1/2}$$

Level 2 [13 channel coupled]

$$4d_{5/2} \rightarrow \epsilon f_{7/2}, \epsilon f_{5/2}, \epsilon p_{3/2}$$
$$4d_{3/2} \rightarrow \epsilon f_{5/2}, \epsilon p_{3/2}, \epsilon p_{1/2}$$
$$5p_{3/2} \rightarrow \epsilon d_{5/2}, \epsilon d_{3/2}, \epsilon s_{1/2}$$
$$5p_{1/2} \rightarrow \epsilon d_{3/2}, \epsilon s_{1/2}$$
$$5s_{1/2} \rightarrow \epsilon p_{3/2}, \epsilon p_{1/2}$$

In level-1 truncation, we include the intra-subshell final-state correlations but exclude interchannel coupling that includes channels from different subshells [independent particle type]. In the level-II truncation scheme, we include both intra-subshell and inter-subshell interchannel coupling for channels originating from the spin–orbit split $4d$ subshell [correlated particle type]. These alternative schemes are very similar, respectively, to the Pseudo-Independent Particle Truncation (PIPT) and Intra Sub-Shell Truncation (ISST) schemes described in [23], and we, therefore, use the same nomenclature.

3   Results and Discussion

The weighted average, as defined in (17), for $4d_{3/2}$ is plotted in Fig. 1. There are four channels which contribute in general to the average with their corresponding weights. The WES time delay from each transition amplitude $\left[T_{10}^{\prime(1\pm)}\right]_{nd_{3/2,1/2}}$ and $\left[T_{10}^{\prime(1\pm)}\right]_{nd_{3/2,3/2}}$ are computed (not shown here). The WES time delay spectra with the corresponding weights are averaged as defined in (19) and plotted in Fig. 1. The results from the coupled calculation (ISST) are shown in bold solid lines for different angles in different colors. Corresponding uncoupled (PIPT) results are shown with thin dashed lines in the same color codes as used in ISST calculation(s).

Looking into each of the transition amplitudes from $4d_{3/2}$, we observe that the interfering channels are $4d_{3/2} \to \epsilon p_{1/2}, \epsilon p_{3/2}$ and $\epsilon f_{5/2}$ along with the respective spherical harmonics of the continuum final states. The WES time delays for $\left[T_{10}^{\prime(1\pm)}\right]_{nd_{3/2,1/2}}$ and $\left[T_{10}^{\prime(1+)}\right]_{nd_{3/2,3/2}}$ are angle dependent owing to the fact that

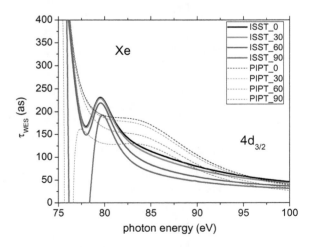

Fig. 1. WES time delay in photoionization of $4d_{3/2}$ subshell of atomic Xe

there are more than one final state with different symmetry (i.e., different angular momenta). The energy and angular variation of the WES time delay in each channel depends on the oscillator strengths of the dipole transitions (indicated as D).

Some sharp structures of the WES time delay spectrum in the specific transition amplitudes are washed out in these averages; but these average results are more amenable by the experiments. The angular variation of time delay is non-monotonous in the near-threshold region. Specially near 60°–90° angle of photoelectron emission, the variation is high and it drops down at the photon energies above 85 eV. The variation with respect to energy is most prominent in the $\sim$10 eV above threshold ($4d_{3/2}$) region. From the threshold, it starts with a very high positive value owing to the dominance of the Coulomb phase and then the correlation-induced phase of the outgoing photoelectron shows up in the WES time delay as a sharp variation near 80eV photon energy. The difference between the PIPT and ISST is very prominent in this energy region, which reflects the impact of SOIAIC on the WES time delay.

Similar to Fig. 1, in Fig. 2 we have shown the WES time delay in photoionization from $4d_{5/2}$ subshell of Xe at different energies and angles. Here, the interfering channels are $4d_{5/2} \rightarrow \epsilon p_{3/2}, \epsilon f_{5/2}$ and $\epsilon f_{7/2}$. The WES time delays for $\left[T_{10}^{\prime(1\pm)}\right]_{nd_{5/2,1/2}}$ and $\left[T_{10}^{\prime(1+)}\right]_{nd_{5/2,3/2}}$ are angle dependent.

The single-hump or no-hump like structure in PIPT level splits into the triple-hump-like structure in ISST. This difference in structure is due to the inclusion of correlation between the channels originating from $4d_{3/2}$ and $4d_{5/2}$ subshells. Up to about 80eV, the spin–orbit effect is much prominent and it decreases as the photon energy increases. Eventually, the PIPT result resembles the results of

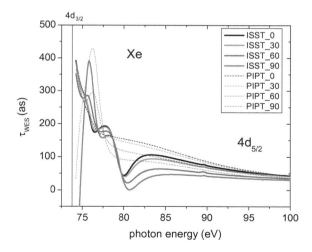

Fig. 2. WES time delay in photoionization of $4d_{5/2}$ subshell of atomic Xe

ISST as also seen in $4d_{3/2}$ case. This is easily understandable, as photoionization becomes mostly one-electron process for high photon energy. This is very clear from the present study that not only the transitions from $l + 1/2$ subshell gets affected by the channels from $l - 1/2$ subshell, the reverse effect also holds. This brings out the result of the general interchannel interaction in which one channel is affected by the other degenerate channels at the same energy. Since the phase and its variations with respect to energy and angle are much sensitive to these changes, the interchannel interactions between the channels from spin–orbit split subshells manifest in the WES time delay in a prominent way, however complex.

4   Conclusions

The angular dependence of WES time delay is present in six out of ten channels, where the interfering continuum final states have different angular momentum quantum numbers. The weighted averages for $4d_{3/2}$ and $4d_{5/2}$ are presented and discussed. The hump-like structure as seen in $4d_{3/2}$ and the triple-hump-like structure in $4d_{5/2}$ time delays in the energy range of 80 ± 5 eV are due to the interchannel coupling between the dipole transitions from $4d_{3/2}$ with those from $4d_{5/2}$.

Two distinct effects play a major role in determining the time delay spectrum and the observable resonance like structures. These are the intra-subshell correlations (interaction between the photoionization channels originating only from the $4d_{5/2}$, PIPT level of truncation) and the other one is the inter-subshell correlation (interaction between the channels originating from $4d_{5/2}$ and $4d_{3/2}$, ISST level of truncation). A part of the latter shows a variation of the structure due to its subset, i.e., intra-subshell correlation (slightly below from 80 eV). The other part (near 80eV) of this interaction causes the structure which is completely absent in PIPT level of calculation and hence attributed to the SOIAIC.

The splitting of ionization thresholds for $l \pm 1/2$ states is important to see the SOIAIC, since the channels originating from lower j state $(l - 1/2)$ dominate over the channels that originate from the higher j state $(l + 1/2)$. This situation happens for the $3d$ subshell in Xe. One may note that the relaxation effects are also important, hence a study of the WES time delay using both the RRPA and the RRPA-R will provide further insight into the effect of SOIAIC in photoionization.

From the previous studies on $d_{5/2}$ subshell's photoionization parameters and from the present study, we may conclude that the SOIAIC affects the WES time delay spectra in the photoionization of both $l \pm 1/2$ subshells. Angle-resolved studies thus play an extremely important role in getting insights into the photoionization dynamics.

References

1. Schultze, M., Fie, M., Karpowicz, N., Gagnon, J., Korbman, M., Hofstetter, M., Neppl, S., Cavalieri, A.L., Komninos, Y., Mercouris, Th., Nicolaides, C.A., Pazourek, R., Nagele, S., Feist, J., Burgdörfer, J., Azzeer, A.M., Ernstorfer, R., Kienberger, R., Kleineberg, U., Goulielmakis, E., Krausz, F., Yakovlev, V.S.: Delay in photoemission. Science **328**, 1658 (2010). https://doi.org/10.1126/science. 1189401
2. Klünder, K., Dahlström, J.M., Gisselbrecht, M., Fordell, T., Swoboda, M., Guénot, D., Johnsson, P., Caillat, J., Mauritsson, J., Maquet, A., Taïeb, R., LHuillier, A.: Probing single-photon ionization on the attosecond time scale. Phys. Rev. Lett. **106**, 143002 (2011). https://doi.org/10.1103/PhysRevLett.106.143002
3. Pazourek, R., Nagele, S., Burgdörfer, J.: Attosecond chronoscopy of photoemission. Rev. Mod. Phys. **87**, 765 (2015). https://doi.org/10.1103/RevModPhys.87.765
4. Gallmann, L., Cirelli, C., Keller, U.: Attosecond science: recent highlights and future trends. Ann. Rev. Phys. Chem. **63**, 447 (2012). https://doi.org/10.1146/annurev-physchem-032511-143702
5. Palatchi, C., Dahlström, J.M., Kheifets, A.S., Ivanov, I.A., Canaday, D.M., Agostini, P., DiMauro, L.F.: Atomic delay in helium, neon, argon and krypton. J. Phys. B **47**, 245003 (2014). https://doi.org/10.1088/0953-4075/47/24/245003
6. Ivanov, I.A., Kheifets, A.S., Serov, V.V.: Attosecond time-delay spectroscopy of the hydrogen molecule. Phys. Rev. A **86**, 063422 (2012). https://doi.org/10.1103/PhysRevA.86.063422
7. Pazourek, R., Nagele, S., Burgdörfer.s, J.: Time-resolved photoemission on the attosecond scale: opportunities and challenges. Faraday Discuss. **163**, 353–376 (2013). https://doi.org/10.1039/C3FD00004D
8. Dahlström, J.M., Guénot, D., Klünder, K., Gisselbrecht, M., Mauritsson, J., LHuillier, A., Maquet, A., Taïeb, R.: Theory of attosecond delays in laser assisted photoionization. Chem. Phys. **414**, 53–64 (2013). https://doi.org/10.1016/j.chemphys. 2012.01.017
9. Deshmukh, P.C., Angom, D., Banik, A.: Symmetry in electron-atom collision and photoionization process. Invited article in DST-SERC-School publication (NAROSA, Nov. 2011); collection of articles based on lecture course given at the DST-SERC School at the Birla Institute of Technology, Pilani, January 9–28, 2011
10. Krausz, F., Ivanov, M.: Attosecond physics. Rev. Mod. Phys. **81**, 163 (2009). https://doi.org/10.1103/RevModPhys.81.163
11. Guénot, D., Klunder, K., Arnold, C.L., Kroon, D., Dahlström, J.M., Miranda, M., Fordell, T., Gisselbrecht, M., Johnsson, P., Mauritsson, J., Lindroth, E., Maquet, A., Taïeb, R., LHuillier, A., Kheifets, A.S.: Photoemission-time-delay measurements and calculations close to the 3s-ionization-cross-section. Phys. Rev. A **85**, 053424 (2012). https://doi.org/10.1103/PhysRevA.85.053424
12. Neppl, S., Ernstorfer, R., Bothschafter, E.M., Cavalieri, A.L., Menzel, D., Barth, J.V., Krausz, F., Kienberger, R., Feulne, P.: Attosecond time-resolved photoemission from core and valence states of magnesium. Phys. Rev. Lett. **109**, 087401 (2012). https://doi.org/10.1103/PhysRevLett.109.087401
13. Yakovlev, V.S., Gagnon, J., Karpowicz, N., Krausz, F.: Attosecond streaking enables the measurement of quantum phase. Phys. Rev. Lett. **105**, 073001 (2010). https://doi.org/10.1103/PhysRevLett.105.073001
14. Kheifets, A.S., Ivanov, I.A.: Delay in atomic photoionization. Phys. Rev. Lett. **105**, 233002 (2010). https://doi.org/10.1103/PhysRevLett.105.233002

15. Moore, L.R., Lysaght, M.A., Parker, J.S., van der Hart, H.W., Taylor, K.T.: Time delay between photoemission from the 2p and 2s subshells of neon. Phys. Rev. A **84**, 061404(R) (2011). https://doi.org/10.1103/PhysRevA.84.061404

16. Deshmukh, P.C.: Relativistic splittings of Cooper minima in radon: an RRPA study. Phys. Lett. A **117**(6), 25, 293–296 (1986). https://doi.org/10.1016/0375-9601(86)90392-0

17. Shanti, N., Deshmukh, P.C.: Xenon 4p photoionization near the 4d Cooper minimum: interchannel coupling effects. Phys. Rev. A **40**, 2400 (1989). https://doi.org/10.1103/PhysRevA.40.2400

18. Ganesan, A., Deshmukh, S., Pradhan, G.B., Radojevic, V., Manson, S.T., Deshmukh, P.C.: Photoionization of the 5s subshell of Ba in the region of the second Cooper minimum: cross sections and angular distributions. J. Phys. B: At. Mol. & Opt. Phys. **46**, 185002 (2013). https://doi.org/10.1088/0953-4075/46/18/185002

19. Saha, S., Mandal, A., Jose, J., Varma, H.R., Deshmukh, P.C., Kheifets, A.S., Dolmatov, V.K., Manson, S.T.: Relativistic effects in photoionization time delay near the Cooper minimum of noble-gas atoms. Phys. Rev. A **90**, 053406 (2014). https://doi.org/10.1103/PhysRevA.90.053406

20. Amusia, M.Ya., Chernysheva, L.V., Manson, S.T., Msezane, A.M., Radojević, V.: Strong electron correlation in photoionization of spin-orbit doublets. Phys. Rev. Lett. **88**, 093002 (2002). https://doi.org/10.1103/PhysRevLett.88.093002

21. Kivimäki, A., Hergenhahn, U., Kempgens, B., Hentges, R., Piancastelli, M.N., Maier, K., Rüdel, A., Tulkki, J. J., Bradshaw, A.M.: Near-threshold study of Xe 3d photoionization. Phys. Rev. A **63**, 012716 (2000). https://doi.org/10.1103/PhysRevA.63.012716

22. Richter, T., Heinecke, E., Zimmermann, P., Godehusen, K., Yalinkaya, M., Cubaynes, D., Meyer, M.: Outstanding spin-orbit-activated interchannel coupling in the Cs and Ba 3d photoemission. Phys. Rev. Lett. **98**, 143002 (2006). https://doi.org/10.1103/PhysRevLett.98.143002

23. Sunil Kumar, S., Banerjee, T., Deshmukh, P.C., Manson, S.T.: Spin-orbit-interaction activated interchannel coupling in dipole and quadrupole photoionization. Phys. Rev. A **79**, 043401 (2009). https://doi.org/10.1103/PhysRevA.79.043401

24. Govil, K., Siji, A.J., Deshmukh, P.C.: Relativistic and confinement effects in photoionization of Xe. J. Phys. B: At. Mol. Opt. Phys. **42**, 065004 (2009). https://doi.org/10.1088/0953-4075/42/6/065004

25. Kheifets, A., Mandal, A., Deshmukh, P.C., Dolmatov, V.K., Keating, D.A., Manson, S.T.: Relativistic calculations of angle-dependent photoemission time delay. Phys. Rev. A **94**, 013423 (2016). https://doi.org/10.1103/PhysRevA.94.013423

26. Watzel, J., Moskalenko, A.S., Pavlyukh, Y., Berakdar, J.: Angular resolved time delay in photoemission. J. Phys. B: At. Mol. Opt. Phys. **48**(2), 025602 (2014). https://doi.org/10.1088/0953-4075/48/2/025602

27. Heuser, S., Galn, J., Cirelli, C. et al.: Angular dependence of photoemission time delay in helium. Phys. Rev. A **94**, 063409 (2016). https://doi.org/10.1103/PhysRevA.94.063409

28. Ivanov, I.A., Dahlström, J.M., Lindroth, E., Kheifets, A.S.: On the angular dependence of the photoemission time delay in helium. arXiv:1605.04539v1 [physics.atom-ph]. https://arxiv.org/abs/1605.04539

29. Deshmukh, P., Mandal, A., Saha, S., Kheifets, A., Dolmatov, V., Manson, S.: Attosecond time delay in the photoionization of endohedral atoms $A@C_{60}$: a probe of confinement resonances. Phys. Rev. A **89**, 053424 (2014). https://doi.org/10.1103/PhysRevA.89.053424

30. Mandal, A., Deshmukh, P.C., Kheifets, A., Dolmatov, V.K., Manson, S.T.: Angle-resolved Wigner time delay in atomic photoionization: the 4d subshell of free and confined Xe. Phys. Rev. A **96**, 053407 (2017). https://doi.org/10.1103/PhysRevA.96.053407

31. Keating, D., Deshmukh, P.C., Manson, S.T.: Wigner time delay and spinorbit activated confinement resonances. J. Phys. B. **50**, 175001 (2017). https://doi.org/10.1088/1361-6455/aa8332

32. Wigner, E.P.: Lower limit for the energy derivative of the scattering phase shift. Phys. Rev. **98**, 145–147 (1955). https://doi.org/10.1103/PhysRev.98.145

33. Eisenbud, L.E.: Ph.D. thesis, Princeton Univ. (1948) (Unpublished)

34. Smith, F.T.: Lifetime matrix in collision theory. Phys. Rev. **118**, 349–356 (1960). https://doi.org/10.1103/PhysRev.118.349

35. de Carvalho, C.A.A., Nussenzveig, H.M.: Time delay. Phys. Rep. **364**, 83–174 (2002). https://doi.org/10.1016/S0370-1573(01)00092-8

36. Johnson, W.R., Lin, C.D.: Multichannel relativistic random-phase approximation for the photoionization of atoms. Phys. Rev. A **20**, 964–977 (1979). https://doi.org/10.1103/PhysRevA.20.964

37. Johnson, W.R.: Atomic Structure Theory. Springer, Berlin, Heidelberg (2007)

38. Johnson, W.R., Lin, C.D., Cheng, K.T., Lee, C.M.: Relativistic random phase approximation. Phys. Scripta **21**, 409 (1980). https://doi.org/10.1088/0031-8949/21/3-4/029

Isotope Separation of Bromine Molecules: A Novel Method

Barun Halder[1], Utpal Roy[1], Jayanta Bera[1], and Suranjana Ghosh[2(✉)]

[1] Indian Institute of Technology of Patna, Bihta, Patna 801103, Bihar, India
[2] Amity University Patna, Rupaspur, Patna 801503, India
sghosh@ptn.amity.edu

Abstract. Wave packets are spatially localized, mesoscopic object in quantum mechanics which are known to manifest interesting phenomena, like revival and fractional revivals, during their course of time evolution when the energy contains nonlinear dependency in the quantum number. Molecular isotope separation from a mixture is experimentally carried out in the literature through different methods: laser isotope separation, wave packet isotope separation, etc. However, the minimum evolution time required for such separation as given by the autocorrelation function is quite long ($\sim 2T_{rev}$), which is not favorable for a quantum system due to its unavoidable interaction with the environment. We propose a novel method to separate isotopes of Bromine molecule, $^{79}Br_2$ and $^{81}Br_2$, in much smaller time compared to that in the existing literature by introducing an initial time delay between the two lasers while preparing the wave packets. The particular observation time is nothing but a chosen fractional revival instant, coincident for both the isotopes.

1 Introduction

The dynamics of a suitably prepared diatomic molecular wave packet have attracted immense attention in the literature [1–3], in both, theory and experiment. Wave packets are created by the short and phase-controlled optical pulses. They are localized in space, hence is a subject of interest in both the classical and quantum domain. Traditional methods for separating isotopes exploit the variation of their masses [4]. Isotope separation by the method of centrifugation utilizes the shift in their spectral lines. Wave packet isotope separation relies on the minute observation of their time evolved wave packets upon appreciable separation due to the difference between their anharmonicity constants.

Wave packet isotope separation method identifies an observation time where the wave packets become distinguishable in their long-time evolution through the auto-correlation function [5]. In this process, the fractional revivals phenomena [6] of the two isotopes do not match at the observation time because of the decoherence. However, a time delay can be introduced between the two lasers while producing the wave packets of the two isotopes. The amount of time delay should be such that the behavior of fractional revival should coincide at the observation time. The observation time is exactly at the chosen fractional revival instant. Hence, by introducing the initial delay, one can get the particular fractional revival for both the isotopes at the same time

© Springer Nature Singapore Pte Ltd. 2019
P. C. Deshmukh et al. (eds.), *Quantum Collisions and Confinement of Atomic and Molecular Species, and Photons*, Springer Proceedings in Physics 230,
https://doi.org/10.1007/978-981-13-9969-5_6

and a careful study of the autocorrelation function offers the separation at much smaller time ($T_{rev}/8$), where the wave packets become quasi-orthogonal, thereby making them distinguishable.

2 Theoretical Model

The phenomena of revival and fractional revival have been experimentally observed in the course of free evolution of the wave packets involving vibrational levels. Here, we consider the Morse potential to study the system as it is capable of capturing the vibrational dynamics of diatomic molecules [7, 8]. For a diatomic molecule with the vibrational motion, the Morse potential has the form

$$V(x) = D\big(e^{-2\beta x} - 2e^{-\beta x}\big) \tag{1}$$

where $x = \frac{r}{r_0} - 1$, r_0 is the internuclear distance at the equilibrium, D is the dissociation energy, and β is a range parameter. For Bromine molecules, we have 30 bound states, with $\beta = 4.482$, reduced mass $\mu = 71933.82$ a.u. for $^{79}Br_2$ and $\mu = 73754.95$ a.u. for $^{81}Br_2$, $r_0 = 4.310654$, and D = 0.073855238 for both the isotopes. Denoting

$$\lambda = \sqrt{\frac{2\mu D r_0^2}{\beta^2 \hbar^2}} \text{ and } s = \sqrt{\frac{-8\mu r_0^2}{\beta^2 \hbar^2}} E, \tag{2}$$

the eigenfunctions of the Morse potential is given by

$$\Psi_n^\lambda(\xi) = N e^{-\frac{\xi}{2}} \xi^{\frac{s}{2}} L_n^s(\xi), \tag{3}$$

where $L_n^s(\xi)$ is the associated Laguerre polynomial and N is the normalization constant given by

$$N = \left[\frac{\beta(2\lambda - 2n - 1)\Gamma(n+1)}{\Gamma(2\lambda - n)r_0} \right]^{1/2} \tag{4}$$

Here, $\xi = 2\lambda e^{-\beta x}$, $0 < \xi < \infty$ and $n = 0, 1, \ldots, [\lambda - 1/2]$. Denoting $[\rho]$ as the largest integer smaller than ρ, the total number of bound states become $[\lambda - 1/2] + 1$. The constraint equation is $s + 2n = 2\lambda - 1$. Here, λ is the potential dependent term and s is related to energy E.

In our analysis, we consider the coherent state (CS) of Morse potential which involves only finite numbers of bound states and is given by

$$\chi(\xi) = \sum_{m=0}^{n'} d_m \psi_m^\lambda(\xi), \tag{5}$$

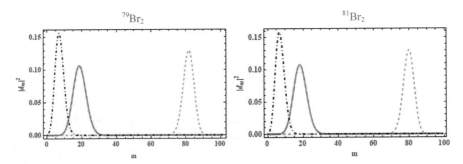

Fig. 1 The $|d_m|^2$ plotted as a function of m for the Morse potential of $^{79}Br_2$ and $^{81}Br_2$ for the values of $\alpha = 0.2$ (Orange, Dashed), $\alpha = 1.4$ (Red, Thick), $\alpha = 2.5$ (Black, Dot Dashed)

where

$$d_m = \frac{(-\alpha)^{n'-m}}{(n'-m)!} \left[\frac{n'!\Gamma(2\lambda - m)}{m'!\Gamma(2\lambda - n')} \right]^{1/2} \tag{6}$$

Figure 1 shows the $|d_m|^2$ distribution of $^{79}Br_2$ and $^{81}Br_2$ for three values of α [9]. When α has a smaller value, $|d_m|^2$ is peaked at higher value of m, where the anharmonicity is larger. The CS wave packet is not well localized due to the presence of an oscillatory tail. As we increase the value of α, the distribution of $|d_m|^2$ moves toward lower levels and the tail disappears. At the higher values of α, the CS wave packet consists only the lower levels, where the anharmonicity is small. Thus, the choice of distribution is important in making the initial wave packet localized.

Temporal evolution of CS state wave packet is given by

$$\chi(\xi) = \sum_m d_m \psi_m^\lambda(\xi) \exp[-iE_m t] \tag{7}$$

with

$$E_m = -(D/\lambda^2)(\lambda - m - 1/2)^2 \tag{8}$$

Due to the quadratic nature of the energy spectrum, the classical and revival times are given by $T_{cl} = T_{rev}/(2\lambda - 1)$ and $T_{rev} = 2\pi\lambda^2/D$, respectively. Figure 2 shows the CS wave packet distribution in the coordinate space for both the isotopes. Here the revival behaviors at $T_{rev}/4$ and $T_{rev}/8$ are not transparent, hence they seem indistinguishable at this moment. By making use of the autocorrelation function, one can find the separation time for the isotopes at around T_{rev}. However, by introducing a time delay between the two lasers and investigating the time evolution of the quantum wave packets we can find that at $t = T_{rev}/8$ the two isotopes are distinguishable.

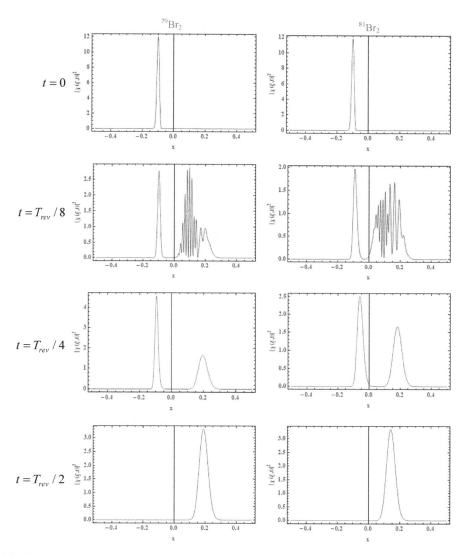

Fig. 2 The wave packet distribution in coordinate space for Bromine molecules at various revival times. Here, $\alpha = 1.4$, $\beta = 4.482$

3 Results and Discussions

The CS wave packet is formed by the coherent superposition of a finite number of energy eigenstates of Morse oscillator. Initially, the CS starts oscillating without much change in its shape. Later, it dissipates and again revives at its revival time T_{rev}. In Fig. 3, the autocorrelation functions for both the isotopes are depicted. At $t \approx T_{rev}$, the CS revives without any phase shift and is called the full revival. The separation in the

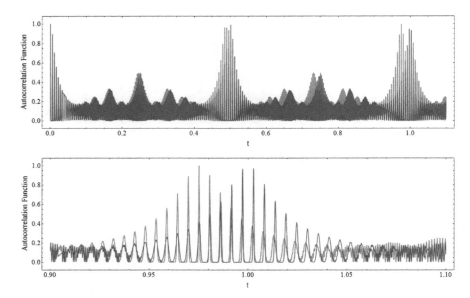

Fig. 3 Above, autocorrelation functions for both $^{79}Br_2$ (red) and $^{81}Br_2$ (blue) molecules are shown. Below is the enlarged view of the functions at the time where the isotopes are getting separated ($\approx T_{rev}$). Time has been scaled by T_{rev} of the first isotope

autocorrelation function is appreciable only after $t \sim T_{rev}$. An enlarged view of the spikes around $t = T_{rev}$ is displayed in the second plot of Fig. 3.

Now, by introducing a time delay between the lasers which excite the initial wave packets, we can get the particular fractional revival for both the isotopes at the same time, at about $t \simeq T_{rev}/8$. Hence, the observation time is exactly at the chosen fractional revival instant. The two quantum states become distinguishable for the two isotopes at this selected observation time (Fig. 4).

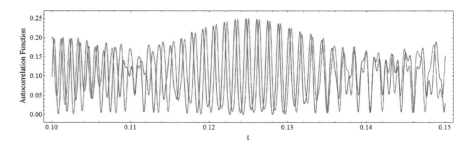

Fig. 4 With the initial time delay, we get the particular fractional revival for both the isotopes at the same time at $t \approx T_{rev}/8$

4 Conclusion

In earlier studies, the autocorrelation functions of $^{79}Br_2$ and $^{81}Br_2$ become distinguishable at $t = T_{rev,}$ which is a very long time as far as experimental feasibility is concerned. By introducing a time delay, we choose the observation time at particular fractional revivals for both the isotopes. The two quantum states become distinguishable for the two isotopes at the selected observation time. In the present method, we are capable of separating isotope in much smaller evolution time, compared to the existing methods in the literature.

References

1. Averbukh, I.S., Vrakking, M.J.J., Villeneuve, D.M.: Wave packet isotope separation. Phys. Rev. Lett. **77**, 3518 (1996). https://doi.org/10.1103/PhysRevLett.77.3518
2. Agarwal, G.S., Pathak, P.K.: Mesoscopic superposition of states with sub-Planck structures in phase space. Phys. Rev. A **70**, 053813 (2004). https://doi.org/10.1103/PhysRevA.70.053813
3. Toscano, F., Dalvit, D.A.R., Davidovich, L., Zurek, W.H.: Sub-Planck phase-space structures and Heisenberg-limited measurements. Phys. Rev. A **73**, 023803 (2006). https://doi.org/10.1103/PhysRevA.73.023803
4. London, H., Phil, D.: Separation of Isotope. George Newsnes Limited, London (1961)
5. Bluhm, R., Kostelecky, V.A., Porter, J.A.: The evolution and revival structure of localized quantum wave packets. Am. J. Phys. **64**, 944–953 (1995). https://doi.org/10.1119/1.18304
6. Eryomin, V.V., Vetchinkin, S.I., Umanskii, I.M.: Manifestations of wave packet fractional revivals in a Morse-like anharmonic system. J. Chem. Phys. **101**, 10730 (1994). https://doi.org/10.1063/1.467885
7. Ghosh, S., Chiruvelli, A., Banerji, J., Panigrahi, P.K.: Mesoscopic superposition and sub-Planck-scale structure in molecular wave packets. Phys. Rev. A **73**, 013411 (2006). https://doi.org/10.1103/PhysRevA.73.013411
8. Averbukh, I.S., Perelman, N.F.: Fractional revivals: Universality in the long-term evolution of quantum wave packets beyond the correspondence principle dynamics. Phys. Rev. A **139**(9), 449–453 (1989). https://doi.org/10.1016/0375-9601(89)90943-2
9. Roy, U., Ghosh, S., Panigrahi, P.K., Vitali, D.: Sub-Planck-scale structures in the Pöschl-Teller potential and their sensitivity to perturbations. Phys. Rev. A **80**, 052115 (2009). https://doi.org/10.1103/PhysRevA.80.052115

Electron Induced Chemistry
of Chlorobenzene

Dineshkumar Prajapati[1](✉), Hitesh Yadav[2], Minaxi Vinodkumar[3],
P. C. Vinodkumar[2], and Chetan Limbachiya[4]

[1] Shree M. R. Arts & Science College, Rajpipla, Gujarat, India
dineshphy13@gmail.com
[2] Department of Physics, Sardar Patel University, Vallabh Vidyanagar,
Gujarat, India
[3] Electronics Department, V. P. & R. P. T. P. Science College, Vallabh Vidyangar,
Gujarat, India
[4] Department of Physics, The M. S. University Baroda, Vadodara, Gujarat, India

Abstract. Cross sections for the electron impact scattering by
chlorobenzene (C_6H_5Cl) are calculated using the R-matrix method and
the spherical complex optical potential (SCOP) method in low energy
(0.1–20 eV) and in high energy (20–5000 eV) regimes, respectively. Using
both the methods, we are able to report data on differential, ionization,
and total cross sections (TCS).

1 Introduction

Electron impact studies with organic targets gained prominence after the study,
that secondary electrons produced by energetic radiations are responsible for
single- and double-strand breaks in DNA. Moreover, systematic and detailed
knowledge of cross sections resulting from electron collisions with simple organic
systems can help us to understand the behaviour of more complex biomolecules.
Chlorobenzene has a spectrum of transient anion state [1] and has very broad
applications, found in the planetary atmosphere [2] and used as radiosensitizer
drugs [3]. This makes the study of (C_6H_5Cl) more appropriate.

Still with the sophisticated techniques available today, there is a scarcity of
cross-sectional data on chlorobenzene. There are a very few works reported on
the cross-sectional study of derivative of benzene molecules. Few efforts are made
by theorist and experimental groups. One of the major contributions comes from
Barbosa et al. [1], they have provided data for DCS and total cross section both
theoretically and experimentally. The other theorist group that gave TCS and ionization cross section is Singh et al. [4]. Lunt et al. [5] gave experimental TCS for
the low-energy region and Makochekanwa et al. [6] provide the TCS from intermediate energy to high energy regions. From the literature survey, it can be observed
that the TCS data available in the literature are scattered. Thus, here we made
an attempt to fill this gap by providing the TCS data from 0.01 to 5000 eV.

© Springer Nature Singapore Pte Ltd. 2019
P. C. Deshmukh et al. (eds.), *Quantum Collisions and Confinement of Atomic
and Molecular Species, and Photons*, Springer Proceedings in Physics 230,
https://doi.org/10.1007/978-981-13-9969-5_7

In this paper, we focus on the calculation of electron collision with chloroben-zene over a broad range of impact energies starting from 0.01 to 5000 eV.

2   Theoretical Methodology

The aim of this study is to compute the electron impact scattering cross-sectional data for the chlorobenzene molecule from low to high energy (0.01–5000 eV). As a single theoretical method is insufficient to calculate these cross sections for such a wide energy range, we have employed two different meth-ods to calculate these cross sections. For the low-energy ($\leq$20 eV) study, we have used the R-matrix ab initio method through the Quantemol-N package [7] and further we have employed the Spherical Complex Optical Potential (SCOP) formalism at energies beyond the ionization threshold of the target. Much details of the theoretical methodologies adopted for such calculations can be found in our earlier publications [8–12]. Here, we briefly explain these two for-malisms for the low-energy as well as for high-energy calculations in the following subsections.

2.1   Low-Energy Formalism (0.01–15 eV)

As described in the earlier publications [8,10,12], the basic principle behind the R-matrix method [13,14] is that the configuration space is divided into two spatial regions. The inner region has a radius of 13 a.u., having short-range potential due to e–e correlation and exchange effects. The electron–chorobenzene R-matrix calculation is done using the quantum chemistry codes [14–16] and the outer region calculation is extended to ~100 a.u. The inner region wave function is described as

$$\Psi_k^{N+1}(x_1...x_{N+1}) = A \sum_{ij} a_{ijk} \Phi_i^N(x_1...x_N) u_{ij}(x_{N+1}) + \sum_i b_{ik} \chi_i^{N+1}(x_1...x_{N+1})$$

(1)

In (1), Φ_i represents the wave function of ith target state. A represents anti-symmetrization operator and u_{ij} denotes the continuum orbitals to represent the scattering electron and a_{ijk} and b_{ik} are the variational parameters. The summa-tion runs over χ_i for N+1 electrons and denotes L^2 function at the boundary of R-matrix with zero amplitude. The continuum orbitals are discussed briefly by Faure et al. [17]. For permanent dipole moment of the target, long-range interaction including $l > 4$ partial waves, we used Born correction for polar tar-gets [18–20]. In the outer region at the boundary of R-matrix, the asymptotic solution is defined by Gailitis [21].

2.2   Intermediate to High-Energy Formalism

For intermediate to high-incident energy for e-(C_6H_5Cl) system, we have used Spherical Complex Optical Potential (SCOP) formalism. The optical potential as given by

$$V_{opt}(r, E_i) = V_{st}(r, E_i) + V_{ex}(r, E_i) + V_{pol}(r, E_i) + iV_{abs}(r, E_i) \qquad (2)$$

where E_i is the incident energy and V_{st}, V_{ex}, V_{pol} and V_{abs} represents static, exchange, polarization, and absorption potential. The absorption potential V_{abs}, is employed with few modifications in the quasi-free model of Staszewska et al. [22,23]. For scattering calculation, we solved Schrödinger equation by partial-wave method. The partial waves' phase-shifts were employed to calculate total (elastic + inelastic) cross sections Q_{TCS}

$$Q_{TCS}(E_i) = Q_{el}(E_i) + \sum Q_{inel}(E_i) \qquad (3)$$

3   Results and Discussion

In Table 1, we present the calculated target properties of the chlorobenzene molecule using cc-pVDZ basis set. The presented properties are in good agreement with the available experimental as well as theoretical data in the literature.

In Fig. 1 we report the differential cross section (DCS) for e-(C_6H_5Cl) scattering at 10 eV. The present data is in good agreement with the data available in the literature qualitatively. While quantitatively, it is in good agreement up to 40° with both the theoretical and experimental data Barbosa et al. [1]. Near around 50° present data shows a shallowness in the cross sections which resembles in experimental data and theoretical data of Barbosa et al. [1]. While in the backward angles ($\geq 130°$), the present data underestimates the available theoretical data of Barbosa et al. [1]. Here we mainly look to reproduce the structures in comparison with the literature. The discrepancy in the case of theoretical results is due to the different methodology and in case of experimental, it depends on the alignment of the experimental setup. The large dipole moment and polarizability plays an important role in the low to intermediate energy electron scattering cross sections.

In Fig. 2, we report the DCS for e-(C_6H_5Cl) scattering at 20 eV. The present result is in very good agreement with the available experimental and theoretical data by Barbosa et al. [1]. The choice of this sample energies is only because resonances can be studied. A lot of discussion about DCS is reported by Barbosa et al. [1].

Table 1. Target properties of chlorobenzene

Target property (unit)		Present	Others
Ground state (Hartree)		−688.64	−689.99 [1]
Ionization potential (eV)		9.200	9.080 [1]
Dipole moment (Debye)		1.689	1.690 [1]
Rotational constants (cm^{-1})	A	0.1893	0.1892 [24]
	B	0.0527	0.0526 [24]
	C	0.0412	0.0411 [24]

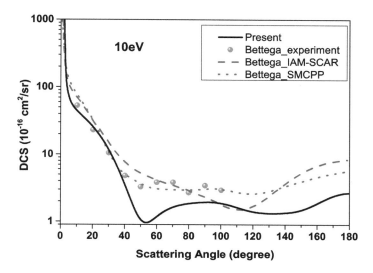

Fig. 1. Elastic differential cross sections (DCS) for incident energy of 10 eV, where solid line represents present calculated data, solid sphere line represents experimental data reported by Barbosa et al. [1], dash line represents calculated data reported for IAM-SCAR Model by Barbosa et al. [1] and dot line represents calculated data reported for SMCPP Model by Barbosa et al. [1]

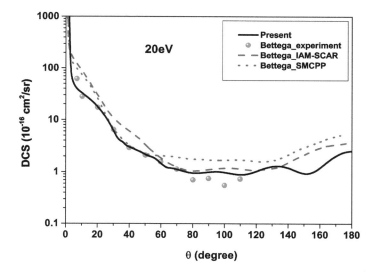

Fig. 2. Elastic differential cross sections (DCS) for incident energy of 10 eV, where solid line represents present calculated data, solid sphere line represents experimental data reported by Barbosa et al. [1], dash line represents calculated data reported for IAM-SCAR Model by Barbosa et al. [1], and dot line represents calculated data reported for SMCPP Model by Barbosa et al. [1]

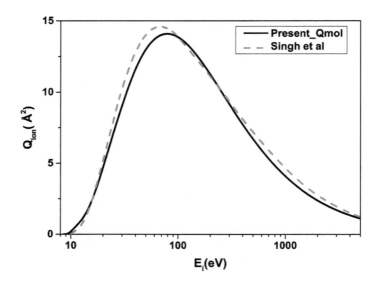

Fig. 3. e-C_6H_5Cl total ionization cross sections, where solid line represents the present calculated data, dash line represents calculated data from Singh [4]

In Fig. 3, we represent the total ionization cross section calculated using the BEB method incorporated with the Quantemol-N package [7]. The present data is in very good agreement with a theoretical data presented by Singh et al. [4].

In Fig. 4, we report the total scattering cross sections (TCS) of chlorobenzene molecule. The present data is in good agreement with available data in the literature [1, 4–6]. The low-energy TCS is in good agreement with Barbosa et al. [1] computed using the IAM model for the cross-sectional calculation. Whereas it overestimates the other theoretical values done by Barbosa et al. [1] using different models. While in the case of experimental comparison the present data is in good agreement with the Makochekanwa et al. [6] from intermediate to high- energy regime. There is a theoretical data in the high energy regime for the comparison given by Singh et al. [4] that nominally under estimates the present data. The difference observed in the data of Singh et al. [4] is due to the difference between the single centre expansion employed in our calculations as against the multi-centre expansion employed by them. The peaks in the region between 1 and 2 eV represent the resonance peak at energy 1.24 and 1.42 eV, which is due to the degeneracy of b1 and a2 symmetry as π^* states. The reason behind this splitting is the weakness of π^* bond in the excitation function, which is quite evident in the cross sections reported in the Fig. 4.

4   Conclusion

In the present study, we report the electron impact scattering cross sections of chlorobenzene molecule. The present DCS data is in good agreement with

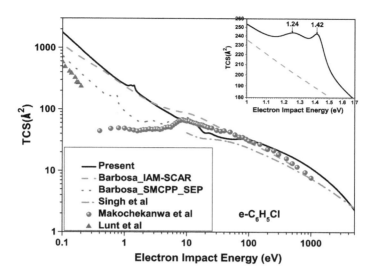

Fig. 4. e-C_6H_5Cl Total scattering cross sections, where solid line represents present calculated data, dash line represents calculated data reported for IAM-SCAR Model by Barbosa et al. [1] and dot line represents calculated data reported for SMCPP Model by Barbosa et al. [1], dash-dot line represents calculated data from Singh [4], solid sphere line represents experimental data reported by Makochekanwa et al. [6], and solid triangle line represents experimental data reported by Lunt et al. [5]

available experimental data reported by Barbosa et al. [1]. The ionization cross section is in very good agreement with the Singh et al. [4] and the TCS is overall in good agreement with available theoretical and experimental data [1,4–6]. The present work will encourage the other experimentalist and theorist to study derivatives of benzene as these are important applied targets.

Acknowledgement. The author acknowledges DST-SERB for the major research project [EMR/2016/000470] for financial support under which part of this work is carried out.

References

1. Barbosa, A.S., Varella, M.T.N., Sanchez, S.A., Ameixa, J., Blanco, F., García, G., Limão-Vieira, P., Ferreira da Silva, F., Bettega, M.H.: Theoretical and experimental study on electron interactions with chlorobenzene: shape resonances and differential cross sections. J. Chem. Phys. **145**, 084311 (2016). https://doi.org/10.1063/1.4961649
2. Freissinet, C., Glavin, D.P., Mahaffy, P.R., Miller, K.E., Eigenbrode, J.L., Summons, R.E., Brunner, A.E., Buch, A., Szopa, C., Archer, P.D.: Organic molecules in the sheepbed mudstone, gale crater, mars. J. Geophys. Res. **120**, 495–514 (2015). American Geophysical Union (AGU). https://doi.org/10.1002/2014JE004737

3. Wardman, P.: Chemical radiosensitizers for use in radiotherapy. Clin. Oncol. **19**, 397–417 (2007). Elsevier BV. https://doi.org/10.1016/j.clon.2007.03.010

4. Singh, S., Naghma, R., Kaur, J., Antony, B.: Calculation of total and ionization cross sections for electron scattering by primary benzene compounds. J. Chem. Phys. **145**, 034309 (2016). https://doi.org/10.1063/1.4955205

5. Lunt, S.L., Field, D., Hoffmann, S.V., Gulley, R.J., Ziesel, J.-P.: Very low energy electron scattering in C_6H_5F, C_6H_5Cl, C_6H_5Br and C_6H_5I. J. Phys. B At. Mol. Opt. Phys. **32**, 2707-2717 (1999). https://doi.org/10.1088/0953-4075/32/11/317

6. Makochekanwa, C., Sueok, O., Kimura, M.: A comparative study of electron and positron scattering from chlorobenze C_6H_5Cl and chloropentafluorobezebe C_6F_5Cl molecules. J. Chem. Phys. **119**, 12257 (2003). https://doi.org/10.1063/1.1626115

7. Tennyson, J., Brown, D.B., Munro, J.J., Rozum, I., Varambhia, H.N., Vinci, N.: Quantemol-N: an expert system for performing electron molecule collision calculations using the R-matrix method. J. Phys. Conf. Ser. **86**, 012001 (2007). https://doi.org/10.1088/1742-6596/86/1/012001

8. Yadav, H., Vinodkumar, M., Limbachiya, C., Vinodkumar, P.C.: Scattering of electrons with formyl radical. Mol. Phys. **115**, 952 (2017). https://doi.org/10.1080/00268976.2017.1293306

9. Swadia, M., Thakar, Y., Vinodkumar, M., Limbachiya, C.: Theoretical electron impact total cross sections for tetrahydrofuran (C_4H_8O). Eur. Phys. J. D **71**, 85 (2017). https://doi.org/10.1140/epjd/e2017-70617-9

10. Vinodkumar, M., Desai, H., Vinodkumar P.C., Mason, N.: Induced chemistry by scattering of electrons from magnesium oxide. Phys. Rev. A **93**, 012702 (2016). https://doi.org/10.1103/PhysRevA.93.012702

11. Yadav, H., Vinodkumar, M., Limbachiya, C., Vinodkumar, P.C.: Study of electron impact inelastic scattering of chlorine molecule (Cl_2). J. Phys. B At. Mol. Opt. Phys. **51**, 045201 (2018). https://doi.org/10.1088/1361-6455/aaa2d6

12. Vinodkumar, M., Limbachiya, C., Chaudhari, A., Desai, H.: Electron induced chemistry of thioformaldehyde. RSC Adv. **5**, 103964 (2015). https://doi.org/10.1039/C5RA19662K

13. Bouchiha, D., Gorfinkiel, J.D., Caron, L.G., Sanche, L.: Low-energy electron collisions with methanol. J. Phys. B At. Mol. Opt. Phys. **40**, 1259 (2007). https://doi.org/10.1088/0953-4075/40/6/016

14. Tennyson, J.: Electron-molecule collision calculations using the R-matrix method. Phys. Rep. **491**, 29 (2010). https://doi.org/10.1016/j.physrep.2010.02.001

15. Mclean, A.D.: Proceedings of Conference on Potential Energy Surfaces in Chemistry. In: Lester, W.A. Jr. (ed.). IBM Reserach Laboratory, San Jose, p. 87 (1971)

16. Mclean, A.D., Yoshimine, M., Lengsfield, B.H., Bagus P.S., Liu, B.: Modern Techniques in Computational Chemistry, MOTECC-91. In: Clementi, E. (ed.) Escom, Leiden (1991)

17. Faure, A., Gorfinkiel, J.D., Morgan, L.A., Tennyson, J.: GTOBAS: fitting continuum functions with Gaussian-type orbitals. Comput. Phys. Commun. **144**, 224 (2002). https://doi.org/10.1016/S0010-4655(02)00141-8

18. Chu, S.-I., Dalgarno, A.: Rotational excitation of CH_+ by electron impact. Phys. Rev. A **10**, 788 (1974). https://doi.org/10.1103/PhysRevA.10.788

19. Itikawa, Y.: Rotational transition in an asymmetric-top molecule by electron collision: applications to H_2O and H_2CO. J. Phys. Soc. Jpn. **32**, 217 (1972). https://doi.org/10.1143/JPSJ.32.217

20. Norcross, D.W., Padial, N.T.: The multipole-extracted adiabatic-nuclei approximation for electron-molecule collisions. Phys. Rev. A **25**, 226 (1982). https://doi.org/10.1103/PhysRevA.25.226

21. Gailitis, M.: New forms of asymptotic expansions for wavefunctions of charged-particle scattering. J. Phys. B At. Mol. Opt. Phys. **9**, 843 (1976). https://doi.org/10.1088/0022-3700/9/5/027

22. Staszewska, G., Schwenke, D.W., Truhlar, D.G.: Complex optical potential model for electron–molecule scattering, elastic scattering, and rotational excitation of H_2 at 10–100 eV. J. Chem. Phys. **81**, 335 (1984). https://doi.org/10.1063/1.447310

23. Staszewska, G., Schwenke, D.W., Truhlar, D.G.: Investigation of the shape of the imaginary part of the optical-model potential for electron scattering by rare gases. Phys. Rev. A **29**, 3078 (1984). https://doi.org/10.1103/PhysRevA.29.3078

24. Computational Chemistry Comparison and Benchmark DataBase. https://cccbdb.nist.gov/

Quantum Dynamics and Frequency Shift of a Periodically Driven Multi-photon Anharmonic Oscillator

Dolan Krishna Bayen and Swapan Mandal$^{(\boxtimes)}$

Department of Physics, Visva-Bharati, Santiniketan 731235, India
swapanvb@rediffmail.com

Abstract. The Hamiltonian and hence the relevant equations of motion involving the dynamics of a driven classical anharmonic oscillator with 2m–th anharmonicity are framed. By neglecting the nonsynchronous energy terms, we derive the model of a driven multi-photon (2m–th) quantum anharmonic oscillator. The dynamical nature of the field operators is expressed in terms of the coupling constant, excitation number and the driven parameter. The solutions presented here are fully analytical and are exact in nature.

PACS 03.65.Ge · 05.45.-a · 02.30.Hq

1   Introduction

The basic physics is easily understood in terms of the physical models. Of course, the model of a harmonic oscillator (HO) is perhaps the most useful one among them. The model of a simple harmonic oscillator is realized when a particle moves under the action of a restoring force. In spite of the wide applications of the SHO model, it is inadequate when we come across with the real physical situations. For a real physical system, the inclusion of damping and/or anharmonicities in the model of HO are inevitable. In addition to these, the oscillator may also be put under the action of an external force and hence the model of a forced (driven) oscillator. It is true that the presence of damping in the model of an SHO is not a serious problem as long as we are interested in the classical regime. On the other hand, the presence of damping in the model of a SHO makes the problem a nontrivial one. Therefore, the damped quantum harmonic oscillator requires special attention. In this short communication, we will ignore the presence of damping if any. Because of the wide range of applications and of the fundamental nature of the problem, the problems of anharmonic oscillator have attracted people from various branches of physics [1–8].

© Springer Nature Singapore Pte Ltd. 2019
P. C. Deshmukh et al. (eds.), *Quantum Collisions and Confinement of Atomic and Molecular Species, and Photons*, Springer Proceedings in Physics 230,
https://doi.org/10.1007/978-981-13-9969-5_8

2   Driven Classical Anharmonic Oscillator

The Hamiltonian of a driven classical anharmonic oscillator of unit mass is given by

$$H = \frac{p^2}{2} + \frac{q^2}{2} + \frac{\lambda}{2m}q^{2m} + fq \tag{1}$$

where q and p are the classical position and momentum, respectively. The parameter λ is a small positive constant and will be termed as anharmonic constant. The frequency of the oscillator governed by the (1) is assumed to be 1. The m ($m \geq 2$) being a positive integer involving the order of anharmonicity. From our choice of the Hamiltonian, it appears that we will be investigating for the anharmonic oscillators of having even orders of anharmonicities. It is attributed because of the fact that most of the time we are interested in the matter–field interaction to discuss the anharmonic effects. The medium with inversion symmetry automatically eliminates the even orders of nonlinear susceptibilities and hence the even orders of nonlinear polarization. Again, the odd orders of nonlinear polarization contribute to the even orders of nonlinear terms in the model Hamiltonian. By virtue of symmetry, the entire gaseous medium is inversion symmetric. Therefore, the choice of the Hamiltonian is quite consistent with the matter–field interaction. Now, the equations of motion corresponding to the Hamiltonian (1) follows from the Hamiltonian's equation. Therefore, we have

$$\begin{aligned}\dot{q} &= p \\ \dot{p} &= -q - \lambda q^{2m-1} - f\end{aligned} \tag{2}$$

The above two equations (2) are combined to obtain a second-order inhomogeneous nonlinear differential equation. The corresponding equation is given by

$$\ddot{q} + q + \lambda q^{2m-1} = -f \tag{3}$$

In absence of driven term (i.e., $f = 0$) and for $m = 2$, the (3) corresponds to the equation of motion of a classical quartic anharmonic oscillator. Unfortunately, the classical quartic anharmonic oscillator does not give a closed-form analytical solution. Of course, for small values of λ, the classical quartic anharmonic oscillator is solved approximately [6]. In absence of anharmonic and driven terms, the Hamiltonian reduces to the Hamiltonian of a simple (one dimensional) harmonic oscillator. Now, the quantum mechanical counterpart of the Hamiltonian (1) is obtained by the replacement of the classically conjugate position $q(t)$ and momentum $p(t)$ by their corresponding operators. During the passage from classical anharmonic oscillator governed by the Hamiltonian (1) to the corresponding quantum mechanical oscillator, the fundamental commutation relation between the position and momentum operators should be respected.

3   Driven Quantum Anharmonic Oscillator

In order to obtain the quantum mechanical counterpart of the Hamiltonian (1), we replace the classical canonical position (q) and momentum (p) by their equivalent operators. Therefore, we have

$$\hat{H} = \frac{\hat{p}^2}{2} + \frac{\hat{q}^2}{2} + \frac{\lambda}{2m}\,\hat{q}^{2m} + f\hat{q} \tag{4}$$

In the Hamiltonian (4), we assumed that the mass and the frequency of the oscillator are unity. During the passage from classical-driven anharmonic oscillator to its quantum mechanical counterpart, we need to the impose the following equal time commutation relation [9,10]:

$$[\hat{q}(t),\,\hat{p}(t)] = [\hat{q}(0),\,\hat{p}(0)] = i \tag{5}$$

where $\hbar = 1$. Now, in order to differentiate between the *c-numbers* and the operators, we use the caret for the operators. Based on the relation (5), we define the usual relations connecting the position and momentum operators with those of the dimensionless annihilation ($\hat{a}$) and creation ($\hat{a}^\dagger$) operators

$$\begin{aligned}
\hat{a}(t) &= \tfrac{1}{\sqrt{2}}\left(\hat{q}(t) + i\hat{p}(t)\right) \\
\hat{a}^\dagger(t) &= \tfrac{1}{\sqrt{2}}\left(\hat{q}(t) - i\hat{p}(t)\right)
\end{aligned} \tag{6}$$

Hence, we have

$$\begin{aligned}
\hat{q}(t) &= \tfrac{1}{\sqrt{2}}\left(\hat{a}(t) + \hat{a}^\dagger(t)\right) \\
\hat{p}(t) &= -\tfrac{i}{\sqrt{2}}\left(\hat{a}(t) - \hat{a}^\dagger(t)\right)
\end{aligned} \tag{7}$$

Therefore, it follows that the annihilation operator $\hat{a}(t)$ and creation operator $\hat{a}^\dagger(t)$ obey the following commutation relation:

$$[\hat{a}(t),\,\hat{a}^\dagger(t)] = 1 \tag{8}$$

In the later part of this article, we shall simply use $\hat{a}$ and $\hat{a}^\dagger$ instead of $\hat{a}(t)$ and $\hat{a}^\dagger(t)$ respectively. Now, the Hamiltonian follows as

$$\hat{H} = \left(a^\dagger a + \frac{1}{2}\right) + \frac{\lambda}{2m}\frac{1}{2^m}\left(\hat{a} + \hat{a}^\dagger\right)^{2m} + f\left(\hat{a} + \hat{a}^\dagger\right) \tag{9}$$

In order to obtain the equations of motion involving the operators $\hat{q}$ and $\hat{p}$, we use the Heisenberg operator equation of motion. Therefore, we have

$$\dot{a} = -ia - \frac{i\lambda}{2m}\left(\hat{a} + \hat{a}^\dagger\right)^{2m-1} - if \tag{10}$$

where, we have made use of the commutation relation $\left[a, g(a, a^\dagger)\right] = \frac{\partial g(a, a^\dagger)}{\partial a^\dagger}$. The equation of motion for the field operator $a^\dagger$ may easily be obtained by taking the Hermitian conjugate of the (10). In order to study the dynamical behavior of the field operators, we need to solve the differential equation (10). Unfortunately,

the differential equation is nonlinear and is involving the noncommuting operators. Therefore, a numerical solution or an approximate analytical solution may
be explored. The differential equation involving the noncommuting operators are
certainly a nontrivial problem if at all possible. Therefore, in the present article, we rely on the so-called rotating wave approximation for solving the above
differential equation. The RWA entails us to remove the nonconserving energy
terms. In order to do so, we express the normal ordered form of $(\hat{a} + \hat{a}^\dagger)^{2m-1}$

$$\left(\hat{a} + \hat{a}^\dagger\right)^{2m-1} = \sum_{r=0}^{m-1} (2r-1)!! \begin{pmatrix} 2m-1 \\ 2r \end{pmatrix} : \left(\hat{a} + \hat{a}^\dagger\right)^{2m-2r-1} : \qquad (11)$$

where the notation :: stands for the normal ordered form, the binomial coefficient
$\begin{pmatrix} x \\ y \end{pmatrix} = \frac{x!}{(x-y)!y!}$ and $(2r-1)!! = (2r-1)(2r-3)....3.1$. Now, we make a binomial
expansion of the term $\left(: \left(\hat{a} + \hat{a}^\dagger\right)^{2m-2r-1} :\right)$ under normal ordered form. We
have $2m-2r$ number of terms. In this expansion, we have two middle terms. Out
of these two terms, the middle term $t_{m-r+1} = \begin{pmatrix} 2m-2r-1 \\ m-r+1 \end{pmatrix} \hat{a}^{\dagger m-r} \hat{a}^{m-r+1}$
will be the synchronous one. Now, we neglect all the terms except the syn-
chronous one in the expansion of $(\hat{a} + \hat{a}^\dagger)^{2m-1}$ in the (10). Hence, the (10)
assumes the following form:

$$\dot{\hat{a}} = -i[\hat{a}, \hat{H}] = -i\hat{O}\hat{a} - if \qquad (12)$$

where, the operator $\hat{O}(t)$ is given by

$$\hat{O}(t) = 1 + \frac{\lambda}{2^m} \sum_{r=0}^{m-1} (2r-1)!! \begin{pmatrix} 2m-1 \\ 2r \end{pmatrix} \begin{pmatrix} 2m-2r-1 \\ m-r+1 \end{pmatrix} \hat{a}^{\dagger m-r} \hat{a}^{m-r} \qquad (13)$$

The equation (12) of the field operator $\hat{a}$ corresponds the equation of motion of
the driven m-photon anharmonic oscillator. As a matter of fact, the m-photon
anharmonic oscillator is found useful for investigating squeezing [11–13], phase
properties [14], Wigner function [15], and the kth power squeezing of the radiation field [16]. However, the driven m-photon anharmonic oscillator is still unexplored. For $f = 0$, the operator $\hat{O}(t)$ is a constant of motion. The operator $\hat{O}(t)$
is a function of number operator $a^\dagger a$. This is quite reasonable since the conser-
vation of number is guaranteed by the use of RWA and hence the removal of
nonsynchronous terms from the equations of motion. The operator $\hat{O}$ has the
dimension of frequency and hence may be called as a frequency operator. In
absence of anharmonicity ($\lambda = 0$), the operator $\hat{O}$ reduces to a unit operator.
Essentially, the anharmonic part of the Hamiltonian (4) is responsible for the
shift of the frequency of the oscillator (here we assumed 1). In other words, the
frequency normalization takes place through the nonlinear term. The complete
solution of the differential equation (12) can only be obtained if the functional

form of $f(t)$ is explicitly known. The corresponding solution will follow as

$$e^{-i\hat{O}t}A - e^{-i\hat{O}t}i\int f(t)e^{i\hat{O}t}dt \tag{14}$$

where A is the integration constant to be evaluated under initial condition. Taking the Hermitian conjugate of the above solution (14), we obtain the the the solution for the creation operator $\hat{a}^\dagger$. Therefore, the dynamical behavior of the driven anharmonic oscillator under the rotating wave approximation are completely solved provided the integral (14) is evaluated for physically acceptable function $f(t)$. As an example, we assume that the oscillator is periodically driven. Therefore, we have

$$f(t) = f(0)\cos\omega t \tag{15}$$

Finally, the analytical expressions for the annihilation operator (14) assumes the following form:

$$\hat{a}(t) = e^{-i\hat{O}t}\hat{a}(0) + f(0)\left(\hat{O}^2 - \omega^2\right)^{-1}\left[\hat{O}e^{-i\hat{O}t} - \hat{O}\cos\omega t + i\omega\sin\omega t\right] \tag{16}$$

It is possible to obtain the creation operator $a^\dagger$ from the (16). Assuming $f(0)$ as real, we have

$$\hat{a}^\dagger(t) = \hat{a}^\dagger(0)e^{i\hat{O}t} + f(0)\left(\hat{O}^2 - \omega^2\right)^{-1}\left[\hat{O}e^{i\hat{O}t} - \hat{O}\cos\omega t - i\omega\sin\omega t\right] \tag{17}$$

where the Hermitian nature of the frequency operator $\hat{O}$ is utilized. Certainly, the solutions for the field operators involving the (16) and (17) correspond the physical model of a multiphoton driven anharmonic oscillator. Interestingly, the dynamics of the driven anharmonic oscillator corresponding to the Hamiltonian (9) is completely known. The exact nature of the solution (16) is certainly useful for investigating quantum statistical properties of the radiation field coupled to the oscillator. It is of interest to calculate the shift of the frequency of the oscillator by calculating the matrix elements of $< n|\hat{a}(t)|n+1 >$ in terms of the number state basis $|n>$. It is clear that the shift of the frequency of the oscillator is governed by the anharmonic parameter Before we go further, we check the equal time commutation relation between the annihilation and creation operators. This can easily be established since the time derivative of the following commutator vanishes:

$$\frac{d}{dt}\left[a(t), a^\dagger(t)\right] = 0 \tag{18}$$

In calculating the above relation (18), we have made use of the (12) and it's Hermitian conjugate. Now, we construct the dimensionless quadrature operators which are happened to be position and momentum operators involving the (7). Therefore, we have

$$\hat{q}(t) = \frac{1}{\sqrt{2}}\left(e^{-i\hat{O}t}\hat{a}(0) + a^\dagger(0)e^{i\hat{O}t} + 2f(0)\hat{O}\left(\hat{O}^2 - \omega^2\right)^{-1}\left\{\cos\hat{O}t - \cos\omega t\right\}\right)$$

$$\hat{p}(t) = -\frac{i}{\sqrt{2}}\left(e^{-i\hat{O}t}\hat{a}(0) - a^\dagger(0)e^{i\hat{O}t} - 2if(0)\left(\hat{O}^2 - \omega^2\right)^{-1}\left\{\hat{O}\sin\hat{O}t - \omega\sin\omega t\right\}\right)$$

$$\tag{19}$$

In addition to the fundamental academic interests, the present solution will find potential applications in the investigation of the quantum statistical properties of the radiation field coupled to a nonlinear medium. However, the present exact solution is obtained at the cost of the sacrifice of nonconserving energy (i.e., RWA) terms in the original Hamiltonian (13).

4   Conclusion

In the present investigation, we provide the solution of the driven anharmonic oscillator. The complete solution of the driven anharmonic oscillator requires the knowledge of the functional form of the driven parameter. The present solution is explored under the RWA and is certainly based on the complete analytical approach. The present solution will find applications in the investigation of quantum statistical properties of the radiation fields and in the dynamics of the trapped ions/atoms in an MOT. These quantum statistical properties include squeezing, higher ordered squeezing, photon antibunching, and photon statistics.

Acknowledgements. The author is thankful to the University Grants Commission, New Delhi for financial support through a major research project (F.No.42-852/2013(SR)). SM is also thankful to the Council of Scientific and Industrial Research (CSIR), New Delhi for financial support (03(1283)/13/EMR-II). DKB thanks the CSIR for awarding the JRF (09/202(0062)/2017-EMR-I).

References

1. Nayfeh, A.H.: Introduction to Perturbation Techniques. Wiley, New York (1981)
2. Nayfeh, A.H., Mook, D.T.: Non-linear Oscillations. Wiley, New York (1979)
3. Bellman, R.: Methods of Nonlinear Analysis, vol. 1, p. 198. Academic Press, New York (1970)
4. Ross, S.L.: Differential Equation, 3rd edn. Wiley, New York (1984)
5. Mandal, S.: Phys. Lett. A **299**, 531–542 (2002)
6. Mandal, S.: J. Phys. A **31**, L501–L505 (1998)
7. Pathak, A., Mandal, S.: Phys. Lett. A **286**, 261–276 (2001)
8. Bender, C.M., Bettencourt, L.M.A.: Phys. Rev. Lett **77**, 4114–4117 (1996)
9. Gasiorowicz, S.: Quantum Physics, p. 271. Wiley, New York (1979)
10. Schiff, L.I.: Quantum Mechanics, vol. 3, p. 176. Mc. Graw Hill Book Company, New York (1987)
11. Gerry, C.C.: Phys. Lett. A **124**, 237–239 (1987)
12. Buzek, V.: Phys. Lett. A **136**, 188–192 (1989)
13. Tanas, R.: Phys. Lett. A **141**, 217–220 (1989)
14. Paprzycka, M., Tanas, R.: Quant. Opt. **4**, 331 (1992)
15. Kheruntsyan, K.V.: J. Opt. B Quantum Semiclass. Opt. **1225** (1999)
16. Buzek, V., Jex, I.: Phys. Rev. A **41**, 4079 (1990)

Diagnostics of Ar/N$_2$ Mixture Plasma with Reliable Electron Impact Argon Excitation Cross Sections

S. Gupta[1], R. K. Gangwar[2], and Rajesh Srivastava[1(✉)]

[1] Department of Physics, Indian Institute of Technology Roorkee,
Roorkee 247667, India
sgupta@ph.iitr.ac.in, rajsrfph@iitr.ac.in
[2] Department of Physics, Visvesvaraya National Institute of Technology,
Nagpur 440010, India
reeteshkr@gmail.com

Abstract. The diagnostic study of Ar/N$_2$ mixture plasma at low temperature has been presented with varying concentrations of N$_2$ gas. For the diagnostics of such mixture plasma, a fine-structure resolved collisional-radiative (CR) model is developed. The CR model requires a large amount of accurate and reliable atomic cross-section data for the proper inclusion of all the relevant processes occurring in the plasma. Thus we incorporated in our model, electron-impact excitation cross-sections of argon calculated by using fully relativistic distorted wave method. Processes which account for the coupling of argon with nitrogen molecules have been further added in the model. We couple our CR model with the OES measurements of Lock et al. [Physics of plasmas 23, 043518 (2016)] and the electron density (n_e) and temperature (T_e) are simultaneously estimated by optimizing the model-simulated intensities with the five emission lines [originates from 3p^{5}4p ($2p_i$) $\rightarrow$ 3p^{5}4s ($1s_i$) transitions] observed in the measurements. On comparison of our results with the values reported by Lock et al. [Physics of plasmas 23, 043518 (2016)], we find that as the concentration of N$_2$ (0–10%) increases in the Ar/N$_2$ mixture, the electron density decreases by three orders of magnitude from 3.0×10^{13}–3.0×10^{10} cm^{-3} while T_e remains approximately close to 1.6 eV.

1 Introduction

Often inert gases are added in trace amounts to different plasmas for the diagnostics purposes, known as Trace-Rare-Gas Optical Emission Spectroscopy (TRG-OES) [1, 2]. In our earlier work, we developed collisional radiative (CR) models for pure Ar and Kr [3, 4] inert gas plasmas for the diagnostics of low temperature (LT) plasmas. Dominant processes in LT plasma are due to electron impact collisions [5]. The inert gases, viz., Ar, Kr, Xe, and Ne show several fine-structure levels due to strong relativistic effect/spin-orbit coupling [6, 7]. Consequently, several fine-structure transitions between the ground state to various fine-structure excited states and among the excited states are involved in the plasma kinetics. Thus, to develop plasma population-kinetic model, i.e., the CR model for inert gas plasmas, required fine-structure electron impact

© Springer Nature Singapore Pte Ltd. 2019
P. C. Deshmukh et al. (eds.), *Quantum Collisions and Confinement of Atomic and Molecular Species, and Photons*, Springer Proceedings in Physics 230,
https://doi.org/10.1007/978-981-13-9969-5_9

excitation cross-section data are calculated using fully relativistic distorted theory. The calculated cross sections are then incorporated into the CR model for diagnostic purposes. In continuation of our earlier work on pure Ar plasma [3], in the present study, we aim to develop a CR model of Ar/N₂ mixture plasma for diagnostic purposes.

The addition of N₂ reactive gas traces in the inert gas plasma is very useful for applications like thin film deposition [8, 9], surface chemistry [10], nitride thin films (viz, GaN and TiN) [10], material modification, etching [11], surface sterilization [12], and lighting [13]. To optimize the plasma processing in these applications, it is essential to understand how the mixing of N₂ in argon influences the plasma parameters [e.g., electron density (n_e) and electron temperature (T_e)] which behave differently depending on the quantity of N₂ introduced in the plasma. Thus, the aim of this work is to develop a reliable CR model for Ar/N₂ plasma and then apply it to understand the behavior of plasma parameters. Here, we applied our model for the diagnostics of an electron beam generated Ar/N₂ plasma. It is worth mentioning here that electron beam generated plasmas are different from the normal discharge plasma and only very few diagnostic studies are available in the literature for these plasmas [14].

In this light, in the present work by coupling our CR model with the OES measurements reported by Lock et al. [14], we performed the diagnostics of an electron beam generated Ar/N₂ plasma. In our study, we obtained the plasma parameters, viz, electron temperature, electron density, and species population which characterize the plasma [15] by optimizing five Ar-I lines with the measurements [14].

2   Collisional–Radiative Model

In the present CR model, we are considering 40 fine-structure levels, viz, $1s_i$ ($i = 2$–5), $2s_i$ ($i = 2$–5), $2p_i$ ($i = 1$–10), $3p_i$ ($i = 1$–10), and $3d_i$ ($i = 1$–12) with the inclusion of ground state and first ionization state of Ar. These states are interconnected through the various collisional and radiative processes occurring in the plasma. The model incorporates different population transfer mechanisms among fine-structure levels such as electron impact excitation, ionization, and radiative decay along with their reverse processes, viz, electron impact de-excitation, three-body recombination. In this work, we follow our earlier CR model on pure Ar and thus more details can be found from our previous publications [3, 16]. However, for Ar/N₂ mixture plasma, the depopulation channels of argon-excited states ($1s_i$ and $2p_i$ manifold) through quenching with N₂ molecule have been added. The particle balance equation for an excited level "u" is given by the following expression,

$$\sum_{\substack{l=1 \\ l \neq u}}^{41} k_{lu}(T_e)n_l n_e + \sum_{l>u} A_{lu}^{eff} n_l + n_e n_+ + n_e\, k_{+u}(T_e) - \sum_{\substack{l=1 \\ l \neq u}}^{41} k_{ul}(T_e)n_u n_e$$

$$- \sum_{l<u} A_{ul}^{eff} n_u - n_u n_e k_{u+}(T_e) - n_u k_u^{diff} - n_u n_{(N_2)} k_u^{quench} = 0 \tag{1}$$

Here, n_l ($l = 1$–41) refers to the density of the lth fine-structure level populations, and n_+ and $n_{(N_2)}$ are the population densities of the Ar^+ and N_2. As defined earlier, n_e and T_e are the plasma parameters. In the (1), positive terms are the population channels and negative terms are the depopulation channels for the level u. The first and fourth terms refer to the population transfer through electron impact excitation and de-excitation. The k_{lu} and k_{l+} are rate coefficients for the electron excitation and ionization processes, respectively [3, 17]. The de-excitation rates (k_{ul}) are obtained following the detailed balance principle as well as the rates for third and sixth terms which denote the three-body recombination and ionization are evaluated as described in our earlier papers [3, 6, 16, 17]. The second and fifth terms are the radiative decay processes, where the effective transition probability is defined as $A_{ul}^{eff} = \Lambda(K_{ul}\rho)A_{ul}$. The modification in the transition probability (A_{ul}) by a factor of $\Lambda(K_{ul}\rho)$ is to incorporate the reabsorption of the radiation. $\Lambda(K_{ul}\rho)$ is known as escape factor, here K_{ul} denotes the reabsorption coefficient and ρ is the source plasma characteristic scale length (2.0 cm. in the present work). We have calculated the escape factor by using the Mewe approximation [18, 19] and calculated it for all the considered transitions and for different N_2 concentrations. The calculated escape factors with the addition of N_2 in the range for 0–10% exhibit only slight changes in the escape factors for the lines originating from $2p_i \rightarrow 1s_i$ transitions. These escape factors are obtained for the same conditions as in the experiment of Lock et al. [14].

The last two terms of the (1) refer, respectively, to the depopulation through diffusion of the excited states via chamber walls and the quenching with N_2 molecules. Diffusion is taken into account mainly for $1s_i$ excited states using the diffusion rate coefficients reported by Kolts and Setser [20]. The rate coefficients for the quenching of $1s_i$ and $2p_i$ levels with N_2 molecules are taken from Scheller et al. [21].

3   Result and Discussion

For the diagnostics of Ar/N_2 mixture plasma, the coupled particle balance equations for all considered states (as represented in (1)) were solved simultaneously. The solution provides the population densities of the various fine-structure excited states of Ar up to $3p_i$ levels considered in the present model as a function of plasma parameters (i.e., electron density (n_e) and electron temperature (T_e)). In the present calculation, the ground state argon population n_{Ar} are calculated from the standard gas law using the pressure and gas temperature reported by the Lock et al. [14] at each concentration of N_2. The ground state argon population was $\sim 8.04 \times 10^{20}$ m^{-3}.

For the diagnostics purpose, we obtain the intensities of the emission lines considered in the present model as a function of n_e and T_e. As the CR model provides the population densities of excited states of an argon atom, thereafter, the line intensities are calculated by using (2). Now in order to fix the particular values of n_e and T_e which really correspond to the plasma parameter, we find the best match of our calculated intensities with the measured intensities of Lock et al. [14]. For this purpose, before doing this, as the intensities reported by Lock et al. [14] are in arbitrary unit, we normalize/scale them and get the value in percentage and also do the same with our calculated intensities. Then the least square method is applied and we obtain the minimum deviation parameter which fixes the plasma parameters (viz., n_e and T_e). The intensity corresponding to these values of n_e and T_e is taken to be the best match with the measured intensity. We calculate the deviation parameter using the minimum scatter approach.

The intensity (I_{ul}) of an emission line (with wavelength λ_{ul}) corresponding to a transition ($u \rightarrow l$) is linked to the population density of the upper excited level (n_u) as [3],

$$I_{ul} \propto n_u \frac{hc}{\lambda_{ul}} A_{ul}^{eff} \tag{2}$$

where h and c are the usual Planck's constant and speed of light, respectively. Thus following (1), the intensities of emission lines have been calculated as a function of T_e and n_e. Here, we have considered five emission lines decaying from $2p_i$ ($i = 1$–10) states to the $1s_i$ ($i = 2$–5) states as observed in the Lock et al. [14] OES measurements.

We individually normalize both the measured and modeled intensities, respectively, as follows,

$$I_{u,OES(Model)}^{normalized} = \frac{I_{u,OES(Model)}}{\sum_{u=1}^{5} I_{u,OES(Model)}} \times 100 \tag{3}$$

In order to check the matching of our simulated intensities with those of OES measurements, we calculated the deviation parameter (Δ) as given below based on the minimum scatter approach [1]. The least magnitude of deviation parameter fixes the plasma parameter.

$$\Delta = \sum_{u=1}^{5} \left(I_{u,OES}^{normalized} - I_{u,Model}^{normalized} \right)^2 \tag{4}$$

In Fig. 1, the deviation parameter is plotted as a function of T_e for different values of n_e at different concentrations of N₂ (0–10%). The minimum value of the deviation parameter fixes the value of T_e and n_e. The T_e and n_e values corresponding to the lowest deviation parameter are given in a table as shown in Fig. 1. We find that the electron temperature does not change much but remains almost close to the value of 1.6 eV while the electron density decreases by three orders of magnitude from 3.0×10^{13} to 3.0×10^{10} cm^{-3} as the concentration of N₂ varies from 0 to 10% in the Ar/N₂ mixture plasma. However, Lock et al. [14] extracted T_e from the EEDF as obtained in their previous work and taken the constant value of $n_e = 2.0 \times 10^{11}$ cm^{-3} for the Ar/N₂ mixture plasma. Using these in their CR model, they obtained the state populations for $1s_i$ and $2p_i$ states which decrease as the concentration of N₂ varies from 0 to 10%. Thus, we note that our results show different and opposite behavior with the modeling results of Lock et al. [14]. In fact, it is important to mention here that the variations of n_e and T_e as obtained by us are similar to those as seen by Kang et al. [10] with their global modeling results carried out in the light of their measurements for Ar/N₂ mixture plasma. Fantz et al. [13] also experimentally studied the Ar/N$_2$ plasma at low pressure and varied the concentration of N$_2$ and from their CR model, they found the similar variations of n_e and T_e as seen by us and Kang et al. [10].

Figure 2 presents our model intensities calculated by using the populations of the $2p_i$ excited levels (which decay to $1s_i$) obtained at the optimized value of n_e and T_e as given in the table with Fig. 1. These are compared with Lock et al. [14] measured intensities at 0, 1, 3, 5, and 10% concentrations of N$_2$. We observe from Fig. 2 that out

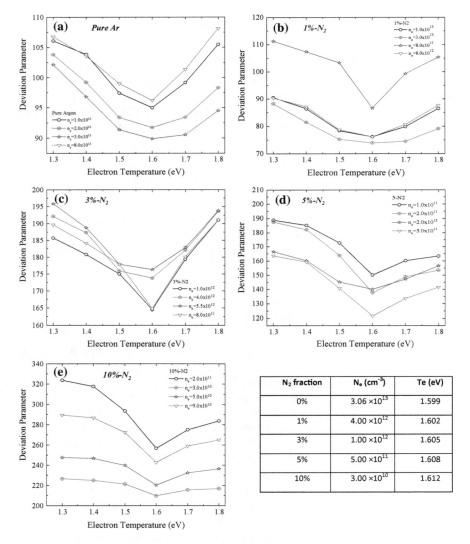

Fig. 1. Deviation parameter as a function of T_e at different n_e for Ar/ N$_2$ (0–10%) mixture are shown in the figures while the table gives the values of T_e and n_e corresponding to the lowest value of the deviation parameter

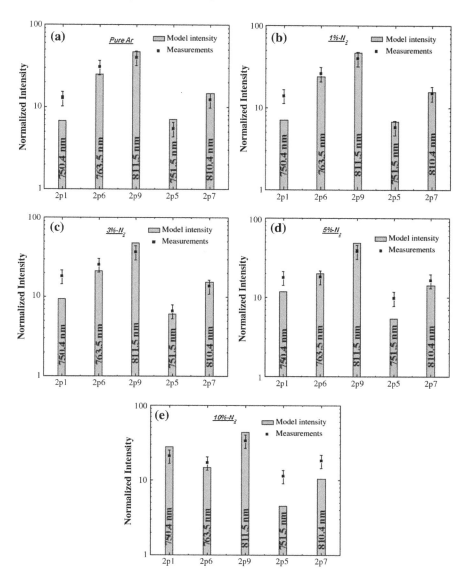

Fig. 2. Comparison of the normalized intensities obtained from the CR model with the OES measurements [14] for Ar ($2p_i \rightarrow 1s_i$) transition lines

of the five lines selected from the measured intensities, the intensities of at least three measured lines are in reasonable agreement with our modeled intensities. We find that the 750.4 nm and 751.5 nm lines whose intensities are not in good agreement, decay from the $2p_1$ and $2p_5$ levels. Both these levels have total angular momentum $J = 0$. Interestingly, we observe a correlation between these two lines with respect to their agreement with the experiment. As can be seen from Fig. 2, for 0% N$_2$ concentration, the 751.5 nm line is relatively in better agreement with the experiment as compared to 750.4 nm line. On the other hand, as the N_2 concentration increases the behavior flips; and for 10% N_2, concentration the 750.4 nm line agrees better with the experiment as compared to the 751.5 nm line. Although the precise reason behind this feature is not clear, it may be due to the fact that the states $2p_1$ and $2p_5$ correspond to the total angular momentum $J = 0$ value and these are also getting excited by electron impact from the Ar ground state ($1s_0$) with $J = 0$. In fact, it is found that for such $J = 0$ to $J' = 0$ transitions, the theoretically calculated electron excitation cross sections are generally larger as compared to the experimentally measured cross sections and possibly this could be one of the reasons. Lock et al. [14] did not report explicitly the intensity obtained from their CR model at the extracted values of n_e and T_e obtained by them; therefore, it is not straightforward to understand the accuracy of their CR model. We believe that our CR model is quite consistent and detailed; therefore, it can improve the diagnostics of the plasma.

4   Conclusions

In the present work, we have developed a CR model for Ar/N_2 mixture plasma. We coupled our model with the OES measurements of Lock et al. [14]. Plasma parameters (T_e and n_e) are obtained by the minimum scatter approach by using experimental intensities of the five intense lines ($2p_i \rightarrow 1s_i$ transitions) at various N_2 fractions (0–10%). We observe that the optimized electron temperature (T_e) does not change and is almost constantly close to 1.6 eV as N_2 fraction changes from 0 to 10% while the electron density decreases by three orders of magnitude. We believe that our CR model is well optimized and demonstrates the importance of the applicability of our reliable RDW cross sections.

Acknowledgements. We are grateful to Dr. Lock for providing us the data of his OES measurement in electronic form. SG is thankful to the Ministry of Human Resources and Development (MHRD), Govt. of India for their fellowship. RS is thankful to the SERB-DST and CSIR, New Delhi, Govt. of India for the sanction of research projects.

References

1. Malyshev, M.V., Donnelly, V.M.: Trace rare gases optical emission spectroscopy: Nonintrusive method for measuring electron temperatures in low-pressure, low-temperature plasmas. Phys. Rev. E **60**, 6016–6029 (1999). https://doi.org/10.1103/PhysRevE.60.6016
2. Donnelly, V.M.: Plasma electron temperatures and electron energy distributions measured by trace rare gases optical emission spectroscopy. J. Phys. D Appl. Phys. **37**, R217–R236 (2004). https://doi.org/10.1088/0022-3727/37/19/R01
3. Gangwar, R.K., Sharma, L., Srivastava, R., Stauffer, A.D.: Argon plasma modeling with detailed fine-structure cross sections. J. Appl. Phys. **111**, 053307 (2012). https://doi.org/10.1063/1.3693043

4. Gangwar, R.K., Dipti, Srivastava, R., Stafford, L.: Spectroscopic diagnostics of low-pressure inductively coupled Kr plasma using a collisional-radiative model with fully relativistic cross sections. Plasma Sources Sci. Technol. **25**, 035025 (2016). https://doi.org/10.1088/0963-0252/25/3/035025

5. Bartschat, K., Kushner, M.J.: Electron collisions with atoms, ions, molecules, and surfaces: fundamental science empowering advances in technology. Proc. Natl. Acad. Sci. **113**, 7026–7034 (2016). https://doi.org/10.1073/pnas.1606132113

6. Gangwar, R.K., Sharma, L., Srivastava, R., Stauffer, A.D.: Electron-impact excitation of argon: cross sections of interest in plasma modeling. Phys. Rev. A **81**, 052707 (2010). https://doi.org/10.1103/PhysRevA.81.052707

7. Gangwar, R.K., Sharma, L., Srivastava, R., Stauffer, A.D.: Electron-impact excitation of krypton: Cross sections of interest in plasma modeling. Phys. Rev. A **82**, 032710 (2010). https://doi.org/10.1103/PhysRevA.82.032710

8. Tao, K., Mao, D., Hopwood, J.: Ionized physical vapor deposition of titanium nitride: a global plasma model. J. Appl. Phys. **91**, 4040–4048 (2002). https://doi.org/10.1063/1.1455139

9. Muratore, C., Leonhardt, D., Walton, S.G., Blackwell, D.D., Fernsler, R.F., Meger, R.A.: Low-temperature nitriding of stainless steel in an electron beam generated plasma. Surf. Coat. Technol. **191**, 255–262 (2005). https://doi.org/10.1016/j.surfcoat.2004.02.026

10. Kang, N., Gaboriau, F., Oh, S.G., Ricard, A.: Modeling and experimental study of molecular nitrogen dissociation in an Ar-N$_2$ ICP discharge. Plasma Sources Sci. Technol. **20**, 045015 (2011). https://doi.org/10.1088/0963-0252/20/4/045015

11. Lock, E.H., Petrovykh, D.Y., Mack, P., Carney, T., White, R.G., Walton, S.G., Fernsler, R. F.: Surface composition, chemistry, and structure of polystyrene modified by electron-beam-generated plasma. Langmuir **26**, 8857–8868 (2010). https://doi.org/10.1021/la9046337

12. Ehlbeck, J., Schnabel, U., Polak, M., Winter, J., Von Woedtke, T., Brandenburg, R., Von Dem Hagen, T., Weltmann, K.D.: Low temperature atmospheric pressure plasma sources for microbial decontamination. J. Phys. D Appl. Phys. **44**, 013002 (2011). https://doi.org/10.1088/0022-3727/44/1/013002

13. Friedl, R., Fantz, U.: Spectral intensity of the N$_2$ emission in argon low-pressure arc discharges for lighting purposes. New J. Phys. **14**, 043016 (2012). https://doi.org/10.1088/1367-2630/14/4/043016

14. Lock, E.H., Petrova, T.B., Petrov, G.M., Boris, D.R., Walton, S.G.: Electron beam-generated Ar/N$_2$ plasmas: the effect of nitrogen addition on the brightest argon emission lines. Phys. Plasmas **23**, 043518 (2016). https://doi.org/10.1063/1.4946880

15. Fantz, U.: Basics of plasma spectroscopy. Plasma Sources Sci. Technol. **15**, S137 (2006). https://doi.org/10.1088/0963-0252/15/4/S01

16. Gangwar, R.K., Srivastava, R., Stauffer, A.D.: Collisional-radiative model for non-Maxwellian inductively coupled argon plasmas using detailed fine-structure relativistic distorted-wave cross sections. Eur. Phys. J. D. **67**, 203 (2013). https://doi.org/10.1140/epjd/e2013-40244-9

17. Srivastava, R.: Electron-impact excitation rate-coefficients and polarization of subsequent emission for Ar$^+$ ion. J. Quant. Spectrosc. Radiat. Transf. **176**, 12–23 (2016). https://doi.org/10.1016/j.jqsrt.2016.02.015

18. Mewe, R.: On the positive column of discharges in helium at intermediate pressures. Physica **47**, 373–397 (1970). https://doi.org/10.1016/0031-8914(70)90272-7

19. Boffard, J.B., Jung, R.O., Lin, C.C., Wendt, A.E.: Measurement of metastable and resonance level densities in rare-gas plasmas by optical emission spectroscopy. Plasma Sources Sci. Technol. **18**, 035017 (2009). https://doi.org/10.1088/0963-0252/18/3/035017

20. Kolts, J.H., Setser, D.W.: Decay rates of Ar(4s,^{3}P$_2$), Ar(4s′,3P0), Kr(5s,^{3}P$_2$), and Xe(6s,^{3}P$_2$) atoms in argon. J. Chem. Phys. **68**, 4848 (1978). https://doi.org/10.1063/1.435638
21. Scheller, G.R., Gottscho, R.A., Intrator, T., Graves, D.B.: Nonlinear excitation and dissociation kinetics in discharges through mixtures of rare and attaching gases. J. Appl. Phys. **64**, 4384–4397 (1988). https://doi.org/10.1063/1.341287

Long Time Evolution of a Bose–Einstein Condensate Under Toroidal Trap

Jayanta Bera[1], Suranjana Ghosh[2], and Utpal Roy[1(✉)]

[1] Indian Institute of Technology of Patna, Bihta, Patna 801103, Bihar, India
uroy@iitp.ac.in
[2] Amity University Patna, Rupaspur, Patna 801503, Bihar, India

Abstract. We consider Bose–Einstein condensates under an experimentally feasible toroidal trap. Temporal expansion of the condensate along the minima of the trap experiences self interference between the counter-propagating constituent parts of the condensate. Condensate density nicely manifests fractional revivals in long time evolution, where the original condensate splits into several subsidiary components at times, rational fractions of the revival time. Upon pointing out this novel phenomenon in this system, we investigate the spatial interference pattern, which requires a careful explanation of the phase distribution among the petals. These features have come out as significantly different from other quantum systems and pave the way to achieve efficient quantum precision measurements in this system.

1 Introduction

The experimental realization of Bose–Einstein condensate (BEC) has gradually made it one of the most appropriate candidates to observe various unexplored quantum phenomena. Although the formation mechanism and dynamics of BEC always remain a nontrivial task, due to its long coherence time and high tunability, it is indeed a favorable candidate for achieving quantum technology. The field is also connected to many important directions: quantum optics, quantum information, weak measurement, higher harmonic generation, negative temperature, condensed matter applications, etc. While dealing with gravimetry, rotation sensing, and interferometric measurements, matter–wave play an important role. Matter–wave interference has drawn huge attention for more than a decade in various aspects like quantum precision measurements [1]. First atomic interference was observed by superposing two independent condensates [2], which can be produced in a compact interferometer. Particularly, a toroidal trap is realized in BEC by tuning the lasers in an ring shaped configuration [3, 4]. Such trapping potential can be created by a blue-detuned laser beam in the middle of the harmonic magnetic trap. BEC loaded in a toroidal trap has got huge attention for the last few years [5–7] and found interesting applications in persistent flow of current [8], solitary waves and vortices [9], and interference [1, 10]. After creating the condensate in a toroidal trap, it radially expands along the minimum of the effective potential with time. The tails of the expanding condensate coming from opposite directions collide, resulting in interference which is observed experimentally [10]. However, the detailed analysis of the dynamics of two condensates in a toroidal trap and its long time

© Springer Nature Singapore Pte Ltd. 2019
P. C. Deshmukh et al. (eds.), *Quantum Collisions and Confinement of Atomic and Molecular Species, and Photons*, Springer Proceedings in Physics 230,
https://doi.org/10.1007/978-981-13-9969-5_10

evolution, which reveals a number of interesting phenomena, has not yet been reported in the literature.

In this work, we are studying the dynamics of two condensates placed in a toroidal trap by numerically solving the Gross–Pitäevskii equation (GPE) under weak interatomic interaction. We, not only report the revival and fractional revivals phenomena in this system but also quantify the time scales and explain the origin of the occurrence of a particular fractional revival situation. When the condensate is allowed to superpose spatially in the central region of the trap, the resulting interference pattern becomes quite different from similar situations in other quantum systems.

We reveal the origin of these structures through a careful phase analysis at time snapshots.

2   Theoretical Model

The dynamics of weakly interacting BEC can be modeled by GPE which is widely used in such nonlinear systems [11]. Incorporating various external potentials make the dynamics richer and tunable to achieve coherent control. The mathematical form of the 3D-GPE can be written, in the presence of some external potential as

$$
\left[-\frac{\hbar^2}{2M} \nabla^2 + V_{ext}(X,Y,Z,T) + g|\Psi(X,Y,Z,T)|^2 \right] \Psi(X,Y,Z,T) = i\hbar \frac{\partial \Psi(X,Y,Z,T)}{\partial T}.
$$

$$(1)$$

The order parameter of the condensate is $\Psi(R,T)$ and the total number of atoms are related as $\int |\Psi(R,T)|^2 dR = N$. The weak coupling constant is given by $g = \frac{4\pi\hbar^2 a}{M}$, where a is the s-wave scattering length for low-energy scattering. It can be positive or negative depending on the repulsive and attractive two-body interactions, respectively. M is the mass of the atom and V_{ext} is the external confining potential. When an external trap is harmonic in nature, i.e., $V_{ext}(X,Y,Z) = \frac{1}{2}M\left(\omega_x^2 X^2 + \omega_y^2 Y^2 + \omega_z^2 Z^2\right)$.

To make the equation in a convenient form, we reduce (1) into a dimensionless form by scaling variables and parameters [3]. If the reference harmonic trap frequency is ω, the corresponding harmonic oscillator length is given by $a_h = \sqrt{\hbar/M\omega}$. The scalings are given by $t = \omega T$, $x = a_h X$, $y = a_h Y$, $z = a_h Z$, $v_{ext}(x,y,t) = \frac{V_{ext}(X,Y,Z)}{\hbar\omega}$, and $\Psi(x,y,z,t) = \sqrt{N/a_h^3}\,\widetilde{\Psi}(x,y,z,t)$. The final form of the 3D-GPE becomes (after removing tilde)

$$
\left[-\frac{1}{2}\left(\frac{\partial^2}{\partial x^2} + \frac{\partial^2}{\partial y^2} + \frac{\partial^2}{\partial z^2} \right) + v_{ext}(x,y,z) + g_{3d}|\Psi(x,y,z,t)|^2 \right] \Psi(x,y,t)
$$
$$
= i\frac{\partial \Psi(x,y,z,t)}{\partial t}
$$

$$(2)$$

where $v_{ext}(x, y, z) = \frac{1}{2}\left(\gamma_x^2 x^2 + \gamma_y^2 y^2 + \gamma_z^2 z^2\right)$ with $\gamma_x = \frac{\omega_x}{\omega}$, $\gamma_y = \frac{\omega_y}{\omega}$, and $\gamma_z = \frac{\omega_z}{\omega}$. Now the weak coupling constant is given by $g_{3d} = \frac{4\pi a N}{a_h}$ and normalization $\int|\Psi(x, y, z, t)|^2 dx\,dy\,dz = 1$.

If we apply strong confinement along any one particular direction, suppose z-direction, then the time scales of the dynamics become quite asymmetric and effectively the condensate will be confined in two-dimension (2D): x and y. This is assuming that the system is in the ground state in the transverse z-axis. In this case, the wave function in (2) can be written as $\Psi(x, y, z, t) = \varphi(x, y, t)\left(\frac{\gamma_z}{\pi}\right)^{\frac{1}{4}} e^{\frac{-\gamma_z z^2}{2}} e^{\frac{-i\gamma_z t}{2}}$. The resulting 3D-GPE reduces to a quasi 2D-GPE:

$$\left[-\frac{1}{2}\left(\frac{\partial^2}{\partial x^2} + \frac{\partial^2}{\partial y^2}\right) + v_{ext}(x, y, t) + g_{2d}|\varphi(x, y, t)|^2\right]\varphi(x, y, t) = i\hbar\frac{\partial\varphi(x, y, t)}{\partial T}. \quad (3)$$

where $v_{ext}(x, y) = \frac{1}{2}\left(\gamma_x^2 x^2 + \gamma_y^2 y^2\right)$. The modified coupling constant for 2D-GPE is given by $g_{2d} = \frac{2\sqrt{2\pi\gamma_z}aN}{a_h}$ and normalization $\int|\varphi(x, y, t)|^2 dx\,dy = 1$.

3 External Confinement and Methodology

The external trapping potential is considered here as a toroidal trap, which can be written as

$$v_{ext}(x, y, t) = \frac{1}{2}\alpha^2\left(x^2 + y^2\right) + U(x, y, t).$$

Here, $\alpha(= \gamma_x = \gamma_y)$ is the harmonic trapping frequency in x- and y-directions and $U(x, y, t)$ is a Gaussian whose peak amplitude, $A(t)$ may be time-dependent, such that one can switch off the potential and control it whenever needed in the experiment. The parameter σ is associated with the width of the Gaussian peak. The expression of the potential peak is given by

$$U(x, y, t) = A(t)\left(e^{-2\left(x^2 + y^2\right)/\sigma^2}\right). \quad (4)$$

The minima of the potential can be calculated and its distance of minima from the center of the harmonic trap is given by (r_0):

$$\left|\frac{\partial v_{ext}}{\partial r}\right|_{r=r_0} = 0,$$

where

$$r_0 \left(= \sqrt{x_0^2 + y_0^2} \right) = \frac{\sigma}{\sqrt{2}} \sqrt{\ln \frac{4A}{\alpha^2 \sigma^2}}. \tag{5}$$

Hence, the radius of the toroid can be controlled by the parameters α, σ, and A. **Numerical method**: We solve the corresponding GPE (3) numerically, which is non-trivial due to the non-linear nature of the second-order partial differential equation of two variables and the inclusion of the space-time-dependent potential function. Bao and Tang haves given a general method to solve the time-dependent GPE to find both stationary and non-stationary states in various types of potential [12]. Runge–Kutta types of method, Adhikari used semi-implicit Crank–Nicolson scheme [13, 14], finite difference method, and Fourier split-step method [15]. Here, we take up the second approach to solve the corresponding dynamics, i.e., split-step Fourier method (SSFM).

To solve the 2D-GP equation (4) using SSFM, we assume the shape of the initial condensate wave function as Gaussian: $\varphi(x, y; t = 0) = \frac{1}{2} \left(\frac{2w}{\pi} \right)^{1/4} \left(e^{-w\left((x - x_0)^2 + (y - y_0)^2 \right)} + e^{-w\left((x + x_0)^2 + (y + y_0)^2 \right)} \right)$, where w is associated with the initial width of the condensate. In our analysis, the amplitude A of the potential is taken as 100. The values of r_0 and α are chosen as 11.0 and 1, respectively. The other parameters are as follows: $a = 4.98 \times 10^{-11}$ m, $a_h = 1$ μm, $N = 10^4$, $\gamma_z = 1$, and $g_{2d} = 2.5$. We discretized the space into 512×512 grids and the step size becomes 0.12.

4   Results and Discussions

We investigate the evolution of the condensate density inside the trap for a longer time. Two Gaussian-shaped condensates are placed at the calculated minima of the effective trap and in the diametrically opposite corners of the toroidal trap. Because of the existence of the Gaussian barrier in the middle of the harmonic trap, the condensate expands along the trap minima. We have chosen the amplitude and width of the Gaussian barrier, such that the tunneling probability becomes negligible through the barrier. During their course of evolution in the ring-shaped dip and the counter-propagating components start overlapping each other, producing self interferences. This spatial density distribution reveals petal formation after a complete mixing. Figure 1a–r depict the petal formation in density at different evolution times. One can observe that at time $t = 375$ (Fig. 1r), the density distribution is the replica of the initial density distribution and this time is known as revival time (T_r). When time t becomes a rational fraction of the revival time, the initial pattern breaks into several parts: the evolution times, $T_r/20$ (Fig. 1b), $T_r/16$ (Fig. 1c), $T_r/14$ (Fig. 1d), $T_r/12$ (Fig. 1e), $T_r/10$ (Fig. 1f), $T_r/8$ (Fig. 1g), $T_r/6$ (Fig. 1h), $T_r/4$ (Fig. 1i), and $T_r/2$ (Fig. 1n), produce 20, 16, 14, 12, 10, 8, 6, 4, and 2, respectively. At time $t = T_r/2$, the half revival time, similar density distribution is observed in the vertical direction with respect to the initial position.

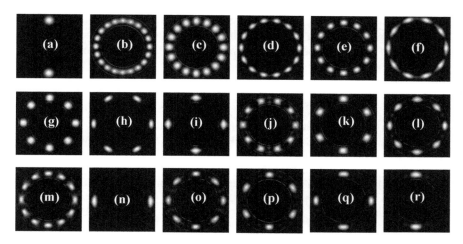

Fig. 1. The density distribution in x-, y space of the condensate at different times. The snapshots at times, $t = 0$, 19.52, 23.42, 27.34, 31.24, 39.06, 46.86, 62.5, 93.74, 113.28, 125, 140.62, 156.24, 187.50, 234.36, 250, 281.24, and 375 (**a–r**), respectively. The bright color manifests maximum density and dark color manifests density minimum. Both x *and* y ranges from -12 to 12

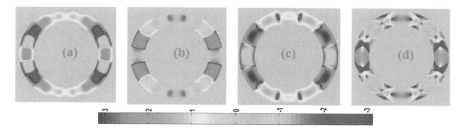

Fig. 2. Phase distribution in x-y space. The snapshots at time $t = 19.52$, 23.42, 31.24, and 46.86, corresponding to figure (**a**), (**b**), (**c**), and (**d**), respectively

To get a clear understanding of the condensate density distribution, phase analysis is also quite important; we have performed the phase analysis which is displayed in Fig. 2. The phase distribution in space is quite random in nature, not having a trivial correlation with the obtained density patterns (Fig. 1). However, the petals at the diametrically opposite points have equal phase. Such phase analysis is important to analyze the spatial interference pattern when the central Gaussian peak is switched off.

5 Conclusion

We have studied the formation of quantum interferences in a toroidal trap. The number of petals formed due to the interference of the counter-propagating condensate probabilities, at different times is studied in details, which clearly reveals fractional revivals.

The phase information during the time evolution as depicted in Fig. 2 made the physics more engrossing and rich. Our model is quite useful towards quantum measurement or quantum metrology. This is a promising system for rotation sensing too.

References

1. Berman, P.R.: Atom Interferometry. Academic, New York (1997)
2. Andrews, R., Townsend, C.G., Miesner, H.-J, Durfee, D.S., Kurn, D.M., Ketterle, W.: Observation of interference between two Bose condensates. Science **275**, 637–641. https://doi.org/10.1126/science.275.5300.637
3. Gupta, S., Murch, K.W., Moore, K.L., Purdy, T.P., Stamper-Kurn, D.M.: Bose–Einstein condensation in a circular waveguide. Phys. Rev. Lett. **95**, 143201 (2005). https://doi.org/10.1103/physrevlett.95.143201
4. Ryu, C., Andersen, M.F., Clade, P., Natarajan, V., Helmerson, K., Phillips, W.D.: Observation of persistent flow of a Bose–Einstein condensate in a toroidal trap. Phys. Rev. Lett. **99**, 260401 (2007). https://doi.org/10.1103/physrevlett.99.260401
5. Murray, N., Krygier, M., Edwards, M., Wright, K.C., Campbell, G.K., Clark, C.W.: Probing the circulation of ring-shaped Bose-Einstein condensates. Phys. Rev. A **88**, 053615 (2013). https://doi.org/10.1103/physreva.88.053615
6. Eckel, S., Lee, J.G., Jendrzejewski, F., Murray, N., Clark, C.W., Lobb, C.J., Phillips, W.D., Edwards, M., Campbell, G.K.: Hysteresis in a quantized, superfluid atomtronic circuit. Nature (London) **506**, 200 (2014)
7. Wood, A.A., McKellar, B.H.J., Martin, A.M.: Persistent superfluid flow arising from the He-McKellar-Wilkens effect in molecular dipolar condensates. Phys. Rev. Lett. **116**, 250403 (2016). https://doi.org/10.1103/PhysRevLett.116.250403
8. Moulder, S., Beattie, S., Smith, R.P., Tammuz, N., Hadzibabic, Z.: Quantized supercurrent decay in an annular Bose–Einstein condensate. Phys. Rev. A **86**, 013629 (2012). https://doi.org/10.1103/PhysRevA.86.013629
9. Mason, P., Berloff, N.G.: Dynamics of quantum vortices in a toroidal trap. Phys. Rev. A **79**, 043620 (2009). https://doi.org/10.1103/PhysRevA.79.043620
10. Bell, T.A., et al.: Bose–Einstein condensation in large time-averaged optical ring potentials. New J. Phys. **18**, 035003 (2016). https://doi.org/10.1088/1367-2630/18/3/035003
11. Pethick, C.J., Smith, H.: Bose–Einstein Condensation in Dilute Gases. Cambridge University Press, Cambridge (2002). Pitevskii, L.P., Stringar, S.: Bose–Einstein Condensation. Oxford University Press, Oxford (2003)
12. Bao, W., Tang, W.: Ground-state solution of Bose–Einstein condensate by directly minimizing the energy functional. J. Comput. Phys. **187**, 230 (2003). https://doi.org/10.1016/S0021-9991(03)00097-4
13. Adhikari, S.K., Muruganandam, P.: Bose–Einstein condensation dynamics from the numerical solution of the Gross–Pitaevskii equation. J. Phys. B: At. Mol. Opt. Phys. **35** 2831 (2002)
14. Muruganandam, P., Adhikari, S.K.: Fortran programs for the time-dependent Gross–Pitaevskii equation in a fully anisotropic trap. Comput. Phys. Commun. **180**, 1888–1912 (2009). https://doi.org/10.1016/j.cpc.2009.04.015
15. Poon, T., Kim, T.: Engineering Optics with Matlab. World Scientific Publishing (2006)

Intriguing Single Photon Induced Processes in Helium Nanodroplets

S. R. Krishnan[1(✉)], Suddhasattwa Mandal[2], Bhas Bapat[2],
Ram Gopal[3], Alessandro D'Elia[4], Hemkumar Srinivas[5],
Robert Richter[6], Marcello Coreno[7], Marcel Mudrich[8],
and Vandana Sharma[9]

[1] Indian Institute of Technology Madras, Chennai 600036, Tamil Nadu, India
srkrishnan@iitm.ac.in

[2] Indian Institute of Science Education and Research Pune, Pune 411008,
Maharashtra, India
suddhasattwa.mandal@students.iiserpune.ac.in

[3] TIFR Centre for Interdisciplinary Sciences, Hyderabad 500107, Telangana,
India
ramgopal@tifrh.res.in

[4] Department of Physics, University of Trieste, 34127 Trieste, Italy
delia@iom.cnr.it

[5] Max-Planck-Institut für Kernphysik, 69117 Heidelberg, Germany
hemkumar.srinivas@mpi-hd.mpg.de

[6] Elettra-Sincrotrone Trieste, Strada Statale 14-km 163.5, 34149 Basovizza,
Trieste, Italy
robert.richter@elettra.eu

[7] Consiglio Nazionale delle Ricerche – Istituto di Struttura della Materia, 34149
Trieste, Italy
marcello.coreno@cnr.it

[8] Department of Physics and Astronomy, Aarhus University, 8000 Aarhus C,
Denmark
mudrich@phys.au.dk

[9] Indian Institute of Technology Hyderabad, Sangareddy 502285, Telangana,
India
vsharma@iith.ac.in

Abstract. Helium nanodroplets are a unique system of quantum fluid clusters possessing several intriguing properties. From the perspective of atomic systems, they have been predominantly viewed as a spectroscopic matrix hosting other molecules and systems of interest. In this report, we draw particular attention to select electronic processes in He aggregates hosting other atoms, in particular, Rb. From this perspective, we present the details of single- and multi-electron processes occurring in this alkali-He system interacting with single extreme ultraviolet and soft X-ray photons. The features brought out in this study are generic and pertinent to systems with similar design on the atomic scale.

© Springer Nature Singapore Pte Ltd. 2019
P. C. Deshmukh et al. (eds.), *Quantum Collisions and Confinement of Atomic and Molecular Species, and Photons*, Springer Proceedings in Physics 230,
https://doi.org/10.1007/978-981-13-9969-5_11

1 Introduction

Helium nanodroplets are being widely used as an "ideal" matrix for spectroscopy of embedded or attached species [1]. Owing to their cold internal temperature due to their formation in supersonic jets, dopant atoms or molecules attached to these droplets are also cooled in the process by the evaporation of He atoms in the host. Combined with finite nanoscale dimensions 1–100 nm and a very large transparency window from the infrared to beyond the ultraviolet, these droplets have served as nanocryostats for hosting individual atoms or molecules as well as dopant aggregates [1, 2]. Dopant species are hosted by the droplets either on the surface or embedded in the interiors, thus providing a universal route for spectroscopy and investigations of cold reactions at sub-Kelvin temperatures. This has been widely exploited to study the rotational, vibration, and electronic spectroscopy of atomic and molecular systems [3]. Since the lambda-point of bosonic ^{4}He in the bulk is 2.1 K, these aggregates with an internal temperature of 400 mK, are evidenced to be in a superfluid state, enabling the realization of a nanoscale quantum fluid. From this perspective, these nanodroplets themselves have become the central object of investigation leading to the observation of irrotational flow, Landau velocity and vortices in these nanometer-sized aggregates [4–6]. Aggregates of bosonic ^{4}He have also been viewed as theoretical testbeds to investigate other such systems which mimic some aspects of atomic nuclei [7]. Hence, these droplets are a prominent candidate for rich and diverse physics and chemistry.

Complementary to their role as hosts for other systems of interest, the electronic properties of He nanodroplets themselves in combination with doped atoms or molecules open a new avenue for investigation when these systems are interrogated with photons and electrons [3, 8–10].

2 Photo-Induced Processes in Helium Droplets

Photodynamics initiated via excitation of He in the droplets with photons of energy >20 eV proves to be intriguing both when these clusters are pristine as well as in weakly doped aggregates. Extreme ultraviolet (EUV) and soft X-ray photoexcitation of doped droplets by the group of Neumark revealed a lowering of the ionization potential of droplets by almost 1.5 eV as compared to atomic He, revealing the role of atomic-like states or bands in their electronic structure [11]; this corroborates earlier fluorescence measurements [12, 13]. Photoelectron spectroscopy revealed that the photoionization of pure nanodroplets is accompanied by the emission of very low energy meV electrons.

Photoexcitation and ionization of doped droplets led to the identification of distinct mechanisms of droplet ionization [14, 15]: (a) at photon energies 20 eV $< h\nu <$ 23 eV alkali doped droplets display a prominent ionization channel. However, droplets doped with rare gas atoms are not ionized when photoexcited at these energies. At these photon energies, the electronic structure of He consists of single-electron excitations from the ground state to the 1s2p state. The photoionization of alkali doped droplets is accompanied by the emission photoelectrons in this *Penning transfer* process. (b) In the photon energy range 23 eV $< hv <$ 24.5 eV below the ionization threshold of atomic He, both rare gas and alkali doped droplets are photoionized. In this case, the accompanying photoelectrons are emitted, and the most prominent photoreaction product is the He dimer ion, He_2^+. In both regimes, (a) and (b) the dynamics is initiated by electronic excitation and the formation of He* and He_2*. In pure droplets in these cases the excited He atom or dimer is ejected from the droplet [16, 17]. (c) At $hv >$ 24.6 eV above the ionization potential of atomic He, the direct formation of He^+ ions in the droplet initiates the photodynamics. While once again, He_2^+ ions are the prominent ionic product of ionization in both doped and pure droplets, characteristic photoelectron with kinetic energies equal to the difference between the ionization potential of atomic He and the photon energy is evidenced. Electron-ion coincidence studies [14] have revealed that even photoelectrons correlated to the dopant ion have these kinetic energies. Thus, a charge transfer between He and the dopant is the dominant ionization channel leading to the formation of dopant photoions. The moderate absorption cross section of bulk He at these energies leads to the incident photon being absorbed in the interiors of the droplet and not just by surface atoms. Hence, a hopping mechanism leading to the migration of the charge which localizes at the droplet center has been proposed [18, 19] (Fig. 1).

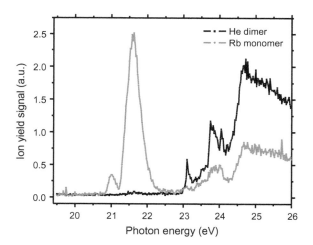

Fig. 1. Photoion yield for He_2^+ and Rb^+ ions emitted from He nanodroplets ($\sim 10^4$ atoms per droplet) doped with Rb atoms. The droplets were photoexcited within the energy range of 19.5–26 eV. Prominent photoion emission peaks reveal the electronic structure of the alkali-He complex. While Rb ions are emitted even below 23 eV corresponding to the 1s3p excitation in atomic He, dimer ions of He are generated only above this energy. The three regimes of photoexcitation of droplets (a)–(c) discussed in the text are enunciated in this plot. The intensities of the ion signals are indicative of relative yields observed

3 Experiment

In this work, we employed a He droplet source attached to a velocity map imaging (VMI) spectrometer with the capability of performing *photoelectron photoion coincidence* (PEPICO) measurements at the GasPhase beamline of the Elettra synchrotron, Trieste. This is similar to the arrangement used in [14, 15]. Depicted in Fig. 2, the set up consists of a cryogenic source for producing He nanodroplets. The beam of droplets is produced by expanding continuously high purity ^{4}He gas at high pressures (50–100 bar) through a cooled (6–30 K) nozzle with an orifice of 5 μm in diameter. To control the size of the droplets the temperature of the nozzle was changed. Using well-established scaling laws [3], the most probable size of the distribution of the droplets emanating from this source in the form of a supersonic beam could be controlled, a log-normal distribution of droplets is produced. We refer to the most probable size of the aggregates as the droplet size. He clusters with 100–10^6 He atoms/droplet could be controllably produced in these jets. This jet then passes through a skimmer into doping chamber, equipped with a gas cell and several doping cells. The doping cell consists of a cylindrical container with two collinear holes which allow the droplet beam to pass and interact with the dopant vapor (in this case Rubidium). Statistical pick-up of atoms (or molecules) by the He droplets leads to their doping in or on droplets. To dope droplets with Rb atoms, the vapor cell was heated between 70 and 120 °C; the conditions are similar to earlier work on Rb doped nanodroplets [20]. The level of doping

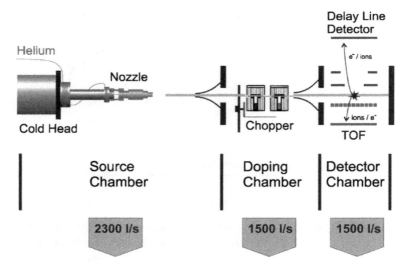

Fig. 2. Experimental set up described in this section consisting of a He nanodroplet source, a doping chamber, and a detector chamber. The supersonic jet of He droplets emanating from the source is doped with the desired atomic species held in a vapor form in the cells in the doping chamber. Thereafter, downstream, this jet of doped droplets intercepts a focused photon beam from the synchrotron. The VMI-PEPICO spectrometer discussed in the text collects and analyzes the electrons and ions produced in the photo interaction. Also shown above are the speeds of the vacuum pump used in these chambers. This set up is similar to [14, 15]

can be controlled by changing the temperature of the doping cell, which is resistively heated using a PID control loop.

The droplet beam which is doped with Rb atoms is expanded into the adjacent chamber. Synchrotron radiation from the U12.5 undulator at the Elettra synchrotron passing through a variable angle monochromator with spherical gratings is delivered with on-demand energy between 13 and 900 eV. The energy resolution of the photons is better than $E/\Delta E > 10^4$, at a pulse repetition rate of 500 MHz, with pulse widths of 150 ps, and maximum intensity in the range of 10 W/m^2 [14, 21]. This photon beam after passing through a pair of slits whose opening can be controlled for attenuating the beam intensity is focused on to the reaction chamber.

The VMI-PEPICO spectrometer in the detector chamber consists of an array of electrodes which accelerates photoelectrons into a position sensitive detector comprised of a microchannel plate with a delay line anode. The position of the impinging photoelectron is deciphered using conventional electronics and data acquisition modules [14, 15, 21]. Photoelectrons up to 30 eV are accepted in this VMI system. The image of photoelectrons from the VMI is converted into kinetic energy spectra using the MEVELER program for Abel inversion [22]. In tandem, the corresponding photoion produced from the same parent as the photoelectron is pushed toward and companion microchannel plate directed opposite to the photoelectron detection element. The flight time of the photoion is recorded using the photoelectron detector pulse as the start for this clock. The mass by charge resolution of this ion spectrometer is ~ 70. Thus, coincidence photoelectron–photoion measurements are possible.

4   Results and Discussion

Figure 3 shows a photoion mass spectrum obtained in the photoionization of Rb doped He nanodroplets. The counts of photoions received at the detector are depicted against their mass-to-charge ratio. In this experiment, a chopper was used to alternatingly interrupt the He droplet jet and signals were acquired in both scenarios—this is similar to earlier studies, see e.g., [14]. The red curve shows the photoion mass spectrum when the chopper blocked the droplet beam, while the blue curve is the signal obtained when the doped He droplet jet passed through. The inset confirms the generation of Rb photoions from the droplet jet as the blue curve shows contrasting peak at m/q values of 85 and 87 atomic units, over the red bottom-line (background). A second set of peaks over background is observed at twice the values above near 170 atomic units from Rb$_2^+$ ions generated from the photoionization of doped droplets in the jet. By varying the droplet size and the heating temperature of the vapor cell containing Rb, the mean number of doped alkali atoms per droplet can be varied [2, 20]. Thus, even Rb trimer doping was observed in other cases in our experiment.

4.1   Multi-electron Processes in Rb Doped He Nanodroplets

In our earlier study [14, 15], we employed similar experimental strategies as described here to investigate single-electron process, i.e., Penning transfer from excited He* in the droplet to hosted alkali atom(s) particularly when the photoexcitation was in the

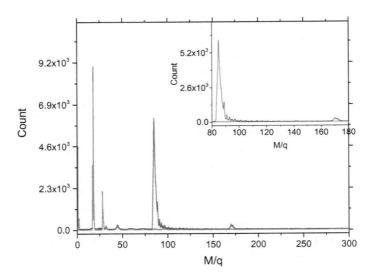

Fig. 3. Photoion mass spectrum: a histogram of the counts of the photoions detected is plotted against the mass-to-charge ratio (m/q) of the ionic species in atomic units. The droplets here contained a mean number of 10,000 He atoms/droplet. *Inset* Mass peak structures at 85 and 87 atomic units in the blue curve corresponding to the *open* condition of the chopper letting the droplet jet through, over red the background curve obtained from the chopper *closed* case which is the configuration when the droplet beam is blocked but the effusive residual gas interacts with the photon beam

energy region of 21.6 eV corresponding to the 1s2p dipole allowed transition of atomic He. In these cases, a prominent ionization of the alkali atoms as well as photoelectron signatures of the excitation transfer induced ionization of the alkali atoms in coincidence measurements was found. Moreover, a strong signal of meV electrons at the same photoexcitation energy was also evidenced. However, multi-electron processes in alkali doped droplets have not yet found their way into the literature. The present study is motivated by this obvious need for investigation.

LaForge et al. [23], studied the double excitation of electrons in pure He nanodroplets observing the Rydberg series in droplets. The observed photoelectron spectra were fit to the Fano formula [24] written in terms of the photoexcitation energy E:

$$\sigma(E) = \sigma_a \frac{(q+\epsilon)^2}{(1+\epsilon^2)} + \sigma_b;$$

where $\epsilon = \frac{E-E_0}{\Gamma/2}$ is the reduced energy, E_0 is the resonance energy, Γ is the resonance linewidth, and q is the Fano parameter [24]. The parameter σ_a is the background cross section associated with fraction of ionization which interferes with the discrete states, whereas σ_b is the fraction which does not interfere with these states [24]. In the sense of a double-slit experiment these two fractions represent the coherent and incoherent aspects of the process. In the investigation with pure droplets, while the line widths were found to increase with the droplet size, quite remarkably, the q parameter and the peak positions varied rather weakly over a change in aggregate size from 10^2 to 10^{11} He atoms/droplet.

We performed similar studies with Rb doped He nanodroplets where most droplets were singly doped with Rb atoms and the droplets had a mean size of 10^4 He atoms/droplet, which corresponds to a geometric size of about 20 nm. Figure 4 depicts the Fano profile of the $2s2p+$ resonance in Rb doped He nanodroplets. Two signals were employed in this case to trace the resonance line—the first is the most prominent He photoion, which is the He dimer ion, and the second is the Rb^+ photoion. Remarkably the ion signals when overlaid follow each other closely, for the $2s2p+$ resonance at a value of $E_0 = 60.39$ eV for both the lines.

Quantum mechanical calculations and simulations performed ab initio on such large systems are challenging and perhaps not feasible [23]. Other formulations may work better in providing a physical model and intuition. However, from the experimental results we can infer the following. The interference process involving two-electron bound states embedded in the continuum of the single-electron excited states is hardly perturbed by the presence of the alkali atom(s) on the droplet surface. We envisage a two-step process of the following kind to explain the observation of the Fano lineshape through both ionic channels, the He_2^+ ion as well as the Rb^+ ion:

$$Rb@He_N + h\nu(>60\,eV) \rightarrow Rb@He_N** \rightarrow Rb@He_N^+ + e^- \rightarrow Rb^+ + He_N + e^-$$

In the first step, the absorption of the photon leads to the formation of the doubly excited state in one of the atoms in the He nanodroplet. This doubly excited state decays either by being autoionized via the quantum interference at the $2s2p+$ state or via the route where the excitation energy is transferred to the attached Rb atom in a

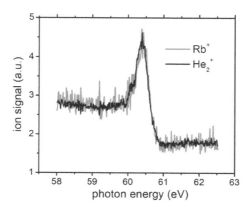

Fig. 4. Fano line shape of the $2s2p+$ Rydberg excitation in Rb doped He nanodroplets: The black curve show He dimer ion signal obtained as a function of incident photon energy in the photoexcitation process. The quantum interference between the electron pathway from the initial state to the continuum and the energetically equivalent route involving the formation of the doubly excited He** state, which leads to the characteristic Fano profile. Subsequent decay of this state leading to the formation of He_2^+ ion and a charge transfer from He_2^+ to Rb leads to the formation of the Rb^+ ion

Penning-like or interatomic decay process leading to its ionization. In the former case, following the formation of the He^+ ion, or very likely the He_2^+, in the droplet, a charge transfer from the Rb atom attached to the droplet leading to an electron hopping from Rb to the dimer ion results in the formation of the Rb^+ ion. Thus, the presence of the Rb atom in the droplet does not significantly change the characteristics of the decay of a doubly excited He** atom in the droplet. Subsequent fast decay of this state which has the characteristic Fano profile leads to efficient ionization of the doped Rb atom by charge transfer. This fast decay is evidenced by studies using attosecond EUV pulses in combination with near-infrared pulses to map the temporal dynamics of the $2s2p+$ state in atomic He [25].

5   Conclusion

In this report, we have presented the development of an experiment to study the doubly excited state of He in He nanodroplets doped with an alkali atom attached to the droplet. We observe that the Fano profile and quantum interference is not significantly perturbed by the environment keeping the coherence of the process intact throughout. This may be surprising considering that alkali atoms do induce structural changes to the surface of the He droplet [2]. However, our study indicates that the fast decay of the doubly excited state may preempt much of the environmental effects. This work kindles again the need for better understanding of multi-electron process in quantum aggregates and should stimulate further experimental and theoretical investigation.

Acknowledgements. S. R. K. thanks the Department of Science and Technology, Govt. of India, and the Max Planck Society, Germany, for partial financial support. All authors declare no competing financial interests.

References

1. Toennies, J.P., Vilesov, A.F.: Angew. Chem. Int. Ed. **43**, 2622 (2004)
2. Tiggesbäumker, J., Stienkemeier, F.: Phys. Chem. Chem. Phys. **9**, 4748 (2007)
3. Stienkemeier, F., Vilesov, A.F.: J. Chem. Phys. **115**, 10119 (2001)
4. Grebenev, S., Toennies, J.P., Vilesov, A.F.: Science **279**, 2083 (1998)
5. Brauer, N.B., et al.: Phys. Rev. Lett. **111**, 153002 (2013)
6. Gomez, L.F., et al.: Science **345**(6199), 906 (2014)
7. Pandharipande, V.R.: Nucl. Phys. A **178**(1), 123 (1971)
8. Kim, J.H., Peterka, D.S., Wang, C.C., Neumark, D.M.: J. Chem. Phys. **124**(21), 214301 (2006)
9. Kuhn, M., et al.: Nat. Commun. **7**, 13550 (2016)
10. Mudrich, M., Stienkemeier, F.: Int. Rev. Phys. Chem. **33**(3), 301 (2014)
11. Peterka, D.S., et al.: Phys. Rev. Lett. **91**(4), 043401 (2003)
12. Joppien, M., Karnbach, R., Möller, T.: Phys. Rev. Lett. **71**(16), 2654 (1993)
13. Wörmer, J., et al.: J. Chem. Phys. **104**(21), 8269 (1996)
14. Buchta, Dominic, Krishnan, Siva R., et al.: J. Phys. Chem. A **117**, 4394 (2013)
15. Buchta, D., Krishnan, S.R., et al.: J. Chem. Phys. **139**, 084301 (2013)

16. Bünermann, O., Kornilov, O., Leone, S.R., Neumark, D.M., Gessner, O.: IEEE J. Sel. Top. Quantum Electron. **18**, 308 (2012)
17. von Haeften, K., Laarmann, T., Wabnitz, H., Möller, T.: J. Phys. B **38**, 373–386 (2005)
18. Lewis, W., Lindsay, C., Bemish, R., Miller, R.: J. Am. Chem. Soc. **127**, 7235 (2005)
19. Atkins, K.: Phys. Rev. **116**, 1339 (1959)
20. Fechner, L., Grüner, B., Sieg, A., Callegari, C., Ancilotto, F., Stienkemeier, F., Mudrich, M.: Phys. Chem. Chem. Phys. **14**, 3843 (2012)
21. O'keeffe, P., et al.: Rev. Sci. Instrum. **82**, 033109 (2011)
22. Dick, B.: Phys. Chem. Chem. Phys. **16**, 570 (2014)
23. LaForge, A.C., et al.: Phys. Rev. A **93**(5), 050502 (2016)
24. Fano, U., Cooper, J.W.: Rev. Mod. Phys. **40**(3), 441 (1968)
25. Kaldun, A., Blättermann, A., Stooß, V., Donsa, S., Wei, H., Pazourek, R., Nagele, S., Ott, C., Lin, C.D., Burgdörfer, J., Pfeifer, T.: Science **354**, 738 (2016)

Two Component Bose–Einstein Condensate in a Pöschl–Teller Potential

N. Kundu and Utpal Roy$^{(\boxtimes)}$

Department of Physics, Indian Institute of Technology Patna, Bihta,
Patna 801103, India
{nilanjan.pph15, uroy}@iitp.ac.in

Abstract. We contribute to the field of rogue waves by providing an exact analytical model of the coupled Bose–Einstein Condensation (BEC) in Pöschl–Teller potential. We start with the dimensionless 1D BEC of a weakly interacting ultra-cold atomic gas with cubic nonlinearity, under the influence of one-dimensional Pöschl–Teller potential. Our main goal is to provide an exact analytical solution of the two-component Gross–Pitäevskii equation which is the modified form of NonLinear Schrödinger Equation (NLSE) for different types of potentials and nonlinearity.

1 Introduction

BEC was experimentally achieved in the year 1995 and its dynamics continue to be an interesting and nontrivial issue to the scientists, both in experiment and theory. It has become a highly emerging area to explore the physics of it and make it applicable to the physically relevant scenarios. Several amazing phenomena like dark and bright solitons, rogue waves, vortices, interference, and Faraday waves are observed by tuning its external potential, nonlinearity, etc. To understand the dynamics of BEC, Gross–Pitäevskii (GP) equation is the most appropriate theoretical model. Our main goal is to solve the two-component GP equation which is the modified form of NLSE for different types of potentials and nonlinearity. Though it is quite difficult using analytical techniques to tackle these kinds of nonlinear systems with the improvement of different methodologies in PDE, it is now possible to map the equation to known localized solutions and observe their dynamics in various potentials. These kind of analytical studies are also helpful in interesting physical phenomena like ultra-cold atoms trapped in BOL [1–3], Anderson localization [4, 5], and subdiffusive transport [6]. Rogue waves have now become a topic of intense research, which are extreme wave events mostly known for its large-scale maritime disasters. This kind of waves is related to an oceanic phenomenon. Apart from the fact that physicist has shown its existence in optics [7, 8], plasmas [9], capillary waves [10], and BEC [11], new studies related to the solution of PDE having special properties of correlated solutions are of severe need. Manakov system [12] is a special kind of system which gives the coupled Rogue wave solutions. One implements this Manakov system in the coupled NLSE to get rogue wave solutions [13–15]. These type of solutions are basically the mathematical description of peregrine solitons [13, 16], which are localized in both the coordinates and is rather a mixture of

© Springer Nature Singapore Pte Ltd. 2019
P. C. Deshmukh et al. (eds.), *Quantum Collisions and Confinement of Atomic and Molecular Species, and Photons*, Springer Proceedings in Physics 230,
https://doi.org/10.1007/978-981-13-9969-5_12

dark-bright solitons [17, 18]. The study of multicomponent scenario on linear waves is a direction of intense research in BEC [19]. These types of multicomponent systems demonstrate multiple coupled condensations. The first experimental realization of this kind of system was in Rubidium atoms [20, 21]. Such multicomponent condensates can manifest unique features in BEC due to its inter-component interactions.

2   Analytical Model and Discussion

At a temperature below the critical temperature, the Bose–Einstein Condensation is described by the self-consistent NLSE known as the GP equation.

$$\left[i\hbar\frac{\partial}{\partial t} + \frac{\hbar^2}{2m}\frac{\partial^2}{\partial z^2} + NU_0(z,t)|\psi(z,t)|^2 - V(z) - i\tau(z,t)\right]\psi(z,t) = 0$$

In order to rescale the equation, we introduce the normalizations as follows:

$$\tilde{t} = \frac{t}{t_s}$$

$$\tilde{x} = \frac{x}{a_0}$$

$$\tilde{\psi}(z,t) = a_0^{3/2}\psi(z,t)$$

We get the dimensionless 1D BEC [22] of weakly interacting ultra-cold atoms with cubic nonlinearity [17, 23], trapped in a one-dimensional spatial potential.

$$\left[i\frac{\partial}{\partial t} + \frac{\partial^2}{\partial z^2} + g(z,t)|\psi(z,t)|^2 - V(z) - i\tau(z,t)\right]\psi(z,t) = 0 \tag{1}$$

Equation (1) can be written as a system of one-dimensional coupled GPE,

$$i\frac{\partial\psi_1}{\partial t} + \frac{1}{2}\frac{\partial^2\psi_1}{\partial z^2} + R(z,t)\left(|\psi_1|^2 + |\psi_2|^2\right)\psi_1(z,t) - i\tau(z,t)\psi_1(z,t) - V(z)\psi_1(z,t) = 0$$

$$i\frac{\partial\psi_2}{\partial t} + \frac{1}{2}\frac{\partial^2\psi_2}{\partial z^2} + R(z,t)\left(|\psi_1|^2 + |\psi_2|^2\right)\psi_2(z,t) - i\tau(z,t)\psi_2(z,t) - V(z)\psi_2(z,t) = 0$$

$$\tag{2}$$

To study the dynamics of BEC construction of the solution to equation (2) is taken of the following ansatz

$$\psi_i(z,t) = A(z,t)F_i(Z(z),t)e^{i\theta(z,t)} \tag{3}$$

where $\theta(z, t)$ and $A(z, t)$ are the space and time-modulated phase and amplitude. $F(Z(z), t)$ is a space and time-dependent variable. Our aim is to construct a set of consistency conditions which will allow us to map (2, 3) to a solvable coupled differential equation of the form:

$$i\frac{\partial F_j}{\partial t} + \frac{\partial^2 F_j}{\partial Z^2} + 2F_j \sum_{k=1}^{2} |F_k|^2 = 0 \tag{4}$$

The above coupled nonlinear differential equation is commonly known as Manakov equation [6], which possesses special forms of solutions such as dark-bright solitons.

We now derive the consistency conditions among the phase, amplitude, nonlinearity, and gain or loss of the system such that the requirements of Manakov system can be fulfilled.

$$\left[A^2(z,t), Z_z(z)\right]_z = 0 \tag{5}$$

$$Z_t(z) + Z_z(z)\theta_z(z,t) = 0 \tag{6}$$

$$Z_z^2(z) - 2A^2(z,t)g(z,t) = 0 \tag{7}$$

$$2A(z,t)A_t(z,t) + \left[A^2(z,t)\theta_z(z,t)\right]_z - 2\tau(z,t)A^2(z,t) = 0 \tag{8}$$

From the above set of consistency conditions, the expression of amplitude, phase, and nonlinearity can be systematically reduced.

$$A(z,t) = \sqrt{\frac{c(t)}{Z_z(z)}} \tag{9}$$

$$\theta_z = -\frac{Z_t(z)}{Z_z(z)} \tag{10}$$

$$g(z,t) = \frac{Z_z^2(z)}{2A^2(z,t)} \tag{11}$$

As we can observe from the above equations, the amplitude is related to the nonlinearity coefficient.

The loss or gain is connected as,

$$c_t(t)Z_z^2(z) - 2c(t)Z_z(z)Z_{z,t}(z) + 2c(t)Z_{zz}(z)Z_t(z) - 2c(t)Z_z^2(z)\tau(z,t) = 0 \tag{12}$$

The Manakov system equation [4] has the following form of solutions:

$$F_1[Z(z),t] = e^{2i\omega T}\left[\left(\frac{L}{B}\right)a_1 + \left(\frac{M}{B}\right)a_2\right]$$
$$F_2[Z(z),t] = e^{2i\omega T}\left[\left(\frac{L}{B}\right)a_2 - \left(\frac{M}{B}\right)a_1\right] \tag{13}$$

where

$$L = \frac{3}{2} - 8\omega^2 t^2 - 2a^2 Z^2(z) + 8i\omega t + |f|^2 e^{2aZ(z)}$$

$$M = 4f \left(aZ(z) - 2i\omega t - \frac{1}{2} \right) e^{aZ(z) + i\omega t}$$

$$B = \frac{1}{2} + 8\omega^2 t^2 + 2a^2 Z^2(z) + |f|^2 e^{2aZ(z)} \tag{14}$$

$$a = \sqrt{a_1^2 + a_2^2}$$

$$\omega = a^2$$

a_1 and a_2 are arbitrary real parameters.

Now from the form of solutions, the ratios L/B and M/B describes asymptotically as a dark soliton and a bright soliton, respectively, as t→(±)∞. Hence,

$$\frac{L}{B} \to \text{Tanh}[Z(z)]$$

$$\frac{M}{B} \to -i\sqrt{2} \left(\frac{ft}{|ft|} \right) \frac{e^{i\omega t}}{\text{Cosh}[Z(z)]} \tag{15}$$

Again Pösch–Teller (PT) potential is one of the most popular quantum mechanical systems with some meticulous property of transmitting linear waves without attenuation. The general form of PT potential is given by

$$V(Z) = -\frac{7s^2}{16} \text{sech}^2(sz) \tag{16}$$

where s controls the width of the potential. With the proper choice of parameters as

$$Z(z) = \gamma \cos ech(sz) \tag{17}$$

the following form of nonlinearity is being observed,

$$g = \frac{G\gamma^3 s^2}{2c(t)} \coth^2(sz) \cos ech^2(sz) \tag{18}$$

The final form of the wave function is as given below:

$$\Psi_1 = \sqrt{\frac{c(t)}{\gamma \text{Coth}(sz)\text{Cosech}(sz)}} \left[e^{2i\omega t} (a_1 \text{Tanh}[\gamma \cos ech(sz)] + a_2 \text{Sech}[\gamma \cos ech(sz)]) e^{-i\theta(z,t)} \right]$$

$$\Psi_2 = \sqrt{\frac{c(t)}{\gamma \text{Coth}(sz)\text{Cosech}(sz)}} \left[e^{2i\omega t} (a_2 \text{Tanh}[\gamma \cos ech(sz)] - a_1 \text{Sech}[\gamma \cos ech(sz)]) e^{-i\theta(z,t)} \right]$$

$$\tag{19}$$

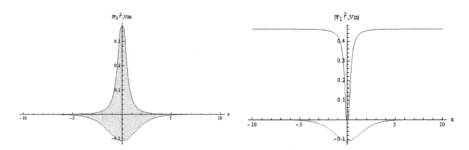

Fig. 1. Variation of condensate density (bright-dark) with potential. The first component (left side) and the second component (right side). The potential is given in the negative y-axis and density is depicted in the positive y-axis

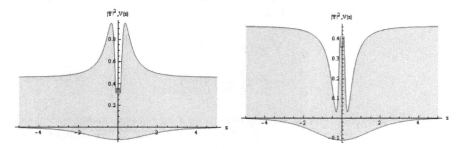

Fig. 2. Variation of condensate density (dark-dark) with potential. The first component (left side) and a second component (right side)

The condensate density variation with Pöschl–Teller potential is given below:

$$s = -0.5$$

As we observe in Fig. 1 for the first component, we switch off the parameter a_2 bright soliton is obtained in the PT potential, whereas we obtain a dark soliton for the second component. In Fig. 2, we observe a mixture of dark and bright soliton for both the two condensate components where the second component condensate density is the inverted condensate density of the first component.

3   Conclusion

In this work, we have constructed a family of solutions for Pöschl–Teller potential for two-component Bose–Einstein Condensation. We have proposed a novel method to find an explicit analytical solution for the coupled GP equation for Pöschl–Teller potential. The present idea can be further extended for higher order nonlinearities and other typical potentials such as Morse potential, Toda Lattice potential, Power law potential, etc.

References

1. Nath, A., Roy, U.: Bose–Einstein condensate in a bichromatic optical lattice: an exact analytical model. Laser Phys. Lett. **11**, 115501 (2014)
2. Sakhel, A.R.: Long-time averaged dynamics of a Bose–Einstein condensate in a bichromatic optical lattice with external harmonic confinement. Phys. B **493**, 72 (2016)
3. Edwards, E.E., et al.: Adiabaticity and localization in one-dimensional incommensurate lattices. Phys. Rev. Lett. **101**, 260402 (2008)
4. Roati, G., et al.: Anderson localization of a non-interacting Bose–Einstein condensate. Nature **453**, 895 (2008)
5. Billy, J., et al.: Direct observation of Anderson localization of matter waves in a controlled disorder. Nature **453**, 891 (2008)
6. Lucioni, E., et al.: Observation of subdiffusion in a disordered interacting system. Phys. Rev. Lett. **106**, 230403 (2011)
7. Solli, D.R., Ropers, C., Koonath, P., Jalali, B.: Optical rogue waves. Nature **450**, 1054 (2007)
8. Erkintalo, M., Genty, G., Dudley, J.M.: Rogue wave like characteristics in femtosecond supercontinuum generation. Opt. Lett. **34**, 2468 (2009)
9. Moslem, W.M., Shukla, P.K., Eliasson, B.: Surface plasma rogue waves. Europhys. Lett. **96**, 25002 (2011)
10. Shats, M., Punzmann, H., Xia, H.: Capillary rogue waves. Phys. Rev. Lett. **104**, 104503 (2010)
11. Bludov, Y.V., Konotop, V.V., Akhmediev, N.: Matter rogue waves. Phys. Rev. A **80**, 033610 (2009)
12. Manakov, S.V.: On the focusing of two-dimensional stationary self-focusing of electromagnetic waves. Sov. Phys. JETP **38**, 248 (1974)
13. Bludov, Y.V., Konotop, V.V., Akhmediev, N.: Vector rogue waves in binary mixtures of Bose–Einstein condensates. Eur. Phys. J. Special Topics **185**, 169 (2010)
14. Yan, Z.: Vector financial rogue waves. Phys. Lett. **375**, 4274 (2011)
15. Guo, B.L., Ling, L.M.: Rogue waves in the three-dimensional Kadomtsev-Petviashvili equation. Chin. Phys. Lett. **28**, 110202 (2011)
16. Peregrine, D.H.: Water waves, nonlinear Schrodinger equations and their solution. J. Aust. Math. Soc. Ser. B, Appl. Math. **25**, 16 (1983)
17. Baronio, F., Degasperis, A., Conforti, M., Wabnitz, S.: Solutions of the vector nonlinear Schrodinger equations: evidence of determinist rogue waves. Phys. Rev. Lett. **109**, 044102 (2012)
18. Guo, B.L., Ling, L.M.: Rogue waves, breathers and bright-dark solutions for the coupled Schrodinger equations. Chin. Phys. Lett. **28**, 110202 (2011)
19. Kevrekidis, P.G., Frantzeskakis, D.J., Carretero-Gonzalez, R.: Emergent nonlinear phenomena in Bose–Einstein condensates. Springer (2008)
20. Hamner, C., Chang, J.J., Engels, P., Hoefer, M.A.: Generation of dark-bright soliton trains in superfluid-superfluid counterflow. Phys. Rev. Lett. **106**, 065302 (2011)
21. Hoefer, M.A., Chang, J.J., Hamner, C., Engels, P.: Dark-dark solitons and modulational instability in miscible two-component Bose–Einstein condensates. Phys. Rev. A **84**, 041605 (R) (2011)

22. Chen, Z., et al.: Waveguides induced by photorefractive screening solitons. J. Opt. Soc. Am. B **14**, 11 (1997)
23. Bronski, J.C., et al.: Bose–Einstein condensates in standing waves: the cubic nonlinear Schrodinger equation with a periodic potential. Phys. Rev. Lett. **86**, 1402 (2001)

Electron Transport Modeling in Biological Tissues: From Water to DNA

Mario E. Alcocer-Ávila[1], Michele A. Quinto[2], Juan M. Monti[2], Roberto D. Rivarola[2], and Christophe Champion[1(✉)]

[1] Université de Bordeaux, CNRS, CEA, CELIA (Centre Lasers Intenses et Applications), UMR 5107, 33405 Talence, France
christophe.champion@u-bordeaux.fr
[2] Instituto de Física Rosario, CONICET – Universidad Nacional de Rosario, S2000 EKF Rosario, Argentina

Abstract. A proper and reliable description of charged particles interactions in the biological matter remains a critical aspect of radiation research. All the more so when new and better methods for treating cancer through the use of ionizing radiation are emerging on the horizon such as targeted alpha therapy. In this context, Monte Carlo track-structure codes are extremely useful tools to study radiation-induced effects at the atomic scale. In the present work, we review the latest version of CELLDOSE, a *homemade* Monte Carlo track-structure code devoted to electron dosimetry in biological matter. We report here some recent results concerning the stopping power and penetration range of electrons in water and DNA for impact energies ranging from 10 eV to 10 keV.

1 Introduction

A complete and accurate description of charged particles' interactions with matter at the nanometric scale is essential for understanding the radio-induced biological effects such as cellular death and chromosomal aberration induction. Computer simulations, particularly those based on Monte Carlo (MC) methods, are a very powerful tool in this field and have been used for decades to tackle particle transport problems. MC radiation transport codes can be grouped into two categories depending on how the particles are tracked: condensed-history (MCCH) codes and track-structure (MCTS) codes. In MCCH codes, also known as general-purpose MC codes, a particle's path is divided into discrete steps, each step encompassing a large number of collision processes and taking into account their combined effects by means of multiple scattering theories. This allows to determine the energy losses and angular changes in the direction of the particle at the end of the step [1]. Some examples of general-purpose MC codes are PENELOPE, MCNP, Geant4, and FLUKA. Although MCCH codes have the obvious advantage of reducing computation time and provide results accurate enough for most applications in radiation therapy, they have also important

© Springer Nature Singapore Pte Ltd. 2019
P. C. Deshmukh et al. (eds.), *Quantum Collisions and Confinement of Atomic and Molecular Species, and Photons*, Springer Proceedings in Physics 230,
https://doi.org/10.1007/978-981-13-9969-5_13

shortcomings, especially when dealing with very small geometries or low-energy transfers [2,3]. Therefore, only MCTS codes are appropriate for microdosimetry applications and remain the only tools able to reproduce in detail the energy deposit pattern in small biological structures such as cell nucleus or DNA subunits (nucleobase, sugar-phosphate backbone) [4]. Contrary to MCCH codes, MCTS codes follow both the primary and the secondary particles in an event-by-event manner, until their energy falls below a cutoff value. Track-structure simulations consist in a series of random samplings, which first determine the distance traveled by the particle, then the type of interaction taking place at the point of arrival and finally the full kinematics of the secondary particles created. To properly simulate the transport of particles in a given medium, MCTS codes require total and differential cross sections for describing the various particle-induced interactions as input data. Most of the time, these cross sections are obtained using a combination of experimental data and theoretical models. It is worth noting that, in the end, the effectiveness and reliability of the code will depend on how accurate and complete the cross sections database is.

The preferred medium for the simulations in most of the existing MCTS codes designed for applications in radiobiology and medical physics is water, since it has often been considered a good surrogate for tissue. A few codes have also included models which allow to explore in more detail the damage to DNA [5,6]. A summary of some well-known MCTS codes and their relevant features may be found in [1,2].

The purpose of this paper is to review the latest version of CELLDOSE, an *homemade* MCTS code for electron dosimetry in biological matter [7]. We present some of our results and discuss the ongoing work devoted to extend the code's capabilities. In Sect. 2, we summarize the main features of CELLDOSE and we provide the theoretical framework used to describe the biological media and the physical processes considered in the code, concluding with a general outline of the future developments. In Sect. 3, we present our calculations regarding the stopping power and the penetration range for electrons in water and DNA and compare them with values found in the literature.

2   CELLDOSE

CELLDOSE is an MCTS code developed in the C++ programming language. In its current version, it is able to simulate the full slowing-down histories of electrons and positrons in gaseous and liquid water, as well as in DNA. The underlying physics of CELLDOSE—needed to describe the elastic as well as the inelastic electron-induced processes—refers to several theoretical models independently developed and mostly within the quantum-mechanical framework. For more details, we refer the reader to some of our previous studies where the ionization treatment for electrons in water is detailed [8], where the elastic scattering of electrons and positrons has been studied in liquid and gaseous water [9], and finally to [10] where the ionization of DNA subunits was investigated. All these models were summarized in [11] and implemented into a *homemade*

Monte Code track-structure code—called EPOTRAN—devoted to electron and positron transport in water for impact energies ranging from 1 MeV down to a predefined energy cutoff of 7.4 eV (i.e. the water excitation threshold) [12]. Thus, CELLDOSE may be seen as a dosimetric extension of EPOTRAN whose current extension includes newly studied electron–DNA interactions. In this context, CELLDOSE has been used previously to assess the electron dose distribution for several radionuclides in simple geometries [7,13] as well as in more complex environments [14].

2.1   Description of the Biological Medium: From Water to DNA

In CELLDOSE, the water target is described following the quantum approach proposed by Moccia for water vapor [15]. The ten bound electrons of the water target are distributed in five molecular orbitals ($j = 5$), which are constructed from a linear combination of atomic orbitals in a self-consistent field. Each molecular orbital wave function is developed in terms of Slater-type-orbital functions centered on the oxygen nucleus. According to this description, the molecular orbital wave functions are written as

$$\psi_j(r) = \sum_{k=1}^{N_j} a_{jk} \phi_{n_{jk}l_{jk}m_{jk}}^{\xi_{jk}}(r) \tag{1}$$

where N_j is the number of Slater atomic orbitals $\phi_{n_{jk}l_{jk}m_{jk}}^{\xi_{jk}}$ and a_{jk} is the corresponding weight. The atomic components are written as

$$\phi_{n_{jk}l_{jk}m_{jk}}^{\xi_{jk}}(r) = R_{n_{jk}}^{\xi_{jk}}(r) S_{l_{jk}m_{jk}}(\hat{r}), \tag{2}$$

with the radial part given by

$$R_{n_{jk}}^{\xi_{jk}}(r) = \frac{(2\xi_{jk})^{2n_{jk}+1/2}}{\sqrt{2n_{jk}!}} r^{n_{jk}-1} e^{-\xi_{jk}r}, \tag{3}$$

and the angular part, expressed by means of real spherical harmonics, is given by:

$$\begin{cases} \text{if } m_{jk} \neq 0 : S_{l_{jk}m_{jk}}(\hat{r}) = \left(\frac{m_{jk}}{2|m_{jk}|}\right)^{1/2} \\ \qquad \times \left\{ Y_{l_{jk}-|m_{jk}|}(\hat{r}) + (-1)^{m_{jk}} \left(\frac{m_{jk}}{|m_{jk}|}\right) Y_{l_{jk}|m_{jk}|}(\hat{r}) \right\}, \\ \text{if } m_{jk} = 0 : S_{l_{jk}0}(\hat{r}) = Y_{l_{jk}0}(\hat{r}). \end{cases} \tag{4}$$

All the necessary coefficients and quantum numbers are reported in [15].

DNA modeling follows a similar ab initio approach in which each component (i.e., nucleobase or sugar-phosphate backbone unit) is described via N molecular subshell wave functions similar to (1); $N = 35, 29, 39, 33, 29$ and 48 for adenine (A), cytosine (C), guanine (G), thymine (T), uracil (U), and sugar-phosphate backbone unit (SP), respectively. Each molecular subshell wave function is expressed as a linear combination of atomic wave functions corresponding to the different atomic components: H_{1s}, C_{1s}, C_{2s}, C_{2p}, N_{1s}, N_{2s}, N_{2p}, O_{1s}, O_{2s}, O_{2p}, P_{1s}, P_{2s}, P_{3s}, P_{2p}, and P_{3p}. Total-energy calculations for all targets were performed in the gas phase with the Gaussian 09 software at the RHF/3-21G level of theory [16]. The computed ionization energies of the occupied molecular orbitals of the targets were scaled so that the ionization energy of their HOMO coincides with the experimental values found in the literature. For each molecular orbital j, the effective number of electrons relative to the atomic component k was derived from a standard Mulliken population analysis and their sum for every occupied molecular orbital is very close to 2, since only atomic shells with very small population were discarded. More details about this model are provided in [17]. Besides, to get insight into the real energy deposit cartography induced by charged particles impact in the biological medium, a DNA molecule composed of a nucleobase-pair plus two SP groups has been considered. Additionally, to fit the realistic composition of living cells, we used the nucleobase repartition percentages reported by Tan et al. [18], namely, 58% (A-T) (adenine–thymine base pair) and 42% (C-G) (cytosine–guanine base pair). Thus, by using the respective molar mass of each DNA component, viz. $M_A = 135.14$ g mol^{-1}, $M_T = 126.12$ g mol^{-1}, $M_C = 111.11$ g mol^{-1}, $M_G = 151.14$ g mol^{-1} and $M_{SP} = 180$ g mol^{-1}, the following mass percentages were obtained: A (12.6%), T (11.8%), C (7.5%), G (10.2%) and SP group (57.9%). However, this description corresponds to dry DNA. In order to obtain an even more realistic biological medium, hydrated DNA was simulated by adding 18 water molecules per nucleotide, which modified the mass percentages as follows: A (8.3%), T (7.7%), C (4.9%), G (6.7%), SP group (38.1%), and water (34.3%). This is consistent with the suggestion made by Birnie et al. [19] who estimated that the total amount of water associated with DNA was of the order of 50 moles per mole of nucleotide, in order to get the expected density of 1.29 g cm^{-3}.

Irrespective of the biomolecular targets investigated, CELLDOSE is based on a cross section database, which refers to *isolated* molecules and to living matter components in *vapor* state. In this context, the present work clearly differs from existing studies on *condensed* matter (water or DNA), where the energy-loss function of realistic biological components was extracted from experimental data and interpolated for use in cross section calculations (see for example [20] and the recent series of works by Abril and coworkers [21,22] and Emfietzoglou et al. [23]) or developed within the independent atom model framework [24]. However, whether it is for water or DNA, the available data—cross sections as well as macroscopic outcomes like ranges and stopping power—are exclusively measured in vapor phase and comparisons with the liquid homologous have been only rarely reported. Consequently, a *vapor-based* approach is chosen here in order to check the suitability of our theoretical models, although comparisons with existing condensed matter models are presented in the Results section of this manuscript. Nevertheless, the theoretical models described in the following section report some information about the extrapolations performed for treating the liquid water phase, in particular for modeling the elastic scattering process [9] and the electron-induced ionization [25].

2.2   Interactions and Cross Sections

CELLDOSE implements a set of theoretical cross sections calculated within the quantum-mechanical framework, using the partial-wave method. The following interactions are taken into account in the code: elastic scattering of the incident particle; ionization and electronic excitation of the target molecule; and Positronium formation, when the primary particle is a positron. We report hereafter some details on the theoretical treatment of electron-induced interactions in water by considering—in some cases—the vapor and liquid environment specificities.

Elastic Scattering. When the selected interaction is elastic scattering, the singly differential cross sections are sampled in order to determine the scattering direction at the electron incident energy E_{inc}, which remains almost unchanged because the energy transfer induced during this process is very small (of the order of meV) [12].

The perturbation potential of the water molecule can be approximated by a spherically symmetric potential $V(r)$ composed of three different terms: the static contribution V_{st}, the correlation-polarization term V_{cp}, and the exchange term V_{ex}. Thus, we can write:

$$\begin{cases} V(r) = V_{\text{st}}(r) + V_{\text{cp}}(r) + V_{\text{ex}}(r) & \text{for electrons,} \\ V(r) = -V_{\text{st}}(r) + V_{\text{cp}}(r) & \text{for positrons.} \end{cases} \tag{5}$$

As we mentioned before, for water *vapor* the description of the water molecule provided by Moccia [15] was used. The static potential $V_{\text{st}}(r)$ was numerically calculated from each target molecular wave function by using the spherical average approximation and is expressed as

$$V_{\text{st}}(r) = \sum_{j=1}^{N_{\text{orb}}} [V_{\text{st}}^j(r)]_{\text{elec}} + [V_{\text{st}}(r)]_{\text{ion}}, \tag{6}$$

where $[V_{\text{st}}^j(r)]_{\text{elec}}$ and $[V_{\text{st}}(r)]_{\text{ion}}$ refer to the electronic and ionic target contribution to the static potential, respectively.

For *liquid* water, the molecular wave functions can only be obtained by theoretical calculations performed in the Dynamic Molecular framework, which rapidly becomes extremely computer time-consuming. To overcome this limitation, the static potential for liquid water has been deduced from the experimental electron density reported by Neuefeind et al. [26]. This latter was then fitted and analytically expressed as [9]:

$$\rho_N(r) = \frac{1}{4\pi}\left(a_1 e^{-r/b_1} + a_2 r e^{-r/b_2}\right), \tag{7}$$

what leads to a static potential given by

$$V_N(r) = -\frac{1}{r}[a_1 b_1^2(r + 2b_1)e^{-r/b_1} + a_2 b_2^2(r^2 + 4b_2 r + 6b_2^2)e^{-r/b_2} + (2\delta - 2)], \tag{8}$$

where $a_1 = 4411$ a.u., $a_2 = 120.435$ a.u., $b_1 = 0.06046$ a.u., $b_2 = 0.3248$ a.u. and the parameter δ is expressed as

$$\delta = \begin{cases} \frac{r}{R_{OH}} & \text{for } r \leq R_{OH} \\ 1 & \text{for } r > R_{OH}, \end{cases} \tag{9}$$

with $R_{OH} = 1.8336$ a.u. for the water molecule in liquid phase [27,28]. The correlation-polarization contribution is based on the recommendations of Salvat [29] and is given by

$$V_{cp}(r) = \begin{cases} \max\{V_{corr}(r), V_p(r)\} & \text{for } r < r_{cr} \\ V_p(r) & \text{for } r > r_{cr}, \end{cases} \tag{10}$$

where r_{cr} is defined as the outer radius at which $V_p(r)$ and $V_{corr}(r)$ cross. The polarization potential $V_p(r)$ is the same for electrons and positrons and is given by

$$V_p(r) = -\frac{\alpha_d}{2(r^2 + r_c^2)^2}, \tag{11}$$

where the static dipole polarizability of the water molecule $\alpha_d = 9.7949$ a.u. [30] and r_c is the cutoff parameter proposed by Mittleman and Watson [31]:

$$r_c = \left[\frac{1}{2} \alpha_d z^{-1/3} b_{pol}^2 \right]^{1/4}, \tag{12}$$

with $z = 10$ and b_{pol} an adjustable parameter given by:

$$b_{pol} = \sqrt{\max\{(E - 0.5)/0.01; 1\}}. \tag{13}$$

The correlation potential differs for electrons and positrons and has been well described by Padial and Norcross [32] for the former and by Jain [33] for the latter. The corresponding expressions $V_{corr}^-(r)$ and $V_{corr}^+(r)$ are reported below:

$$\begin{cases} V_{corr}^-(r_s) = 0.0311 \ln r_s - 0.0584 + 0.006 r_s \ln r_s \\ \qquad\qquad -0.015 r_s & \text{for } r_s \leq 0.7, \\ V_{corr}^-(r_s) = -0.07356 + 0.02224 \ln r_s & \text{for } 0.7 \leq r_s \leq 10, \\ V_{corr}^-(r_s) = -0.584 r_s^{-1} + 1.988 r_s^{-3/2} - 2.450 r_s^{-2} \\ \qquad\qquad -0.733 r_s^{-5/2} & \text{for } r_s \geq 10, \end{cases} \tag{14}$$

with $r_s = \left[\frac{3}{4\pi\rho(r)} \right]^{1/3}$ and:

$$\begin{cases} V_{corr}^+(r_s) = \frac{1}{2} \left\{ -\frac{1.82}{r_s^{1/2}} + (0.051 \ln r_s - 0.115) \ln r_s \right. \\ \qquad\qquad \left. +1.167 \right\} & \text{for } r_s \leq 0.302, \\ V_{corr}^+(r_s) = \frac{1}{2} \left\{ -0.92305 - \frac{0.09098}{r_s^2} \right\} & \text{for } 0.302 \leq r_s \leq 0.56, \\ V_{corr}^+(r_s) = \frac{1}{2} \left\{ -\frac{8.7674 r_s}{(r_s+2.5)^3} + \frac{-13.151+0.9552 r_s}{(r_s+2.5)^2} \right. \\ \qquad\qquad \left. +\frac{2.8655}{r_s+2.5} - 0.6298 \right\} & \text{for } 0.56 \leq r_s \leq 8.0. \end{cases} \tag{15}$$

Finally, the exchange effect (only used for electrons) was treated via the phenomenological potential $V_{ex}(r)$ proposed by Riley and Truhlar [34], and is expressed as

$$V_{ex}(r) = \frac{1}{2}\left\{ \begin{array}{l} E_{inc} - (V_{st}(r) + V_{cp}(r)) \\ -\left[(E_{inc} - (V_{st}(r) + V_{cp}(r)))^2 + 4\pi\rho(r)\right]^{1/2} \end{array} \right\}. \tag{16}$$

Ionization. In the case of ionization, the kinetic energy of the ejected electron E_e is first determined by random sampling among the singly differential cross sections. Then by applying a direct MC sampling according to the relative magnitude of the partial total ionization cross sections (at E_e), the ionization potential IP_j is determined. The particle energy is reduced by $E_e + IP_j$, whereas the ejected and scattered directions are determined according to the triply and double differential cross sections, respectively. The particular ionization potential IP_j is stored as locally deposited energy, except when inner-shell ionization occurs, in which case an Auger electron is produced and assumed to be isotropically emitted with a kinetic energy E_{Auger} [7,12].

Regarding the theoretical approach, the ionization process is described within the frozen-core first Born approximation that consists in reducing the ten-electron problem to a one active electron problem (for more details see [35]). Thus, the triply differential cross sections (TDCS) are defined as

$$\sigma^{(3)}(\Omega_s, \Omega_e, E_e) \equiv \frac{d^3\sigma_j}{d\Omega_s d\Omega_e dE_e} = \frac{k_e k_s}{k_i}|M|^2, \tag{17}$$

where $d\Omega_s = \sin\theta_s d\theta_s d\phi_s$ and $d\Omega_e = \sin\theta_e d\theta_e d\phi_e$ correspond to the scattered and ejected direction, respectively. The momenta $\mathbf{k_i}, \mathbf{k_s}$ and $\mathbf{k_e}$ are related to the incident, the scattered and the ejected electron, respectively. The transition amplitude M is given by

$$M = \frac{1}{2\pi}\langle \Psi_f|V|\Psi_i\rangle, \tag{18}$$

where V represents the interaction between the incident electron and the target and is written as

$$V = -\frac{8}{r_0} - \frac{1}{|\mathbf{r_0} - \mathbf{R_1}|} - \frac{1}{|\mathbf{r_0} - \mathbf{R_2}|} + \sum_{i=1}^{10}\frac{1}{|\mathbf{r_0} - \mathbf{r_i}|}, \tag{19}$$

with $R_1 = R_2 = R_{OH} = 1.814$ a.u., while $\mathbf{r_i}$ is the position of the ith bound electron of the target with respect to the oxygen nucleus [36]. The initial state is written as the product of two wave functions: a first one $\phi(\mathbf{k_i}, \mathbf{r_0})$ describing the incident electron by a plane wave and a second one $\varphi_i(\mathbf{r_1}, \mathbf{r_2}, \ldots, \mathbf{r_{10}})$ for the ten bound electrons described by the Slater functions given by Moccia. Thus, the initial state is expressed as

$$|\Psi_i\rangle = |\phi(\mathbf{k_i}, \mathbf{r_0})\varphi_i(\mathbf{r_1}, \mathbf{r_2}, \ldots, \mathbf{r_{10}})\rangle. \tag{20}$$

Considering the final state of the collisional system, it was successively described within five different models: two first models where the incident and the scattered electrons are both described by plane waves, whereas the ejected one is described by a Coulomb wave (FBA-CW) or a distorted wave (DWBA), respectively; a two-Coulomb wave (2CW) model where the scattered and ejected electrons are both described by a Coulomb wave; the Branner, Briggs, and Klar (BBK) model whose main characteristic consists of exhibiting a correct asymptotical Coulomb three-body wave function for the ejected and scattered electrons in the residual ion field and finally the DS3C—for dynamic screening of the three two-body Coulomb model. For more details about these models as well as a fine analysis of the respective improvements, we refer the reader to our previous work [36]. However, on the basis of TDCS calculations and comparisons with available experimental measurements, we concluded that the two first-order models, namely, FBA-CW and DWBA, reproduced quite well the shape of the TDCS, the 2CW model giving nevertheless a better agreement with experiments than the first cited ones, even considering that the recoil peak remains still underestimated for some molecular orbitals. Considering the double structure of the binary region, it is generally well reproduced for all models. However, we observed that none of the models—even the rather sophisticated DS3C approach—reproduces correctly the data at low ejection angles. To conclude, for computational time reasons, we first focused our efforts to the building and the implementation in CELLDOSE of a DWBA database including triply, doubly, singly differential, and total cross sections.

Let us remind that the water target wave function used here refers to an *isolated* water molecule, namely, in *vapor* phase. For a more realistic modeling of the electron tracking in biological matter, it is worth noting that water should be considered in its liquid phase. To that end, various semi-empirical models were developed since the pioneer works of the Oak Ridge group [37,38]. For more details, we refer the reader to the works of Emfietzoglou and coworkers [39–41]. In this context, we reported in a recent work [25] a unified methodology to express the molecular wave functions of water in the liquid phase by means of a single center approach and reported a low influence of the water phase on the cross section calculations, limited to the low-incident energy regime. Indeed, in this energy range, the contribution of the outermost target subshells to the ionization cross sections is very important, which leads to an increasing influence of the water phase as the incident electron energy decreases.

Excitation. When the selected interaction is excitation, direct MC sampling is applied according to the relative magnitude of all partial excitation cross sections for determining the excitation level n. The particle energy is reduced by the potential energy W_n, assumed as locally deposited, whereas the incident direction remains unaltered [7].

Excitation includes all the processes which modify the internal state of the molecule without emission of electrons. These processes include, in particular, the following:

1. Electronic transitions toward Rydberg states or degenerate states ($\widetilde{A}_1 B_1$, $\widetilde{B}_1 A_1$, diffuse band);
2. The dissociative excitation processes leading to excited radicals (H*, O*, and OH*);
3. The vibrational and rotational channels;
4. The dissociative attachment leading to the formation of negative ions.

It is worth noting that describing the electronic states of molecular targets remains a challenging task in a quantum-mechanical approach. In this context, we first privileged semi-empirical models for treating both the electron- and positron-induced excitations. For the first three processes listed above, the total excitation cross sections are given by [11, 42, 43]

$$\sigma_n(E_{\text{inc}}) = \frac{4\pi a_0^2 Ry^2 f_0^{(n)} c_0^{(n)}}{W_n^2} \left(\frac{W_n}{E_{\text{inc}}} \right)^\Omega \left[1 - \left(\frac{W_n}{E_{\text{inc}}} \right)^\beta \right]^\nu, \tag{21}$$

where W_n is the deposited energy during the excitation channel labeled n, and Ω, β, ν are fitting parameters; $f_0^{(n)}$ and $c_0^{(n)}$ being the oscillator strength and a constant of normalization deduced from experimental total cross sections $\sigma_n(E_{\text{inc}})$, respectively. All these parameters are listed in Olivero et al. [42]. For the dissociative attachment process, the following expression is used [44]:

$$\sigma_n(E_{\text{inc}}) = \frac{A e^t}{U(1 + e^t)^2} \tag{22}$$

with:

$$t = \frac{E_{\text{inc}} - W_n}{U}, \tag{23}$$

where A and U are fitting parameters. Moreover, let us add that according to experimental results [44], we assume that impact excitation induces no angular deflection.

Positronium Formation. When the incident particle is a positron, a target electron capture process can take place, leading in the final channel of the reaction to the formation of Positronium atoms (Ps), a bound system consisting of an electron and a positron. In its ground state (triplet state), this exotic system has a mean lifetime of approximately 1.47×10^{-7} s (ortho-Positronium) annihilating in three photons, whereas in the singlet state (para-Positronium) its mean lifetime is approximately 1.25×10^{-10} s with annihilation in two photons of 511 keV each, the rest mass of the electron. The knowledge of this process is important for many domains of physics as well as in medicine, a field where it is nowadays common the use of positron emission tomography (PET) for both diagnostic and treatment monitoring.

The description of the Positronium formation process when positrons collide with a water target is based on the work by Hervieux et al. [45], who developed a continuum distorted-wave final state (CDW-FS) approximation. Within this approach, the final state of the collision is distorted by two-Coulomb wave functions associated with the interaction of both the positron and the active electron (the captured one) with the residual ionic target. Since the collision of a positron with a vapor water molecule is a many-electron problem, the independent electron model (IEM) was adopted. This means that at the high impact energies considered here, it is possible to assume that the electron is, before being captured, suddenly ejected into an intermediate state corresponding to a high-velocity continuum of the target (with a velocity close to the projectile one). Therefore, it is considered that the other electrons remain as frozen in their initial orbitals due to the fact that the relaxation time is much larger than the collision one. The total cross sections are calculated by using a partial-wave technique. More details are given in [45].

2.3 Future Developments

The next step in the development of CELLDOSE is to extend the code by including the interactions of heavier charged particles, namely protons and alpha particles, with water and DNA. This will be achieved by coupling CELLDOSE with TILDA-V (an acronym for "Transport d'Ions Lourds Dans l'Aqua & Vivo"), a newly developed MCTS code [6,46]. In its current version, TILDA-V is able to simulate the full transport of protons and its secondaries in water and DNA components. Thus, the main task remains the modeling of alpha particles, taking into account all the charge states of the helium atom [47,48]. With the growing interest in targeted alpha therapy (TAT) for treating microscopic or small-volume disease [49], an accurate nano-dosimetric description of the pattern of energy deposit of alpha particles could be a valuable input to assess and compare the potential therapeutical advantages of several alpha-emitting radionuclides. This is even more true given that microdosimetry seems essential to investigate the therapeutic efficacy of TAT [50].

3 Results

We present here some recent results obtained with the latest version of CELLDOSE. We have computed the stopping power and penetration range for electrons in water and DNA for incident energies, E_{inc} in the range 10 eV to 10^4 eV. To compute the stopping power one million histories of stationary projectiles, i.e., primary particles for which the initial energy remains unchanged, were simulated. In this mode, each primary particle is followed until it experiences an interaction with the medium. A sum over the energy loss and the distance traveled by each particle is carried out and the linear stopping power—expressed in keV μm^{-1}—is then defined as

$$SP = \frac{E_{tot}}{L}, \tag{24}$$

where E_{tot} is the total energy lost by the stationary projectiles and L is the track length. The results were later normalized using the mass densities for

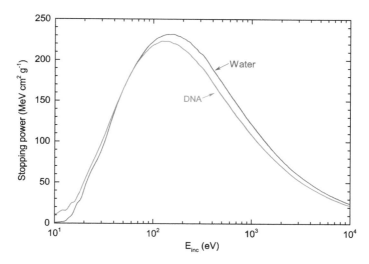

Fig. 1. Total stopping power for electrons in water versus DNA as provided by CELL-DOSE

each medium ($\rho_{H_2O} = 1$ g cm^{-3}, $\rho_{DNA} = 1.29$ g cm^{-3} [19]) to obtain the mass stopping power, expressed in MeV cm^2 g^{-1}. Figure 1 represents the total stopping power for electrons in water versus DNA. The results display a similar behavior for both media, with a maximum stopping power of 232 MeV cm^2 g^{-1} at 160 eV and 223 MeV cm^2 g^{-1} at 120 eV for water and DNA, respectively. It is worth noting that the linear stopping power of DNA is greater than for water; however, when normalizing with respect to the mass density of the medium to obtain the mass stopping power as mentioned before, the DNA curve initially above the one from water is shifted downwards, as shown in Fig. 1.

Figure 2 shows the contribution of ionization as well as excitation processes to the total stopping power of electrons in both media. In panel (a) we report our results for water and compare them with values found in the literature. A good agreement can be seen with the data provided in the IAEA TECDOC 799 [51] and the ICRU Report 16 [52] for $E_{inc} > 300$ eV, as well as with the calculations by Garcia-Molina et al. [53] and Tan et al. [54] for $E_{inc} > 500$ eV. At lower energies, however, important discrepancies are observed, especially with respect to the maximum value of the stopping power, since there is a difference of about 12% and 19% between our values and the predictions of Garcia-Molina et al. [53] and Tan et al. [54], respectively. This can be explained by the fact that both authors used the dielectric formalism for their calculations, which without all the necessary corrections is known to overestimate the inelastic cross sections at the peak region, as pointed out in the work by Tan et al. [54]. Finally, our predictions agree with the results obtained by Paretzke [55] in most of the energy range here considered, although a slight underestimation of the maximum is still noticeable. Panel (b) shows our results for DNA. As in the case of water, significant dif-

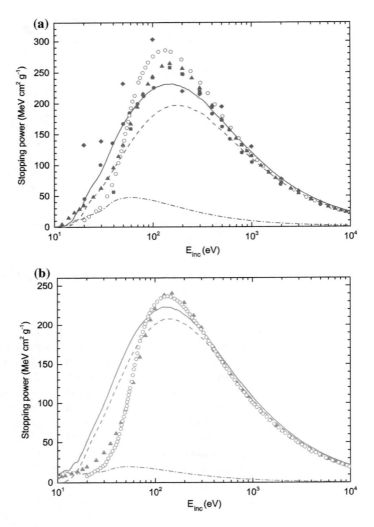

Fig. 2. Contribution of ionization and excitation processes (dashed and dash-dotted line, respectively) to the total stopping power for electrons in **a** water and **b** DNA (blue and red solid lines, respectively), as provided by CELLDOSE. Total stopping power data are taken from the IAEA TECDOC 799 [51] (squares), the ICRU report 16 [52] (diamonds), Garcia-Molina et al. [53] (triangles), Tan et al. [54] (open circles) and Paretzke [55] (solid circles)

ferences are observed between our values and those of Garcia-Molina et al. [53] and Tan et al. [54] at low energies and near the peak region. Nevertheless, now the maximum stopping power values predicted by Garcia-Molina et al. [53] and Tan et al. [54] are only 7% and 6% greater than our result, respectively. On the other hand, a good agreement between the three calculations can be observed for $E_{inc} > 300$ eV.

Figure 3 presents our results for the penetration range of electrons in water and DNA (blue and red solid line, respectively). We apply here the same definition of penetration range used by Meesungnoen et al. [56], namely, the length of the vector $|\mathbf{R_f} - \mathbf{R_i}|$ from the point of departure $(\mathbf{R_i})$ to the final position $(\mathbf{R_f})$ of the electron after thermalization. As shown in Fig. 3, at energies below $\sim$700 eV our results for water fall between those of Meesungnoen et al. [56] (open circles) and the values computed with the default version of Geant4-DNA [57] (solid diamonds). For incident energies above $\sim$700 eV the three predictions overlap perfectly. The penetration range for electrons in DNA is represented in Fig. 3 by the red solid line. It can be observed that in the whole energy range here considered the penetration range values for DNA are lower than those for water. There is a maximum difference of 80% in the penetration range for both media at 10 eV, while for energies higher than 2 keV the difference falls to about 15%. In order to confirm that this behavior is not only due to the different densities considered for water and DNA, we have plotted in Fig. 4 the penetration range in water with the density rescaled to a value of 1.29 g cm^{-3} (blue dashed line). The results show that even after the density correction the difference in the penetration range for electrons in water and DNA is still remarkable at low incident energies, reaching 75% at 10 eV. On the other hand, it can be seen that at about 500 eV the penetration range is almost the same for both media. At

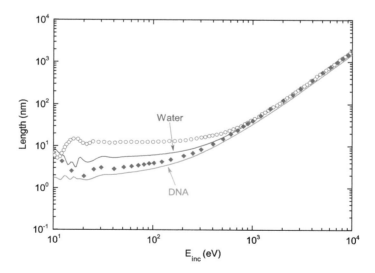

Fig. 3. Penetration range for electrons in water (blue solid line) and DNA (red solid line), as provided by CELLDOSE. Calculations provided by Meesungnoen et al. [56] (open circles) and results obtained with the default version of Geant4-DNA, as reported by Bordage et al. [57] (solid diamonds), are shown for comparison

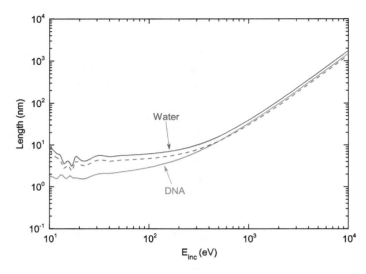

Fig. 4. Penetration range for electrons in water (blue solid line), water with the density rescaled to $1.29\,\mathrm{g\ cm^{-3}}$ (blue dashed line) and DNA (red solid line), as provided by CELLDOSE

higher impact energies ($E_{inc} > 500$ eV), the penetration range in DNA becomes greater than in density-rescaled water, with a difference of about 9% at 10 keV.

4 Conclusion

We have reviewed in this paper the main features and physical models implemented in the CELLDOSE MCTS code. We have also presented our latest calculations regarding the stopping power and penetration range for electrons in water and DNA, showing that for the former medium our results are in agreement with the values found in the literature. In the case of DNA we have provided here the predictions of our code and compared our stopping power values with the calculations performed by other authors, observing differences at low incident energies and a better agreement for $E_{inc} > 300$ eV. Our results could be taken as a reference for future comparisons as more experimental data and theoretical calculations on this medium become available. Finally, we have briefly discussed the objectives of the ongoing work to improve and extend our code, underlining the significant potential applications in ionizing radiation microdosimetry and targeted alpha therapy.

References

1. Nikjoo, H., Uehara, S., Emfietzoglou, D., Cucinotta, F.A.: Track-structure codes in radiation research. Radiat. Meas. **41**, 1052–1074 (2006). https://doi.org/10.1016/j.radmeas.2006.02.001

2. Nikjoo, H., Emfietzoglou, D., Liamsuwan, T., Taleei, R., Liljequist, D., Uehara, S.: Radiation track, DNA damage and response-a review. Rep. Prog. Phys. **79**, 116601 (2016). https://doi.org/10.1088/0034-4885/79/11/116601

3. Kyriakou, I., Emfietzoglou, D., Ivanchenko, V., Bordage, M.C., Guatelli, S., Lazarakis, P., Tran, H.N., Incerti, S.: Microdosimetry of electrons in liquid water using the low-energy models of Geant4. J. Appl. Phys. **122**, 024303 (2017). https://doi.org/10.1063/1.4992076

4. Nikjoo, H., Taleei, R., Liamsuwan, T., Liljequist, D., Emfietzoglou, D.: Perspectives in radiation biophysics: from radiation track structure simulation to mechanistic models of DNA damage and repair. Radiat. Phys. Chem. **128**, 3–10 (2016). https://doi.org/10.1016/j.radphyschem.2016.05.005

5. Friedland, W., Dingfelder, M., Kundrát, P., Jacob, P.: Track structures, DNA targets and radiation effects in the biophysical Monte Carlo simulation code PAR-TRAC. Mutat. Res. Fund Mol. Mech. Mutagen. **711**, 28–40 (2011). https://doi.org/10.1016/j.mrfmmm.2011.01.003

6. Quinto, M.A., Monti, J.M., Weck, P.F., Fojón, O.A., Hanssen, J., Rivarola, R.D., Senot, P., Champion, C.: Monte Carlo simulation of proton track structure in biological matter. Eur. Phys. J. D **71**, 130 (2017). https://doi.org/10.1140/epjd/e2017-70709-6

7. Champion, C., Zanotti-Fregonara, P., Hindié, E.: CELLDOSE: a Monte Carlo code to assess electron dose distribution S values for ^{131}I in spheres of various sizes. J. Nucl. Med. **49**, 151–157 (2007). https://doi.org/10.2967/jnumed.107.045179

8. Champion, C., Hanssen, J., Hervieux, P.A.: Electron impact ionization of water molecule. J. Chem. Phys. **117**, 197–204 (2002). https://doi.org/10.1063/1.1472513

9. Aouchiche, H., Champion, C., Oubaziz, D.: Electron and positron elastic scattering in gaseous and liquid water: a comparative study. Radiat. Phys. Chem. **77**, 107–114 (2008). https://doi.org/10.1016/j.radphyschem.2007.09.004

10. Champion, C.: Quantum-mechanical predictions of electron-induced ionization cross sections of DNA components. J. Chem. Phys. **138**, 184306 (2013). https://doi.org/10.1063/1.4802962

11. Champion, C.: Theoretical cross sections for electron collisions in water: structure of electron tracks. Phys. Med. Biol. **48**, 2147–2168 (2003). https://doi.org/10.1088/0031-9155/48/14/308

12. Champion, C., Le Loirec, C., Stosic, B.: EPOTRAN: a full-differential Monte Carlo code for electron and positron transport in liquid and gaseous water. Int. J. Radiat. Biol. **88**, 54–61 (2012). https://doi.org/10.3109/09553002.2011.641451

13. Champion, C., Quinto, M.A., Morgat, C., Zanotti-Fregonara, P., Hindié, E.: Comparison between three promising β-emitting radionuclides, ^{67}Cu, ^{47}Sc and ^{161}Tb, with emphasis on doses delivered to minimal residual disease. Theranostics **6**, 1611–1618 (2016). https://doi.org/10.7150/thno.15132

14. Hindié, E., Champion, C., Zanotti-Fregonara, P., Rubello, D., Colas-Linhart, N., Ravasi, L., Moretti, J.-L.: Calculation of electron dose to target cells in a complex environment by Monte Carlo code "CELLDOSE". Eur. J. Nucl. Med. Mol. Imaging **36**, 130–136 (2009). https://doi.org/10.1007/s00259-008-0893-z

15. Moccia, R.: One-center basis set SCF MO's. III. H_2O, H_2S, and HCL. J. Chem. Phys. **40**, 2186–2192 (1964). https://doi.org/10.1063/1.1725491

16. Frisch, M.J., et al.: Gaussian 09, Revision A.02. Gaussian, Inc., Wallingford, CT (2009)

17. Galassi, M.E., Champion, C., Weck, P.F., Rivarola, R.D., Fojón, O., Hanssen, J.: Quantum-mechanical predictions of DNA and RNA ionization by energetic

proton beams. Phys. Med. Biol. **57**, 2081–2099 (2012). https://doi.org/10.1088/0031-9155/57/7/2081

18. Tan, Z., Xia, Y., Zhao, M., Liu, X.: Proton stopping power in a group of bioorganic compounds over the energy range of 0.05–10 MeV. Nucl. Instrum. Methods B **248**, 1–6 (2006). https://doi.org/10.1016/j.nimb.2006.04.073

19. Birnie, G.D., Rickwood, D., Hell, A.: Buoyant densities and hydration of nucleic acids, proteins and nucleoprotein complexes in metrizamide. Biochim. Biophys. Acta **331**, 283–294 (1973). https://doi.org/10.1016/0005-2787(73)90441-3

20. Dingfelder, M., Inokuti, M., Paretzke, H.G.: Inelastic-collision cross sections of liquid water for interactions of energetic protons. Radiat. Phys. Chem. **59**, 255–275 (2000). https://doi.org/10.1016/S0969-806X(00)00263-2

21. Abril, I., Garcia-Molina, R., Denton, C.D., Kyriakou, I., Emfietzoglou, D.: Energy loss of hydrogen- and helium-ion beams in DNA: calculations based on a realistic energy-loss function of the target. Radiat. Res. **175**, 247–255 (2011)

22. de Vera, P., Garcia-Molina, R., Abril, I.: Angular and energy distributions of electrons produced in arbitrary biomaterials by proton impact. Phys. Rev. Lett. **114**, 018101 (2015). https://doi.org/10.1103/PhysRevLett.114.018101

23. Emfietzoglou, D., Karava, K., Papamichael, G., Moscovitch, M.: Monte Carlo simulation of the energy loss of low-energy electrons in liquid water. Phys. Med. Biol. **48**, 2355–2371 (2003). https://doi.org/10.1088/0031-9155/48/15/308

24. Blanco, F., Roldán, A.M., Krupa, K., McEachran, R.P., White, R.D., Marjanović, S., Petrović, Z.L., Brunger, M.J., Machacek, J.R., Buckman, S.J., Sullivan, J.P., Chiari, L., Limão-Vieira, P., García, G.: Scattering data for modelling positron tracks in gaseous and liquid water. J. Phys. B At. Mol. Opt. Phys. **49**, 145001 (2016). https://doi.org/10.1088/0953-4075/49/14/145001

25. Champion, C.: Electron impact ionization of liquid and gaseous water: a single-center partial-wave approach. Phys. Med. Biol. **55**, 11–32 (2010). https://doi.org/10.1088/0031-9155/55/1/002

26. Neuefeind, J., Benmore, C.J., Tomberli, B., Egelstaff, P.A.: Experimental determination of the electron density of liquid H_2O and D_2O. J. Phys. Condens. Matter **14**, L429–L433 (2002). https://doi.org/10.1088/0953-8984/14/23/104

27. Ichikawa, K., Kameda, Y., Yamaguchi, T., Wakita, H., Misawa, M.: Neutron-diffraction investigation of the intramolecular structure of a water molecule in the liquid phase at high temperatures. Mol. Phys. **73**, 79–86 (1991). https://doi.org/10.1080/00268979100101071

28. Moriarty, N.W., Karlström, G.: Geometry optimization of a water molecule in water. A combined quantum chemical and statistical mechanical treatment. J. Chem. Phys. **106**, 6470–6474 (1997). https://doi.org/10.1063/1.473637

29. Salvat, F.: Optical-model potential for electron and positron elastic scattering by atoms. Phys. Rev. A **68**, 012708 (2003). https://doi.org/10.1103/PhysRevA.68.012708

30. Jain, A., Thompson, D.G.: Elastic scattering of slow electrons by CH_4 and H_2O using a local exchange potential and new polarisation potential. J. Phys. B At. Mol. Phys. **15**, L631–L637 (1982). https://doi.org/10.1088/0022-3700/15/17/012

31. Mittleman, M.H., Watson, K.M.: Effects of the Pauli principle on the scattering of high-energy electrons by atoms. Ann. Phys. **10**, 268–279 (1960) (New York). https://doi.org/10.1016/0003-4916(60)90024-5

32. Padial, N.T., Norcross, D.W.: Parameter-free model of the correlation-polarization potential for electron-molecule collisions. Phys. Rev. A **29**, 1742–1748 (1984). https://doi.org/10.1103/PhysRevA.29.1742

33. Jain, A.: Low-energy positron-argon collisions by using parameter-free positron correlation polarization potentials. Phys. Rev. A **41**, 2437–2444 (1990). https://doi.org/10.1103/PhysRevA.41.2437

34. Riley, M.E., Truhlar, D.G.: Approximations for the exchange potential in electron scattering. J. Chem. Phys. **63**, 2182–2191 (1975). https://doi.org/10.1063/1.431598

35. Champion, C., Hanssen, J., Rivarola, R.D.: The first born approximation for ionization and charge transfer in energetic collisions of multiply charged ions with water. In: Advances in Quantum Chemistry. Elsevier, pp. 269–313 (2013). https://doi.org/10.1016/B978-0-12-396455-7.00010-8

36. Champion, C., Cappello, C.D., Houamer, S., Mansouri, A.: Single ionization of the water molecule by electron impact: angular distributions at low incident energy. Phys. Rev. A **73**, 012717 (2006). https://doi.org/10.1103/PhysRevA.73.012717

37. Ritchie, R.H., Hamm, R.N., Turner, J.E., Wright, H.A.: The interaction of swift electrons with liquid water. In: 6th Symposium on Microdosimetry, pp. 345–354. Harwood Academic, London (1978)

38. Ashley, J.C.: Stopping power of liquid water for low-energy electrons. Radiat. Res. **89**, 25–31 (1982). https://doi.org/10.2307/3575681

39. Emfietzoglou, D.: Inelastic cross-sections for electron transport in liquid water: a comparison of dielectric models. Radiat. Phys. Chem. **66**, 373–385 (2003). https://doi.org/10.1016/S0969-806X(02)00504-2

40. Emfietzoglou, D., Cucinotta, F.A., Nikjoo, H.: A complete dielectric response model for liquid water: a solution of the bethe ridge problem. Radiat. Res. **164**, 202–211 (2005). https://doi.org/10.1667/RR3399

41. Emfietzoglou, D., Nikjoo, H.: Accurate electron inelastic cross sections and stopping powers for liquid water over the 0.1–10 keV range based on an improved dielectric description of the Bethe surface. Radiat. Res. **167**, 110–120 (2007). https://doi.org/10.1667/RR0551.1

42. Olivero, J.J., Stagat, R.W., Green, A.E.S.: Electron deposition in water vapor, with atmospheric applications. J. Geophys. Res. **77**, 4797–4811 (1972). https://doi.org/10.1029/JA077i025p04797

43. Green, A.E.S., Dutta, S.K.: Semi-empirical cross sections for electron impacts. J. Geophys. Res. **72**, 3933–3941 (1967). https://doi.org/10.1029/JZ072i015p03933

44. Compton, R.N., Christophorou, L.G.: Negative-ion formation in H_2O and D_2O. Phys. Rev. **154**, 110–116 (1967). https://doi.org/10.1103/PhysRev.154.110

45. Hervieux, P.-A., Fojón, O.A., Champion, C., Rivarola, R.D., Hanssen, J.: Positronium formation in collisions of fast positrons impacting on vapour water molecules. J. Phys. B. At. Mol. Opt. **39**, 409–419 (2006). https://doi.org/10.1088/0953-4075/39/2/015

46. Quinto, M.A., Monti, J.M., Galassi, M.E., Weck, P.F., Fojón, O.A., Hanssen, J., Rivarola, R.D., Champion, C.: Proton track structure code in biological matter. J. Phys. Conf. Ser. **583**, 012049 (2015). https://doi.org/10.1088/1742-6596/583/1/012049

47. Garcia-Molina, R., Abril, I., Heredia-Avalos, S., Kyriakou, I., Emfietzoglou, D.: A combined molecular dynamics and Monte Carlo simulation of the spatial distribution of energy deposition by proton beams in liquid water. Phys. Med. Biol. **56**, 6475–6493 (2011). https://doi.org/10.1088/0031-9155/56/19/019

48. Liamsuwan, T., Uehara, S., Emfietzoglou, D., Nikjoo, H.: Physical and biophysical properties of proton tracks of energies 1 keV to 300 MeV in water. Int. J. Radiat. Biol. **87**, 141–160 (2011). https://doi.org/10.3109/09553002.2010.518204

49. Mulford, D.A., Scheinberg, D.A., Jurcic, J.G.: The promise of targeted α-particle therapy. J. Nucl. Med. **46**, 199S–204S (2005)
50. Huang, C.-Y., Guatelli, S., Oborn, B.M., Allen, B.J.: Microdosimetry for targeted alpha therapy of cancer. Comput. Math. Methods Med. **2012**, 1–6 (2012). https://doi.org/10.1155/2012/153212
51. International Atomic Energy Agency: Atomic and Molecular Data for Radiotherapy and Radiation Research IAEA-TECDOC-799. IAEA, Vienna (1995)
52. International Commission on Radiation Units and Measurements: ICRU Report 16 Linear Energy Transfer. ICRU, Bethesda, MD (1970)
53. Garcia-Molina, R., Abril, I., Kyriakou, I., Emfietzoglou, D.: Inelastic scattering and energy loss of swift electron beams in biologically relevant materials: energy loss of electron beams in biomaterials. Surf. Interface Anal. **49**, 11–17 (2017). https://doi.org/10.1002/sia.5947
54. Tan, Z., Xia, Y., Zhao, M., Liu, X.: Electron stopping power and inelastic mean free path in amino acids and protein over the energy range of 20–20,000 eV. Radiat. Environ. Biophys. **45**, 135–143 (2006). https://doi.org/10.1007/s00411-006-0049-0
55. Paretzke, H.G.: Simulation von Elektronenspuren im Energiebereich 0.01–10 keV in Wasserdampf. Report No. 24/88, GSF—National Research Center for Health and Environment (1988)
56. Meesungnoen, J., Filali-Mouhim, A., Mankhetkorn, S.: Low-energy electron penetration range in liquid water. Radiat. Res. **158**, 657–660 (2002)
57. Bordage, M.C., Bordes, J., Edel, S., Terrissol, M., Franceries, X., Bardiès, M., Lampe, N., Incerti, S.: Implementation of new physics models for low energy electrons in liquid water in Geant4-DNA. Phys. Med. **32**, 1833–1840 (2016). https://doi.org/10.1016/j.ejmp.2016.10.006

Dissociative Electron Attachment Study of Di- & Triatomic Molecule

Minaxi Vinodkumar[1(✉)], Hitesh Yadav[1,2], and P. C. Vinodkumar[2]

[1] Electronics Department, V. P. & R. P. T. P. Science College, Vallabh Vidyangar, Gujarat, India
minaxivinod@yahoo.co.in
[2] Department of Physics, Sardar Patel University, Vallabh Vidyanagar, Gujarat, India

Abstract. In this article, we report Dissociative Electron Attachment (DEA) cross sections. DEA is a resonant process and happens at low energy electron scattering on electron collision with simple molecules ($\leq 20\,\text{eV}$). Using the well-established ab inito R-matrix method via. Quantemol-N DEA estimator we compute the cross sections for the diatomic molecules and applied the same for a triatomic molecule.

1 Introduction

From the perspective of fundamental study, it becomes more important to understand different properties of atoms and molecules including their dynamics and different processes involving bond formation, bond breaking, and other chemical processes. And the easiest and simple technique for understanding the system is to perturb the system, and one such method is electron impact scattering study.

Variety of atomic and molecular processes are widely studied using electron collision. And the important part of the electron impact collision study at low energy is that it is not governed by the dipole selection rule and contains all the angular momentum components [1,2]. Thus, the study helps in understanding the molecular structure more precisely and their relative dynamics. With this, it underlines the importance of scattering study with electron impact. The scattering study was the first phenomenal change, that helped in understanding the structural symmetries very well. Rutherford's experiment with the alpha particle and the gold foil is the base for this scattering studies. With the advancement in technology and availability of sophisticated and well-equipped instruments, through quantum scattering elastic and inelastic processes were well studied. The inelastic process includes rotational excitation, vibrational excitation, dissociative electron attachment/recombination, neutral dissociation, electronic excitation, and ionizations [3,4].

© Springer Nature Singapore Pte Ltd. 2019
P. C. Deshmukh et al. (eds.), *Quantum Collisions and Confinement of Atomic and Molecular Species, and Photons*, Springer Proceedings in Physics 230,
https://doi.org/10.1007/978-981-13-9969-5_14

In this article, we focus on the study of dissociative electron attachment (DEA) and the study of resonance phenomena that leads to dissociation of neutral target.

2   Resonance

Resonance is a temporary formation of a compound state between the incident electron and the target molecule i.e a transient state that decays by emitting electron [5]. And these resonances have a lifetime appreciably longer than the electron passage time, i.e., a/v where a is the linear dimension of the target and v is electron velocity [5]. In collision process, the formation of a resonance from the target system and incident electron will occur at some energies as it generally leads to a severe distortion of the incident electron wave function. This only happens when the incident electron falls in one of the discrete energy bands, where the incident electron finds quasi-stationary orbit of the target molecule.

The resonance process is not only the effect of temporary capture of the incident electron to quasi-stationary state, but also effects the potential of the target system [5]. The signature of inelastic processes such as rotational, vibrational, and electronic excitations can be found in resonances. In this process, the target is excited by incident electron energy and also the electron loses its energy. One of the inelastic process closely related to the resonance is the dissociative electron attachment (DEA).

2.1   Theoretical Methodology

To calculate the resonance for our electron–target interaction, we have employed the ab initio R-matrix via Quantemol-N [7]. The principle behind the R-matrix method [6] relies on the division of the configuration space into two spatial regions, viz., inner and outer regions. The inner region is chosen generally to have a R-matrix radius of 10–15 au, which accounts for the short-range potentials arising due to electron–electron correlation and exchange effects. It is solved using the quantum chemistry codes [7]. All required target wave functions must vanish at the boundary of the R-matrix. The outer boundary is extended to $\sim$100 au. The trial wave function in the inner region is expressed as

$$\Psi_k^{N+1}(x_1\ldots x_{N+1}) = A \sum_{ij} a_{ijk}\Phi_i^N(x_1\ldots x_N)u_{ij}(x_{N+1}) + \sum_i b_{ik}\chi_i^{N+1}(x_1\ldots x_{N+1})$$

(1)

In (1), the first summation Φ_i represents the wave function of ith target state. A is the anti-symmetrization operator taking care of exchange effect and u_{ij} is the continuum orbitals to represent the scattered electron, whereas a_{ijk} and b_{ik} are the variational parameters determined by the diagonalization of $N+1$ Hamiltonian matrix. In the second term, summation runs over χ_i for $N+1$ electrons and represents L^2 function with zero amplitude at the R-matrix boundary [4].

The continuum orbitals are represented by the orbitals given by Faure et al. [8]. The asymptotic solution is obtained in the outer region by expanding the R-matrix on the boundary by the expression given by Gailitis [9]. The K-matrix obtained used to derive all the physical quantities which are used to calculate the cross sections. Further, the eigenphase sum obtained on diagonalizing the K-matrix is used to obtain the position and width of the resonance by fitting them to the Breit–Wigner profile [10] using the RESON program [10]. The details of the methodology can be obtained from our earlier publications [3,4,11].

3   Dissociative Electron Attachment (DEA)

As discussed in the earlier section of resonance, DEA is one of the process which is closely related to resonance phenomena. In DEA, electron is temporally captured by the molecule to form a transient negative ion (TNI) or a resonant state [4,12,13]. This can be understood as one of the inelastic channel being introduced, i.e., electronic excitations from ground state of the neutral molecule to the ground state or any other accessible excited states of the anion. A graphical representation of the potential energy curve of diatomic target, anion, and excited transient negative ion is given in Fig. 1. The DEA process in general for electron scattered by the diatomic molecule AB is given by

$$e^- + AB \longrightarrow AB^{*-} \longrightarrow A + B^- \, or \, B + A^- \tag{2}$$

The dissociative electron attachment process is active below the ionization threshold of the molecule and it is very important in understanding the local chemistry induced by the incident electron. The necessary condition for target to fragment is that the energy of the incident electron must be greater than the dissociation energy of the bond. The cross sections for a given DEA process is defined by the initial attachment cross sections resulting from attachment of electron with the target molecule multiplied by the survival probability of TNI beyond the cross point (r_c in the Fig. 1) of the respective potential energy curves. In dissociation, excitation is the initial step involved either in the direct or exchange scattering. Dissociation is certain once the interatomic distance of the resonance state anion exceeds cross point r_c (see Fig. 1).

3.1   Theoretical Methodology

In general, the DEA cross sections is calculated as a product of resonance cross section $\sigma_r(E_i)$ arising from electron collision with ground state of the target and the survival probability "S" of these resonances that occur during the interaction. Thus, DEA cross section $\sigma_T(E_i)$ is represented as

$$\sigma_T(E_i) = C \sum_r S_i \sigma_{ri} \tag{3}$$

where E_i is the incident energy of electron and the summation runs over the different resonances being considered. Whereas C accounts for energy-dependent

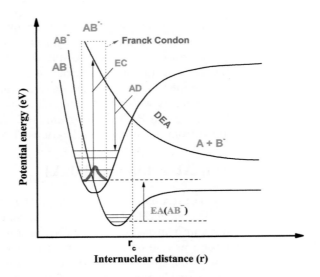

Fig. 1. A typical interatomic potential energy curve of a diatomic molecule. Here AB represents the neutral diatomic molecule, AB^{*-} represents the excited diatomic molecule, A is the neutral fragmented atom, B^- represents the negative ion formed after fragmentation and EC is the electron-captured process for the Franck–Condon region, AD represents the auto-detachment process, DEA represents the dissociative electron attachment process, and EA represents Electron affinity of the AB molecule

widths for the resonances. To perform the DEA calculations the inputs required are the dissociation energy, the ground state vibrational energy, and the electron affinity E_a of the target molecule. Further, the resonance cross sections, σ_{ri} are calculated by computing resonance width and position using Breit-Wigner profile using RESON program [10]. Resonance cross-sections (σ_{ri}) is as follows

$$\sigma_r(E_i) = \int_0^\infty \psi^*(r)\sigma_{BW}(E_i, r)\psi(r)dr \tag{4}$$

where $\psi^*(r)$, is the wave function of the vibrational ground state is assumed to have a simple harmonic form given by

$$\psi(r) = (\frac{rm_rE_v}{\pi})^{\frac{1}{4}}e^{(-m_rE_v)(r-R_e)^2} \tag{5}$$

The Breit–Wigner [10] cross section is expressed as

$$\sigma_{BW}(E_i, r) = \frac{2\pi}{E_i}\frac{\frac{1}{4}\Gamma^2}{(E_i - V_r(r))^2 + \frac{1}{4}\Gamma^2} \tag{6}$$

where V_r is the resonance potential. Using the molecular geometry, the equilibrium distance for the molecule in the direction of the dissociating coordinate (r_e)

is computed. This is approximately the average distance of each atom from the center of mass multiplied by a dimensional correction factor of $\sqrt{2}$. Further, the dissociation energy is computed as the difference between the target dissociation energy and the electron affinity of the dissociating fragment, $D_r = D_e - E_a$. A Morse potential is used to approximate the ground state electronic target potential. The details regarding the Morse potential can be found in our earlier publication [3,4].

From Fig. 1, the anion reaches r_c point without autoionizing, then the molecule must dissociate. A survival probability (S) for target to dissociate is then computed. When $r_c > r_e$, the survival probability S is estimated by considering the classical time to reach r_c given by

$$t = \sqrt{\frac{2m_e(r_c - r_e)^2}{V_r(r_e) - V_t(r_c)}} \tag{7}$$

and survival probability S is given by

$$S = e^{-\Gamma t} \tag{8}$$

where Γ is resonance width. When the crossing point r_c is less than the target equilibrium geometry r_e the survival probability is assumed to be unity (S = 1).

4   Results and Discussion

Target selected for the present study, HCl, ClO, and FNO molecules are of varying importance in different fields of applied physics.

In Fig. 2, we report the result of DEA process of H^- anion formation from HCl. The present theoretical data is in good agreement with the experimental data of Orient and Srivastava [14] above 10 eV. Below 10 eV present results are lower compared to experimental data of Orient and Srivastava [14]. The calculated resonance by Quantemol-N is at 9.4 eV. This shift in the DEA cross sections computed here is attributed by an energy shift of resonances, due to the formulation through which we calculated position and width of resonance using R-matrix (Quantemol-N) [3,4].

Chlorine-containing molecules play a particularly important role in the removal of ozone and atomic oxygen at Earth's upper atmosphere due to which its importance in global stratospheric ozone is lost. The open shell nature of many of the Cl_xO_y compounds makes them difficult systems to treat theoretically. Electron correlation plays an important role in these systems, not only for characterizing the ground state wavefunction, but also in the excitation of the many low lying excited states which significantly affect low energy electron collisions. The DEA-estimator [4] method is based on electron–molecule calculations performed with the R-matrix method [6] and is implemented as part of the Quantemol-N [3] expert system. In Fig. 3 it is our maiden effort to report Cl^- anion on dissociation of ClO molecule. The two prominent shape resonances are

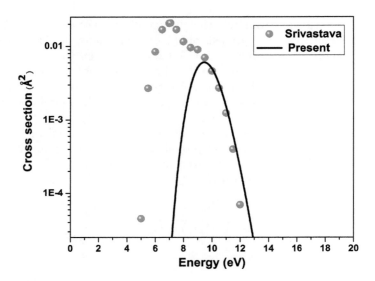

Fig. 2. HCl dissociative to Cl + H^-, where solid line represents the present calculated result of DEA cross section of H^-, solid sphere represents the measured data by Orient and Srivastava [14]

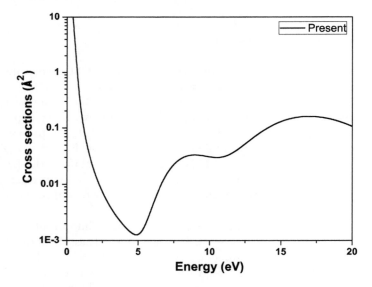

Fig. 3. ClO dissociative to O + Cl^-, where solid line represents the present calculated result of DEA cross section of Cl^-

at 6.7 and 12.6 eV in Fig. 3. The contribution to these peaks comes from symmetry state $^3A_2$. ClO belongs to $C_{\infty v}$ symmetry but we have done our calculations using C_{2v} symmetry as Quantemol-N uses only Abelian point group symmetry. These resonances can be easily recognized from the nature of Fig. 3.

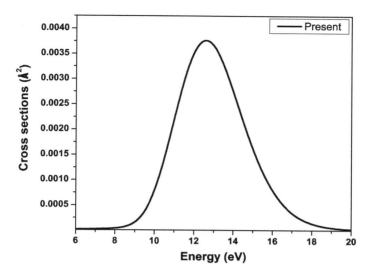

Fig. 4. FNO dissociative to F + NO^-, where solid line represents the present calculated result of DEA cross section of NO^-

FNO molecule is a reactive agent which is mainly used for converting the metal to their respective fluorides and release of nitric oxide during this process. It is also considered as a source to have NO during the organic processes and also been proposed as an oxidizer for the rocket propellants. Nitric oxide (NO) is a radical and has a probability of NO^- ion on the dissociation of FNO molecule. In Fig. 4, we report the DEA cross sections for the NO^- anion on dissociation of FNO molecule. Electron collisions with FNO gives different resonance structures. Our DEA calculation finds resonances at 9.4 and 13.07 eV due to $^2A'$ symmetric state and at 5.3 eV due to $^2A''$ symmetric state.

5   Conclusion

In the present study, we report the DEA cross sections of HCl (H^-), ClO (Cl^-), and FNO (NO^-) molecules. The H^- of HCl molecule is in good agreement with available experimental data reported by Orient and Srivastava [14]. The Cl^- of ClO and NO^- of FNO are reported here for the first time to the best of our knowledge. The present work will encourage the other experimentalist and theorist to study DEA of these targets as these are important applied targets.

Acknowledgements. Minaxi Vinodkumar acknowledges DST-SERB, New Delhi for the Major research project [EMR/2016/000470] for financial support under which part of this work is carried out. Authors are grateful to the referees for their constructive input.

References

1. Nag, P., Nandi, D.: Identification of overlapping resonances in dissociative electron attachment to chlorine molecules. Phys. Rev. A. **93**, 012701 (2016). https://doi.org/10.1103/PhysRevA.93.012701
2. Prabhudesai, V.S., Kelkar, A.H., Nandi, D., Krishnakumar, E.: Functional group dependent site specific fragmentation of molecules by low energy electrons. PRL **95**, 143202 (2005). https://doi.org/10.1103/PhysRevLett.95.143202
3. Yadav, H., Vinodkumar, M., Limbachiya, C., Vinodkumar, P.C.: Scattering of electrons with formyl radical. Mol. Phys. **115**, 952 (2017). https://doi.org/10.1080/00268976.2017.1293306
4. Yadav, H., Vinodkumar, M., Limbachiya, C., Vinodkumar, P.C.: Study of electron impact inelastic scattering of chlorine molecule (Cl_2). J. Phys. B: At. Mol. Opt. Phys. **51**, 045201 (2018). https://doi.org/10.1088/1361-6455/aaa2d6
5. Shimamura, I., Takayanagi, K.: Electron-Molecule Collisions. Plenum Press (1984). https://doi.org/10.1007/978-1-4613-2357-0
6. Jonathan, T.: Electron-molecule collision calculations using the R-matrix method. Phys. Rep. **491**, 29 (2010). https://doi.org/10.1016/j.physrep.2010.02.001
7. Tennyson, J., Brown, D.B., Munro, J.J., Rozum, I., Varambhia, H.N., Vinci, N.: Quantemol-N: an expert system for performing electron molecule collision calculations using the R-matrix method. J. Phys.: Conf. Ser. **86**, 012001 (2007). https://doi.org/10.1088/1742-6596/86/1/012001
8. Faure, A., Gorfinkiel, J.D., Morgan, L.A., Tennyson, J.: GTOBAS: fitting continuum functions with Gaussian-type orbitals. Comput. Phys. Commun. **144**, 224 (2002). https://doi.org/10.1016/S0010-4655(02)00141-8
9. Gailitis, M.: New forms of asymptotic expansions for wavefunctions of charged-particle scattering. J. Phys. B: At. Mol. Opt. Phys. **9**, 843 (1976). https://doi.org/10.1088/0022-3700/9/5/027
10. Tennyson, J., Noble, C.J.: RESONA program for the detection and fitting of Breit-Wigner resonances. Comput. Phys. Commun. **33**, 421 (1984). https://doi.org/10.1016/0010-4655(84)90147-4
11. Vinodkumar, M., Desai, H., Vinodkumar, P.C., Mason, N.: Induced chemistry by scattering of electrons from magnesium oxide. Phys. Rev. A **93**, 012702 (2016). https://doi.org/10.1103/PhysRevA.93.012702
12. Thorman, R.M., Ragesh Kumar, T.P., Howard Fairbrother, D., Inglfsson, O.: The role of low-energy electrons in focused electron beam induced deposition: four case studies of representative precursors. Beilstein J. Nanotechnol. **6**, 1904 (2015). https://doi.org/10.3762/bjnano.6.194
13. Christophorou, L.G.: Negative ions of polyatomic molecules. Environ. Health Perspect. **36**, 3 (1980). PMC1637732
14. Orient, O.J., Srivastava, S.K.: Cross sections for H and Cl production from HCl by dissociative electron attachment. Phys. Rev. A **32**, 2678 (1985). https://doi.org/10.1103/PhysRevA.32.2678

Single-Photon Decay Processes in Atomic Systems with Empty K Shell

L. Natarajan$^{(\boxtimes)}$

Department of Physics, University of Mumbai, Mumbai, India
ln@physics.mu.ac.in

Abstract. Two extreme cases of atomic configurations and the resulting radiative decay are considered: (a) only two electrons knocked out and (b) only two electrons present. In the first case, an empty K shell with an otherwise normal atomic configuration (hollow atom) is considered and the decay properties of the two possible modes of radiative decay of empty K shell resulting in one-electron one-photon (OEOP) or two-electron one-photon (TEOP) processes are analyzed to study the impact of correlation on the coupling schemes as the coupling moves from LS at low Z to jj at high Z. In the second case, the correlation induced intense TEOP transitions from He-like ions with empty K shell and only two electrons in the $2s$ subshell are investigated. The radiative rates of transitions between $2s^2$ and $1snp$ ($n = 2-4$) groups have been studied to evaluate the contributions from each group to the total TEOP rates. The dependence of the estimated line intensities of X-ray photons from $2s^2$-$1s2p$ on the nature of orthogonalization of the spin orbitals is analyzed.

1 Introduction

The excitation and de-excitation of hollow atoms where the inner shell is completely empty have been experimentally investigated using different processes since it was first produced by Briand et al. [1] and theoretically analyzed by many workers using Dirac-Hartree-Slater (DHS) and multi-configuration Dirac-Fock (MCDF) models with the inclusion of Breit interaction and quantum electrodynamics corrections. An electron jump from the $2p$ shell to the doubly ionized $1s$ shell leads to one-electron one-photon (OEOP) transition and an alternate simultaneous two electron jumps from the L shell (one from the $2s$ and the other from the $2p$ subshells) give rise to two-electron one-photon (TEOP) transition. The former radiation is termed as Kα X-ray hypersatellite and the latter which is exclusively due to the correlation between $2s$ and $2p$ electrons is known as K$\alpha\alpha$ X-ray. Though K$\alpha\alpha$ X-rays are less intense than Kα X-ray hypersatellites, accurate evaluation of measurable quantities like lifetimes, fluorescence yields, ionization cross sections does require the X-ray rates from both OEOP and TEOP transitions. While Kα X-ray hypersatellites have been extensively analyzed both theoretically [2–5] and experimentally [1,6–12], experimental data

© Springer Nature Singapore Pte Ltd. 2019
P. C. Deshmukh et al. (eds.), *Quantum Collisions and Confinement of Atomic and Molecular Species, and Photons*, Springer Proceedings in Physics 230,
https://doi.org/10.1007/978-981-13-9969-5_15

on K$\alpha\alpha$ X-rays are scarce [13] placing added importance on the systematic evaluation of theoretical data. The double ionization process of K shell is a sensitive tool to investigate the relativistic and QED effects in atoms [4] and is probably the only testing ground to study the variation of coupling scheme from almost pure LS at low Z to jj coupling at high Z. In spectroscopy, the nomenclatures employed to specify the initial and final configurations leading to radiative transitions are usually of two types. In the first type, the holes characterize the states and the configurations are indicated by the missing electrons in the initial and final states. In the second type, the initial and final configurations are indicated by the occupation numbers of electrons in the different shells. In this work, the first nomenclature is used to describe the Kα X-ray hypersatellites and K$\alpha\alpha$ X-rays and the second one is used to describe He-like ionic configurations. The relative intensities of the Kα_1^h and Kα_2^h lines from $1s^{-2} - 1s^{-1}2p^{-1}$ fine structure transitions have been used to prove the LS and jj coupling schemes. Just like Kα_1^h and Kα_2^h lines, the relative intensities of K$\alpha_1\alpha_3$ and K$\alpha_2\alpha_3$ lines from $1s^{-2} - 2s^{-1}2p^{-1}$ are also highly sensitive to the degree of intermediacy of the coupling scheme (LS vs. jj) and an attempt has been made by us for the first time to analyze the impact of correlation on the variation in the coupling with Z [14].

The K X-ray spectra from singly excited states of He-like ions have been investigated both experimentally and theoretically for some selected configurations. However, experimental results on the OEOP transition from the doubly excited $2s^2$ $^1S_0$ state of He-like ion are scarce due to the low probability of emission of the only possible magnetic dipole (M1) line. This situation places an added importance on the systematic evaluation of other possible alternative modes of decay. While the normal two-electron one-photon (TEOP) transitions where both the electrons jump to the same inner shell (K$\alpha\alpha$) are weak, special type of TEOP transitions in which the two electrons jump to different shells are highly intense [15]. As this unusual type of transition is due exclusively to the correlated motion of the two electrons and hence very sensitive to the nature of the spin orbitals of the initial and final states, in our previous work [15] we had carried out two different versions of Optimal level scheme. The first method was based on bi-orthogonalization in which the initial and final state functions are orthogonal to each other and the second method considered a common set where the orbital functions of the two states are not completely orthogonal to each other. The transition data from these two methods were analyzed to test the dependence of the calculated line intensities on the nature of the spin orbitals. In [15], we had considered $2s^2 - 1s2p$ TEOP and $1s2s$ M1 transitions. In this work we have extended our calculations to $2s^2 - 1snp$ ($n = 3, 4$) groups of transitions to evaluate all nonnegligible contributions to the total transition rates. All the calculations have been computed in the relativistic configuration interaction formalism (RCI) which uses correlated MCDF wavefunctions and include Breit interaction and QED effects [16].

2   Numerical Procedure

In a multi-configuration relativistic calculation, the configuration state functions (CSFs) are symmetry- adapted linear combinations of Slater determinants constructed from a set of one-electron Dirac spinors. A linear combination of these configuration state functions (CSFs) is then used in the construction of atomic state functions (ASFs) with the same J and parity.

$$\Psi_i(J^P) = \sum_{\alpha=1}^{n_{CSF}} c_{i\alpha}\Phi(\Gamma_\alpha J^P) \tag{1}$$

where $c_{i\alpha}$ are the mixing coefficients for the state i and n_{CSF} are the number of CSFs included in the evaluation of ASF. The Γ_α represents all the one-electron and intermediate quantum numbers needed to define the CSFs and the configuration mixing coefficients are obtained through the diagonalization of the Dirac-Coulomb Hamiltonian

$$\mathbf{H}_{DC} = \sum_i [c\alpha_i.\mathbf{p}_i + (\beta_i - 1)c^2 - \frac{Z}{r_i}] + \sum_{i<j} \frac{1}{r_{ij}} \tag{2}$$

Once a set of radial orbitals and the expansion coefficients are optimized for self-consistency, RCI calculations can be performed by including higher order interactions in the Hamiltonian. The most important of these is the transverse photon interaction

$$\mathbf{H}_{trans} = -\sum_{i<j}^{N} [\frac{\alpha_i.\alpha_j\cos(\omega_{ij}r_{ij})}{r_{ij}} + (\alpha_i.\nabla_i)(\alpha_j.\nabla_j)\frac{\cos(\omega_{ij}r_{ij}) - 1}{\omega_{ij}^2 r_{ij}}] \tag{3}$$

where ω_{ij} is the wavenumber of the exchanged virtual photon and is obtained as the difference between the diagonal Lagrange multipliers associated with the orbitals. However, this is valid only when the shells are singly occupied and the diagonal energy parameters may not represent the correct binding energies of the orbitals in a variously ionized atomic system. Hence in the present work, the low-frequency limit $\omega_{ij} \to 0$ is considered and only the mixing coefficients are recalculated by diagonalizing the Dirac-Coulomb-Breit Hamiltonian matrix. The dominant QED corrections comprise of self energy and vacuum polarization. While the former contribution is evaluated in the hydrogen-like approximation, the latter correction is treated perturbatively. The theoretical background using relativistic wavefunctions and higher order corrections is fully described in [16–19].

In the active space approximation method [17], the electrons from the occupied orbitals are excited to unoccupied orbitals in the active set. Since the orbitals with the same principal quantum number n have near similar energies, the active set is expanded in layers of n. The correlation contribution was evaluated by considering SD excitations of electrons from the occupied shells to unoccupied virtual shells. In the first step, as a zeroth-order minimal basis set, we generated mono-configuration DF wavefunctions separately for the initial and final reference sets in Extended Optimal Level (EOL) scheme using variational method [18]. In EOL calculations, the radial functions and the mixing coefficients are determined by optimizing the energy functional, which is the weighted sum of the energy values corresponding to a set of (2j + 1) eigenstates. In the subsequent steps, we included higher order correlation corrections by gradually expanding the size of the CSF sets in layers and repeated the computations for each step by step multi-configuration expansion until the convergence and stability of the observable is obtained. Finally in the relativistic configuration interaction (RCI) calculations, we included Breit interaction and QED corrections and evaluated the transition parameters. As pointed out earlier, the TEOP transitions from $2s^2$ ($^1S_0$) are due exclusively to the correlated motion of the two electrons and hence to test the influence of relaxed and frozen spin orbitals on the transition amplitudes, in our earlier work [15] we generated two sets of spin orbitals. In the first method (Method I), we used a set of separately generated initial and final state spin orbitals which were subsequently made mutually orthogonal to each other. In the second method (Method II), we opted for a common orthonormal basis set incorporating some amount of non-orthogonality in the radial functions. In this work, only Method I is considered in the study of transitions between $2s^2$ and $1snp$ ($n = 3, 4$) groups in He-like ions.

3   Results and Discussion

The $K\alpha_2^h$ and $K\alpha_1^h$ fine structure lines are due to $1s^{-2}$ $^1S_0 \rightarrow 1s^{-1}2p^{-1}$ $^1P_1$ and $1s^{-2}$ $^1S_0 \rightarrow 1s^{-1}2p^{-1}$ $^3P_1$ transitions respectively. Similarly the respective $K\alpha_2\alpha_3$ and $K\alpha_1\alpha_3$ lines represent the $1s^{-2}$ $^1S_0 \rightarrow 2s^{-1}2p^{-1}$ $^1P_1$ and $1s^{-2}$ $^1S_0 \rightarrow 2s^{-1}2p^{-1}$ $^3P_1$ transitions. For closed shell atoms with an initial empty K shell, the number of $K\alpha^h$ or $K\alpha\alpha$ fine structure transitions will be only two whereas atoms with open shell ground configurations will yield a lot more transitions. In our work, we have neglected the interaction of the electrons in the unfilled outer shells with the inner shells. While this simplification might lead to a marginal change in the transition data, the intensity ratios of the fine structure lines will not be affected.

The intensity ratio of the $K\alpha_1^h$ to the $K\alpha_2^h$ lines in terms of the spin–orbit and electrostatic interactions matrix elements can be expressed as [10]

$$I(K\alpha_1^h)/I(K\alpha_2^h) = (8/9)\chi^2/[1 - \chi/3 + (1 - (2/3)\chi^\dagger\chi^2)^{1/2}]^2 \qquad (4)$$

where $\chi = (3/4)\zeta(2p)/G_1(1s,2p)$. The symbols $\zeta(2p)$ and $G_1(1s,2p)$ represent the spin–orbit and electrostatic interaction matrix elements respectively. For low Z ions, $\zeta(2p)$ is negligible in comparison to $G_1(1s,2p)$. With increasing Z, contribution from $\zeta(2p)$ increases and for high Z ions, the intensity ratio approaches 2, the corresponding limit observed for the $K\alpha$ diagram lines. In Fig. 1, the intensity ratios of $K\alpha_1^h$ to $K\alpha_2^h$ lines are compared with other relativistic and available experimental data [3,4,6–9]. In pure LS coupling, only $K\alpha_2^h$ and $K\alpha_2\alpha_3$ transitions are dipole allowed and $K\alpha_1^h$ and $K\alpha_1\alpha_3$ are spin forbidden where as the

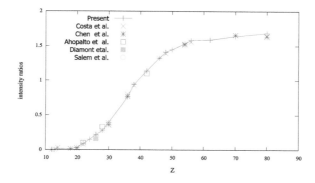

Fig. 1. Comparison of intensity ratios of $K\alpha$ hypersatellite lines with other available data. Costa et al. [3], Chen et al. [4], Diamont et al. [6,7], Salem et al. [8,9], Ahopelto et al. [12]

later two transitions are fully allowed in jj coupling due to mixing between the two final states.

As $2p_{1/2}$ is close to the nucleus than $2p_{3/2}$, $K\alpha_1^h$ line intensity increases with Z due to relativistic effects [3]. As shown in our earlier paper [5], our calculated ratios are in good agreement with the existing relativistic and experimental data. The ratio for low Z atoms is nearly zero showing the prominence of pure LS coupling. As Z increases, spin -orbit interaction starts contributing to the transition parameters and at Z around 40, $I(K\alpha_h^1)/I(K\alpha_2^h) \simeq 1$, thereby proving that the strengths of e–e interaction and spin–orbit interaction are nearly the same in this Z range. With increasing Z, spin–orbit interaction dominates and the coupling moves from intermediate to jj scheme. Thus, as pointed out by earlier workers [3,4,6,7], the intensity ratio $I(K\alpha_1^h)/I(K\alpha_2^h)$ is strongly dependent on the coupling scheme. However, while the intensity ratio is nearly 0 for low Z, it does not approach the jj coupling limit of 2 for high Z as predicted by the angular momentum coupling scheme [2,10]. For example, our present calculation for Z = 80 gives a ratio of 1.69. As experimental measurements for Z > 45 where the smooth variation of coupling from intermediate to jj scheme occurs are limited, the accuracy of the theoretical predictions can be verified only when experimental data on high Z elements are made available.

In Fig. 2, $I(K\alpha_1\alpha_3)/I(K\alpha_2\alpha_3)$ values reported by us using relativistic configuration interaction method [14] are plotted for $12 \leq Z \leq 80$. For the sake of comparison, our calculated $K\alpha_h^1/K\alpha_2^h$ ratios [5] are also shown in the figure. While the same trend in moving from pure LS to pure jj scheme is reflected in both decay modes, TEOP intensity ratios are consistently lower than OEOP ratios for all Z. We also note that while OEOP intensity ratio is unity around Z = 40, TEOP gives unit value around Z = 50. It may be pointed out here that the transition rates of $K\alpha$ hypersatellites calculated in the length and velocity forms are nearly the name for most of the ions considered in our work whereas

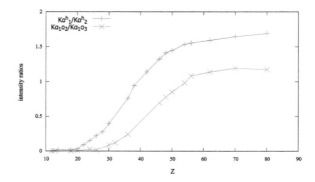

Fig. 2. Comparison of intensity ratios of Kα hypersatellites and Kαα lines

the intensity ratios of Kαα fines structure lines calculated in length and velocity gauges vary substantially [14].

Precise experimental measurements are needed to know how well pure jj coupling applies for high Z elements and also the influence of correlation on the coupling scheme, especially at the regime intermediate between LS and jj couplings.

The configurations mixing between $2s^2$ $^1S_0$ and $2p^2$ $^1S_0$, $^3P_0$ is responsible for the E1 transitions from $^1S_0$ state and $^3P_1 - ^1S_0$ is a result of configuration interaction and spin–orbit mixing with the final term [21]. Detailed investigations on the influence of correlation and higher order corrections from Breit interaction and QED effects to the X-ray energies and rates and also on the effects of fully and partially relaxed spin orbitals of the initial and final states on the correlation sensitive radiative rates are reported in our earlier paper [15]. In Table 1, to analyze the influence of spin-orbitals on the unusually intense TEOP transitions from $2s^2$ $^1S_0$ with empty K shell, we reproduce the energies of E1 and M1 transitions from [15] along with available data [20–23] for some select ions (Z = 26, 27, 46, 66 and 92) calculated using relaxed (Method I) and frozen orbital (Method II) sets. The calculated energies from Method I are in excellent agreement with earlier data [20–22] and differ by 50 eV for U^{90+} from the values Ref. [23]. The Method II energies agree reasonably well with the previous values only for ions with Z ≤ 50. The present M1 energies compare well with those of Ref. [21] except for U^{90+}.

Table 2 reproduces the transition rates from Ref. [15] calculated using Methods I and II along with other available data for ions with Z = 26, 27, 46, 66 and 92. The transition rates from both the methods are in good agreement with each other for ions with Z ≤ 46. However, the differences increase with increasing Z. The present rates are in excellent agreement with the NIST compilation [20] and MZ rates [22]. While the rates from Method I compare well with REOS rates of Zhang et al. [21] for low and medium Z ions, the two rates deviate considerably for ions with Z > 50 whereas the rates from Method II are in better agreement

Table 1. Comparison of E1 and M1 energies in eV [15] from $2s^2\,^1S_0$ state with other relativistic data. The energies listed in the second row under columns 2, 4, and 6 are values calculated with frozen set of orbitals

Z	$^1P_1$- $^1S_0$		$^3P_1$ - $^1S_0$		$^3S_1$ - $^1S_0$	
	Present	Others	Present	Others	Present	Others
26	6845.41	6845.93[c]	6878.76	6879.37[c]	6910.02	6910.47[c]
	6848.82	6846.0[a]	6882.11	6878.9[a]	6913.00	
		6845.44 [b]		6878.49[b]		
27	7388.92	7389.6[a]	7425.58	7425.8[a]	7458.2	
	7392.78	7389.42[b]	7429.39	7425.60[b]	7461.33	
46	21799.49	21810.66[c]	22021.66	22023.99[c]	22085.23	22087.49[c]
	21822.79		22044.49		22107.29	
66	45845.82	45852.44[c]	46832.55	46839.27[c]	46931.89	46938.84[c]
	45924.38		46909.98			
92	92897.39	92940.43[c]	97338.71	97381.78[c]	97482.21	97527.60[c]
	93177.7	92950[d]	97619. 7	97392[d]		97536[d]

[a]Reference [20]
[b]Reference [22]
[c]Reference [21]
[d]Reference [23]

with REOS rates than the rates obtained from Method I. The M1 rates from Method I compare well with Ref. [21] except for heavy elements.

It is clear from Table 2 that the chosen optimization approach does influence the transition rates for high Z elements.

The length to velocity gauges rates (A_l/A_v) listed in the last two columns show that the ratio is nearly unity for both dipole allowed and spin forbidden transitions thereby providing the reliability of the spin orbitals used in the evaluation of transition parameters.

Table 3 lists the rates of $2s^2- 1snp$ ($^1P_1,^3P_1$) ($n=3,4$) transitions calculated in this work for some select ions. Unlike the $2s^2- 1s2p$ transitions the agreement between the TEOP rates in the two gauges for $2s^2- 1s3p$ is only 60–76% for allowed transitions.The spin-forbidden rates agree well by 90% only for ions with Z < 30 and differ largely for heavy ions. For the $2s^2- 1s4p$ weak lines, the differences are more pronounced. As the agreement between the rates calculated in the two gauges is a measure of the accuracy of the atomis state function (ASF) representation, these less intense transitions need improved ASFs. However, a comparison of this table with those listed in Table III of Ref. [15], a part of which is reproduced in Table 2 of this paper, shows that the decay of $2s^2\,^1S_0$ is mainly governed by transitions to $1s2$p $^1P_1$ and $^3P_1$ states and the contributions from the other groups of transitions to the total TEOP rates is negligible. The transition parameters for $2s^2- 1snp$ ($^1P_1,^3 P_1$) ($n=3,4$) transitions are calculated using only Method I as the contributions from these groups to the total X-ray rates is

Table 2. Comparison of length gauge E1 and M1 rates in s^{-1} [15] from $2s^2\,^1S_0$ with other relativistic data. The rates listed in the second row under columns 2, 4, and 6 are values calculated with common set of orbitals. The length to velocity (A_l/A_v) values for the E1 transitions are listed in the last two columns. The numbers within the parentheses are powers of ten

Z	$^1P_1$- $^1S_0$		$^3P_1$ -$^1S_0$		$^3S_1$ -$^1S_0$		$^1P_1$- $^1S_0$	$^3P_1$ -$^1S_0$
	Present	Others	Present	Others	Present	Others	(A_l/A_v)	(A_l/A_v)
26	5.891(13)	5.78(13)[b]	4.997(13)	4.75(13)[b]	5.711(8)	5.46(8)[b]	1.00	1.01
	5.837(13)	5.9(13)[a]	4.908(13)	4.9(13)[a]				
		5.9(13)[c]		4.1(13)[c]				
27	6.267(13)	6.26(13)[a]	6.743(13)	6.38(13)[a]	1.186(9)		1.00	1.01
	6.133(13)	6.2(13)[c]		6.4(13)[c]				
46	5.114(13)	5.02(13)[b]	1.359(15)	1.04(15)[b]	1.74(11)	1.8(11)[b]	1.00	1.01
	5.151(13)		1.156(15)		1.822(11)			
66	2.178(13)	2.37(13)[b]	7.508(15)	4.07(15)[b]	6.787(12)	7.79(12)[b]	0.991	1.02
	2.469(13)		5.209(15)		7.203(12)			
92	4.342(12)	8.88(12)[b]	5.005(16)	1.38(16)[b]	1.742(14)	2.82(14)[b]	0.941	1.42
	8.264(12)		2.345(16)		2.135(14)			

[a]Reference [22]
[b]Reference [21]
[c]Reference [20]

Table 3. E1 rates in length gauge in s^{-1} from $2s^2$-$1snp$ $(n = 3, 4]$ transitions

Z	1s3p		1s4p		Z	1s3p		1s4p	
	$^1P_1$ -$^1S_0$	$^3P_1$ -$^1S_0$	$^1P_1$ -$^1S_0$	$^3P_1$ -$^1S_0$		$^1P_1$ -$^1S_0$	$^3P_1$ -$^1S_0$	$^1P_1$ -$^1S_0$	$^3P_1$ -$^1S_0$
18	2.195(10)	1.460(9)	3.815(9)	3.332(8)	20	3.379(10)	2.981(9)	2.204(10)	1.222(9)
26	2.288(10)	1.023(10)	8.717(9)	2.685(9)	28	4.654(10)	2.072(10)	9.572(9)	3.568(9)
36	4.564(10)	5.4791(10)	1.901(10)	1.410(10)	46	4.554(9)	5.689(10)	1.959(10)	1.363(10)
54	2.131(11)	3.221(11)	1.280(10)	3.054(10)	74	9.881(10)	4.985(11)	1.173(9)	4.675(10)

minimal. However, we expect that the impact of relaxed and frozen orbitals on the energies and rates will follow the same trend as was observed for $2s^2-1s2p$ transitions.

4   Conclusion

The intensity ratios of the Kα^h and K$\alpha\alpha$ fine structure lines prove the well known angular momentum coupling schemes, namely pure LS coupling at low Z, moving from intermediate coupling to jj coupling at high Z. We also note that while the intensity ratio is highly sensitive to the degree of intermediacy of the coupling scheme, it does not approach the jj coupling limit of 2 for high Z as predicted by the angular momentum coupling scheme. Also, a comparison of the intensity ratios of Kα^h and K$\alpha\alpha$ fine structure lines clearly indicates the influence of correlation on the intermediate and jj coupling schemes. Precise experimental measurements are needed to know how well pure jj coupling applies for high Z elements and also the influence of correlation on the coupling schemes. The sequential decay of the doubly excited $2s^2$ $^1S_0$ state of He-like ion to the ground state is possible only through this special type of intense TEOP transitions to states of $1snp$ with maximum contribution from $2s^2-1s2p$ and hence the decay of such a system will provide valuable information on the transition dynamics of such hollow ions. Future precise experimental measurements/observations can help in judging the dependence of line intensities on the nature of spin-orbitals as noticed in this work.

Acknowledgements. The author acknowledges the support from Science and Engineering Research Board (SERB), Department of Science and Technology, Government of India under the project EMR/2016/005501.

References

1. Briand, J.P., et al.: Phys. Rev. Lett. **65**, 159 (1990)
2. Aberg, T.: Advances in X-Ray Spectroscopy. In: Bonnelle, C., Mande, C. (eds.), p. 1. Pergamon, New York (1982)
3. Costa, A.M., Martins, M.C., Santos, J.P., Indelicato, P., Parente, F.: J. Phys. B: At. Mol. Phys. **40**, 57 (2007)
4. Chen, M.H., Crasemann, B., Mark, H.: Phys. Rev. A **25**, 391 (1982)
5. Natarajan, L.: Phy. Rev. A **78**, 052505 (2008)
6. Diamant, R., Huotari, S., Hamalainen, K., Sharon, R., Kao, C.C., Deutsch, M.: Phys. Rev. Lett. **91**, 193001 (2003)
7. Diamant, R., Huotari, S., Hamalainen, K., Sharon, R., Kao, C.C., Honkimäki, V., Buslaps, T., Deutsch, M.: Phys. Rev. A **79**, 062512 (2013) and references therein
8. Salem, S.I., Kumar, A., Scott, B.L.: Phys. Rev. A **29**, 2634 (1984)
9. Salem, S.I.: Phys. Rev. A **21**, 858 (1980)
10. Cue, N., Schulz, W., Li-Schulz, A.: Phys. Lett. **63A**, 54 (1977)
11. Mikkola, E., Keski-Rahkonen, O., Kuoppala, R.: Phys. Scr. **19**, 29 (1979)
12. Ahopelto, J., Rantavouri, E., Keski-Rahkonen, O.: Phys. Scr. **20**, 71 (1979)
13. Hoszowska, et al.: Phys. Rev. Lett. **107**, 053001 (2011)
14. Kadrekar, R., Natarajan, L.: J. Phys. B: At. Mol. Phys. **43**, 155001 (2010)
15. Natarajan, L.: Phys. Rev. A **90**, 032509 (2014)
16. Jonsson, P., Gaigalas, G., Bieron, J.B., Froese Fischer, C., Grant, I.P.: Comput. Phys. Commun. **184**, 2197 (2013)
17. Olsen, J., Godefroid, M., Jonsson, P., Malmqvist, A., Froese Fischer, C.: Phys. Rev. E **52**, 4499 (1995)
18. Dyall, K., Grant, I.P., Johnson, P., Parpia, F.A., Plummer, E.: Comput. Phys. Commun. **55**, 425 (1989)
19. Grant, I.P.: Relativistic Quantum Theory of Atoms and Molecules: Theory and Computation. Springer, New York (2007) and references therein
20. Online bibliographic database at NIST. http://www.physics.nist.gov/asd3
21. Zhang, D.H., Dong, C.Z., Koike, Fumihiro: Chin. Phys. Lett. **23**, 2059 (2006)
22. Goryayev, Y.Y., Urnov, A.M., Vainshtein, L.A.: At. Data Nucl. Data Tables **113**, 117 (2017)
23. Zackowicz, S., Harman, Z., Grun, N., Scheid, W.: Phys. Rev. A **68**, 042711 (2003)

Electron Excitation Cross Sections of Fine-Structure ($5p^5 6s$–$5p^5 6p$) Transitions in Xenon

Priti[1], R. K. Gangwar[2], and Rajesh Srivastava[1(✉)]

[1] Department of Physics, Indian Institute of Technology Roorkee, Roorkee
247667, India
{pritidph, rajsrfph}@iitr.ac.in
[2] Department of Physics and Astronomy, National Institute of Technology
Rourkela, Rourkela 769008, India
reeteshkr@gmail.com

Abstract. Electron impact excitation cross sections of xenon atoms from its two metastable states $(5p^5 6s)_{J=2,0}$ to the ten fine-structure $(5p^5 6p)_{J=0,1,2,3}$ levels have been calculated using relativistic distorted wave (RDW) theory. The results are reported in the wide range of incident electron energies (excitation threshold to 1 keV). Fitting of the obtained cross sections of each transition with suitable analytic expression is also provided for plasma modeling purposes.

1 Introduction

In recent years, trace-rare-gas (TRG) optical emission spectroscopy (OES) is widely used for plasma diagnostics [1, 2]. In this technique, a small amount of a rare-gas (Ne, Ar, Kr, and Xe) atom is introduced to the plasma and the optical emission intensities coming from it are recorded. By coupling the OES measured intensities with an appropriate kinetics plasma model [mostly a collisional radiative (C-R) model] one can obtain the plasma parameters, such as electron temperature and electron density of the plasma. Moreover, the most commonly used lines of an inert gas for diagnostics are from the $2p_i$ ($i = 1$–10) $\rightarrow$ $1s_i$ ($i = 1$–4) transitions (the levels are written in Paschan notation). In this context, the knowledge of the populations of the $2p_i$ states is very important as the emitted intensities are directly proportional to the upper-level populations. Further, to develop a C-R model and get accurate plasma parameters, we need to utilize reliable cross-section results in the model for the dominant electron impact processes which populate the $2p_i$ levels [3–7]. These levels can be excited by the electrons from the ground $1s_0$ as well as the metastable $1s_3$ and $1s_5$ states. Since the metastable states, as compared to the ground state, are closer to the $2p_i$ energy levels, their cross sections are expected to be larger [8, 9].

Most of the earlier studies have focused on the electron impact $1s_i$-$2p_i$ transitions in Ar and Kr inert gases and reported their cross sections in a detailed manner. Not much attention has been paid to study similar transitions in Xe which are also prominent in Xe plasma having many applications such as in thrusters [10]. However, for these transitions, experimental and theoretical cross-section results are very limited and available for a few

© Springer Nature Singapore Pte Ltd. 2019
P. C. Deshmukh et al. (eds.), *Quantum Collisions and Confinement of Atomic
and Molecular Species, and Photons*, Springer Proceedings in Physics 230,
https://doi.org/10.1007/978-981-13-9969-5_16

selected electron energies which are often fine-structure unresolved [11–13]. For $1s_5$-$2p_i$ ($i = 5, 7, 9, 10$) and $1s_5$-$2p_i$ ($i = 6, 8$) transitions, Jung et al. [12] reported cross sections for electron incident energy only up to 10 eV and 300 eV, respectively. To compare with their results, Srivastava et al. [13] performed the RDW calculations for the same transitions and in a similar incident energy range. Therefore, in the light of the required detailed complete set of cross section data, in the present study, we obtain these cross sections for all the excitations from both the $1s_5$ and $1s_3$ metastable states to all ten $2p_i$ levels in the wide range of energy up to 1 KeV using the RDW theory.

2   Theoretical Background

For the electron impact excitation of an atom having N electrons from an initial state a to final state b, the distorted wave T-matrix [14] can be written as

$$T_{a \to b}^{RDW}(J_b M_b, k_b \mu_b; J_a M_a, k_a \mu_a, \theta) = \left\langle \Phi_b^{rel}(\mathbf{1, 2, \dots, N}) \mathbf{F}_{b, \mu_b}^{DW-}(\mathbf{k}_b, \mathbf{N} + \mathbf{1}) | V - U(N+1) \right.$$
$$\left. \times | A \left\{ \Phi_a^{rel}(\mathbf{1, 2, \dots, N}) \mathbf{F}_{a, \mu_a}^{DW+}(\mathbf{k}_a, \mathbf{N+1}) \right\} \right\rangle$$

$$(1)$$

here, $J_{a(b)}$ and $M_{a(b)}$ represent, respectively, the total angular momentum and the corresponding magnetic component of the atom in the initial (final) state. $K_{a(b)}$ and $\mu_{a(b)}$ denote, respectively, the wave vectors and the spin projection of the incident (scattered) electron. θ is the scattering angle between the wave vectors of the incident and scattered electron. $\Phi_{a(b)}^{rel}(\mathbf{1, 2, \dots, N})$ represents relativistic N-electron wave function for the target atom in the initial (final) state. $\mathbf{F}_{a(b), \mu_{a(b)}}^{DW+(-)}$ is the relativistic distorted projectile electron wave function in the incident (scattered) channel where "+" denotes an outgoing wave and "−" refers to an incoming wave. V is the interaction potential between projectile electron and target atom and U is the distortion potential which is assumed to be a function of $\mathbf{r}_{N+1}$ coordinate only. The electron exchange has been taken into account by using the antisymmetrization operator A.

In order to evaluate T-matrix, one requires accurate initial and final state wave function of the xenon atom as well as of projectile electrons distorted waves. The bound state wave functions in initial and final states of the atom are calculated within multiconfiguration Dirac–Fock (MCDF) approach by GRASP2K code [15]. These atomic target wave functions (initial/final) are expressed as a linear combination of other configuration state functions (CSFs) having the same J value and parity. In addition to the ground state ($5p^6$), the $5p^5 6s^1$, $5p^5 6p^1$, $5p^5 5d^1$, $5p^5 7s^1$, and $5p^5 7p^1$ are the various CSFs used to represent the xenon atom wave functions in GRASP2K [15]. To insure the reliability of the obtained wave function, we have compared our calculated transition energies and oscillator strengths with the available results from the NIST database [16] as well as other experimental [12, 17–19] and theoretical [20, 21] reports and we found good agreement with them. To obtain the wave function for the projectile electrons, Dirac equations are solved in the field of the spherically averaged static potential of the final state of the atom.

The excitation cross section to its magnetic sublevel M_b can be expressed in terms of the T-Matrix as follows:

$$\sigma_{M_b} = (2\pi)^4 \frac{k_b}{2(2J_a + 1)k_a} \int \left| T_{a \to b}^{RDW} \left(J_a, M_a, \mu_a; J_b, M_b, \mu_b, \theta \right) \right|^2 d\Omega \qquad (2)$$

where the integration is performed over the solid angle of the scattered electron. We get the total excitation cross section by summing σ_{M_b} for all the magnetic sublevels of the final state of the atom.

3   Results and Discussion

In Fig. 1, 2a and b, we have shown electron impact cross section results for the excitation from the metastable states $1s_3$ and $1s_5$ to the ten $5p^5 6p$ ($2p_i$, $i = 1$–10) fine-structure levels of xenon in the incident energy range up to 1 keV. Only experimental cross section results of Jung et al. [12] are available for the $1s_5 \to 2p_6$ and $1s_5 \to 2p_8$ excitation up to higher impact energies. In Fig. 1, we have compared our calculated cross sections with the experimental results and find excellent agreement in the energy range above the 80 eV. However, below this energy, our results are slightly higher than the experimental results. Being a perturbative approach, RDW method sometimes slightly overestimates the cross sections for the low incident electron energies. In Fig. 2b and c, we have presented our complete set of results for the $2p_i$ level excitations, respectively, from the $1s_5$ and $1s_3$ states. We find from these figures that the dipole allowed transitions ($\Delta J = 0, \pm1$), have lager cross sections as compare to the forbidden

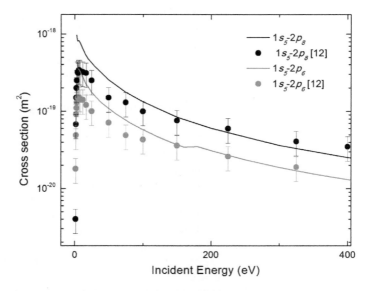

Fig. 1. Electron impact excitation cross section for the transitions $1s_5$-$2p_6$ and $1s_5$-$2p_8$, solid lines represent the present calculation and solid circles represent the experimental results of Jung et al. [12]

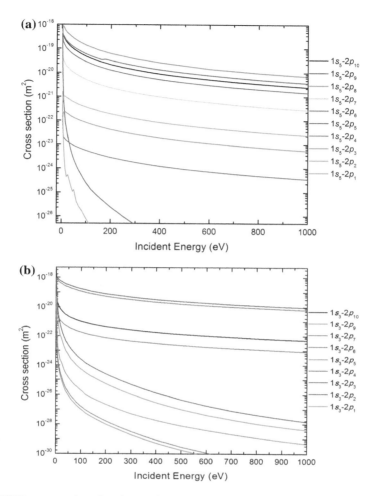

Fig. 2. RDW cross sections for electron impact excitation from the metastable level $1s_5$ ($J = 2$) and $1s_3$ ($J = 0$) to the $2pi$ fine-structure levels of the $5p^56p$ configuration

transitions which are quite expected. The cross sections for the forbidden transitions decay faster at higher impact energies as E^{-3}. Further, cross sections for core preserving excitations, is larger than core changing excitations. Therefore, cross sections for $(5\bar{p}^25p^36s) \rightarrow (5\bar{p}^25p^36p)$ transitions are larger than the $(5\bar{p}^25p^36s) \rightarrow (5\bar{p}5p^46p)$. For example, cross section for $1s_5$-$2p_{10}$ is large than $1s_5$-$2p_2$ while both are dipole allowed transitions. Finally, for plasma modeling purposes, we have fitted our calculated cross sections of all the transitions up to 1 keV by the following expression:

$$\sigma = \frac{\sum_{i=0}^{n} b_i E^i}{c_0 + c_1 E + c_2 E^2} \tag{3}$$

Table 1. Coefficients for the fitting function for cross sections for electron impact excitation for transitions from $1si$ and $2pi$ levels

Transition	Energy interval (eV)	b_0	b_1	b_2	b_3	c_0	c_1	c_2
$1s_5 \rightarrow 2p_{10}$	E_{th}–1000	2.6098e+1				1.5121e-1	3.2918e-1	1.6700e-2
$1s_5 \rightarrow 2p_9$	E_{th}–10	3.9755e−4	7.2257e+5			−3.0221e+3	2.5272e+4	8.0032e+4
	10–1000	1.0122e+1				1.0684e−1	2.312e−1	7.4400e−2
$1s_5 \rightarrow 2p_8$	E_{th}–1000	3.6499e+2	1.2666e−1	−3.9480e+4		6.1273e+0	6.5354e+1	1.0580e+2
$1s_5 \rightarrow 2p_7$	E_{th}–10	7.0814e+0				1.9165e−1	2.3946e+0	−2.5172e+0
	10–1000	2.5173e+0				1.5313e−1	3.1449e−1	8.6500e−3
$1s_5 \rightarrow 2p_6$	E_{th}–10	7.3764e−6	1.8704e+8			−2.4360e+5	3.6945e+6	6.2914e+6
	10–1000	2.3404e+1				1.3008e−1	2.5480e−1	4.5500e−3
$1s_5 \rightarrow 2p_5$	E_{th}–10	1.2579e+1	−3.1775e+1			4.0798e+1	−5.4764e+2	2.0071e+3
	10–200	4.9200e−3	−4.1720e+1					
	200–1000	3.4700e−2						
$1s_5 \rightarrow 2p_4$	E_{th}–10	4.8510e−2				−2.6937e+0	2.42388e+1	2.8147e+1
	10–200	2.0074e+1				2.2740e+2	2.37348e+2	−1.1196e+0
	200–1000	1.4891e+2				7.0555e+3	1.64423e+4	4.3993e+2
$1s_5 \rightarrow 2p_3$	E_{th}–10	1.6970e−2	1.6418e−1	1.2860e−2		1.1657e+0	−9.5431e+0	2.0322e+1
	10–1000	−2.1300e−3	7.9129e+4			4.8117e+4	6.0709e+5	5.5174e+5
$1s_5 \rightarrow 2p_2$	E_{th}–10	9.0160e−2	6.7293e−1	6.7626e−1		1.0771e+1	−8.7902e+1	1.8948e+2
	10–1000	8.2549e−1				1.8115e+0	1.1182e+0	4.6650e−2
$1s_3 \rightarrow 2p_{10}$	E_{th}–10	5.2976e+1	1.1100e−3	7.1165e+2		−4.4000e−2	4.0975e+0	−3.9228e+0
	10–1000	1.4071e+1				−1.9707e+0	−5.1316e+0	−5.0370e+1
$1s_3 \rightarrow 2p_9$	E_{th}–20	3.9048e+0	−3.3033e+0			9.0941e+2	−1.9226e+4	1.0925e+5
	20–300	2.0734e−5	−4.5519e+0					
	300–1000	3.7034e−4						
$1s_3 \rightarrow 2p_7$	E_{th}–10	1.5408e+0	1.6700e−3	5.4206e+1		9.3580e−2	8.1152e+0	−6.9688e+0
	10–200	1.4604e+0				−1.9582e+0	−7.1655e+0	−3.6481e+1
	200–1000	1.4436e+0				−5.9308e+0	7.0601e+0	1.6550e−1

(continued)

Table 1. (continued)

Transition	Energy interval (eV)	b_0	b_1	b_2	b_3	c_0	c_1	c_2
$1s_3 \to 2p_6$	E_{th}–10	3.2290e−2	−2.9783e+0			4.2454e+0	−9.0377e+1	5.1407e+2
	10–300	3.7715e−5	−4.5518e+0					
	300–1000	6.1656e−4						
$1s_3 \to 2p_5$	E_{th}–10	4.5790e−5	−4.1147e+0					
	10–200	2.8912e−4	−2.7052e+0					
	200–1000	9.3760e−4	−3.5701e+0					
$1s_3 \to 2p_4$	E_{th}–20	3.4357e−6	4.8063e−7	2.2660e−6	2.3146e+7	4.9307e+2	−6.6931e+2	−4.7232e+4
	20–300	2.9420e+2				9.9270e−2	1.4493e−1	7.1800e−3
	300–1000	6.5692e+3				−3.6586e+1	4.5181e+1	9.8724e−1
$1s_3 \to 2p_3$	E_{th}–10	7.3481e−1	−3.0874e+0			1.7417e−1	−2.4551e+0	9.4552e+0
	10–200	6.0840e−2	−4.3205e+0					
	200–1000	5.3314e−1						
$1s_3 \to 2p_2$	E_{th}–20	2.9483e−4	2.0089e−3	6.5456e−4	−2.0241e+4	−2.0241e+4	1.31256e+0	6.15596e+0
	20–1000	4.8482e+1				0.12252	0.16453	0.00765
$1s_3 \to 2p_1$	E_{th}–10	−7.5782e−1	1.1885e+1			−1.1100e+0	5.0443e+0	2.43598e+1
	10–300	1.0480e−2	−3.2194e+0					
	300–1000	1.8262e−1	−4.5049e+0					

Here, σ is the electron impact excitation cross section and E is the incident electron energy. Both σ and E are in the atomic units. The b_0, b_1, b_2, b_3, c_0, c_1, and c_2 are the fitting parameters which are given in Table 1. The fitted cross sections are accurate to 5%.

4   Conclusions

We have studied electron impact excitation of xenon from the lowest lying metastable states to the ten fine-structure states of the $5p^5 6p$ using RDW method. Cross sections are calculated in the wide electron incident energy range from excitation threshold to 1 KeV. We have compared our results with the available experimental results of Jung et al. [12] and found good agreement. Our present study is an effort to provide a complete set of reliable cross-section data for the most prominent transition lines of the Xe plasma.

References

1. Donnelly, V.M.: Plasma electron temperatures and electron energy distributions measured by trace rare gases optical emission spectroscopy. J. Phys. D Appl. Phys. **37**, R217–R236 (2004). https://doi.org/10.1088/0022-3727/37/19/R01
2. Zhu, X.-M., Pu, Y.-K.: A simple collisional-radiative model for low-temperature argon discharges with pressure ranging from 1 Pa to atmospheric pressure: kinetics of Paschen 1s and 2p levels. J. Phys. D Appl. Phys. **43**, 15204 (2010)
3. Boffard, J.B., Lin, C.C., Wendt, A.E.: Application of excitation cross-section measurements to optical plasma diagnostics, 1st edn. Elsevier Inc.
4. Dipti, Gangwar R.K., Srivastava, R., Stauffer, A.D.: Collisional-radiative model for non-maxwellian inductively coupled argon plasmas using detailed fine-structure relativistic distorted-wave cross sections. Eur. Phys. J. D **67**, 203 (2013). https://doi.org/10.1140/epjd/e2013-40244-9
5. Gangwar, R.K., Dipti, Srivastava, R., Stafford, L.: Spectroscopic diagnostics of low-pressure inductively coupled Kr plasma using a collisional-radiative model with fully relativistic cross sections. Plasma Sour. Sci. Technol. **25**, 35025 (2016). https://doi.org/10.1088/0963-0252/25/3/035025
6. Gangwar, R.K., Sharma, L., Srivastava, R., Stauffer, A.D.: Argon plasma modeling with detailed fine-structure cross sections. J. Appl. Phys. **111**, 53307 (2012). https://doi.org/10.1063/1.3693043
7. Dressler, R.A., Chiu, Y., Zatsarinny, O., et al.: Near-infrared collisional radiative model for Xe plasma electrostatic thrusters: the role of metastable atoms. J. Phys. D Appl. Phys. **42**, 185203 (2009)
8. Gangwar, R.K., Sharma, L., Srivastava, R., Stauffer, A.D.: Electron-impact excitation of argon: cross sections of interest in plasma modeling. Phys. Rev. A **81**, 52707 (2010). https://doi.org/10.1103/PhysRevA.81.052707
9. Gangwar, R.K., Sharma, L., Srivastava, R., Stauffer, A.D.: Electron-impact excitation of krypton: cross sections of interest in plasma modeling. Phys. Rev. A At. Mol. Opt. Phys. **82**, 32710 (2010). https://doi.org/10.1103/PhysRevA.82.032710

10. Karabadzhak, G.F., Chiu, Y.H., Dressler, R.A.: Passive optical diagnostic of Xe propelled Hall thrusters. II. collisional-radiative model. J. Appl. Phys. **99**. https://doi.org/10.1063/1.2195019 (2006)

11. Hyman, H.A.: Electronimpact excitation cross section for the transition $(n-1)p^5ns \rightarrow (n-1)p^5np$ in the rare gases. Phys. Rev. A **24**, 1094 (1981). https://doi.org/10.1103/PhysRevA.24.1094

12. Jung, R.O., Boffard, J.B., Anderson, L.W., Lin, C.C.: Electron-impact excitation cross sections from the xenon J = 2 metastable level. Phys. Rev. A At. Mol. Opt. Phys. **72**, 1–9 (2005). https://doi.org/10.1103/PhysRevA.72.022723

13. Srivastava, R., Stauffer, A.D., Sharma, L.: Excitation of the metastable states of the noble gases. Phys. Rev. A At. Mol. Opt. Phys. **74**, 1–10 (2006). https://doi.org/10.1103/PhysRevA.74.012715

14. Zuo, T., McEachran, R.P., Stauffer, A.D.: Relativistic distorted-wave calculation of electron impact excitation of xenon. J. Phys. B At. Mol. Opt. Phys. **24**, 2853–2870 (1991). https://doi.org/10.1088/0953-4075/24/12/008

15. Jönsson, P., Gaigalas, G., Bieroń, J., et al.: New version: {Grasp 2 K} relativistic atomic structure package. Comput. Phys. Commun. **184**, 2197–2203 (2013)

16. Kramida, A., Ralchenko, Y., Reader, J.: TNNASD (version 5. 4). NASD (version 54) atomic spectra database NIST. https://www.nist.gov/pml/atomic-spectra-database. Last accessed 2 Dec 2017

17. Cabrera, J.A., Ortiz, M., Campos, J. Transition probabilities of 6s-6p lines and lifetimes of 6p configuration levels of XeI. Phys. B+C **104**, 416–422 (1981). https://doi.org/10.1016/0378-4363(81)90190-x

18. Sabbagh, J., Sadeghi, N.: Experimental transition probabilities of some Xe(I) lines. J. Quant. Spectrosc. Radiat. Transf. **17**, 297–301 (1977). https://doi.org/10.1016/0022-4073(77)90108-x

19. Miller, M.H., Roig, R.A.: Transition probabilities of XeI and XeII. Phys. Rev. A **8**, 480–486 (1973)

20. Chen, C.J., Garstang, R.H.: Transition probabilities for Xe I. J. Quant. Spectrosc. Radiat. Transf. **10**, 1347–1348 (1970). https://doi.org/10.1016/0022-4073(70)90017-8

21. Aymar, M., Coulombe, M.: Theoretical transition probabilities and lifetimes in Kr I and Xe I spectra. At. Data Nucl. Data Tables **21**, 537–566 (1978). https://doi.org/10.1016/0092-640X(78)90007-4

Ionisation of Nanoclusters at Relativistic Laser Intensities

R. Rajeev and M. Krishnamurthy[(✉)]

TIFR Centre for Interdisciplinary Sciences, Hyderabad 500 075, India
mkrism@tifr.res.in

Abstract. Nanoclusters exposed to intense ultrashort pulses have attracted immense attention due to higher absorption and high charge state ion emission. Insofar, most of the studies in Coulomb explosion of clusters have been at non-relativistic laser intensities. The question of how the ionisation of the clusters change at relativistic intensities ($>2 \times 10^{18}$ Wcm^{-2}) is little explored. In this article, we present charge resolved ion spectroscopic measurement of ion emission for two different cluster sizes of Ar exposed to relativistic intensities. Low charge states ($<8+$) dominate for smaller cluster sizes and high charges states ($>8+$) dominate the spectrum at larger cluster sizes. Electron impacting ionisation seem to be dominantly important to comprehend the charge propensity spectrum. Detailed simulations are important to decipher the role of over the barrier ionisation and electron impact ionisation.

1 Introduction

At low intensities, light interaction with the matter can be understood by the induced polarisation that is linear to the electric field of the electromagnetic radiation. The intensity of light is proportional to the square of the electric field amplitude of the electromagnetic radiation. At high intensities with larger electric field amplitude, the induced polarisation does not remain linear to the field [1]. Nonlinear interaction brings in multi-photon processes and typically at intensities beyond 10^{10} Wcm^{-2} most atomic systems would undergo multi-photon ionisation. At larger intensities, such nonlinear interactions are non-perturbative and electromagnetic fields severely distort the atomic coulomb potential. Tunnel-ionisation becomes increasingly dominant at intensities larger than 10^{12} Wcm^{-2} [2]. At a 100 times larger intensity, the coulomb potential is so severely distorted that a barrier-less ionisation sets in most atoms [3]. All atoms exposed to the over-the-barrier ionisation fields would ionise. Free electrons generated from any atomic system continue to interact with the field after becoming free from the atomic potential. Electrons can gain energy in the field when they oscillate under the electron-magnetic field. The equation of the motion of free electrons under of the interaction of the light is characterised by:

$$m_e \ddot{x}_e(t) = eE_0 \sin \omega t, \tag{1}$$

© Springer Nature Singapore Pte Ltd. 2019
P. C. Deshmukh et al. (eds.), *Quantum Collisions and Confinement of Atomic and Molecular Species, and Photons*, Springer Proceedings in Physics 230,
https://doi.org/10.1007/978-981-13-9969-5_17

where E_0 is the field amplitude of the electric field of the electromagnetic radiation, ω is the laser frequency, m_e and x_e are the mass and instantaneous position of the electron. If we solve for the velocity of the electron and obtain the cycle average kinetic energy, it is

$$U_p(eV) = \frac{1}{2}m_e \langle V(t)^2 \rangle = \frac{e^2 E_0^2}{4m_e \omega^2} \quad (2)$$

This is called as the ponderomotive energy and can be simplified for useful evaluation to be $\approx 9.33 \times 10^{-14}\, I(W/cm^2)(\lambda(\mu m)^2)$, where I is the intensity in W/cm^2 and λ is the light wavelength in μm units [4]. Even at an intensity of 10^{16} W/cm^2 for a laser of wavelength $\lambda = 800$ nm, the electron energy is ≈ 600 eV, which is more than 40 times the ionisation energy of a Hydrogen atom and is more than 380 times larger than the photon energy. Non-perturbative phenomenon dominates at this regime and the classical motion of the electron in the field dominate the interaction physics. At even higher intensity, the velocity of the electrons becomes sufficiently large to appreciate the interaction of the electrons with the magnetic field of the radiation. In such a case it becomes more relevant to use the Lorentz force equation

$$\mathbf{F} = e(\mathbf{E} + \frac{\mathbf{V}}{c} \times B) \quad (3)$$

The motion of the electron in this regime is very different. A simple harmonic electron motion of the dipole regime changes to a figure eight motion in the high-intensity regime [5]. An intensity regime is considered relativistic once the laser field E is so high that [6,7] ponderomotive potential given as $m_e c^2(\sqrt{(a_o^2 + 1)})$ becomes close to the electron rest mass, where a_0 is the normalised vector potential representing the amplitude of the laser radiation given by $a_0 = eA/m_e c$ and A is the vector potential defining the electron magnetic radiation. At these intensities, the ponderomotive energy of the electron becomes comparable with its rest mass (~ 500 keV) and the motion of the electron is governed by the relativistic mass correction of the electron. For $a_0 \sim 1$, the intensity corresponds to about 2×10^{18} Wcm^{-2} with 800 nm light. This gives $U_p \sim 200$ keV.

When there is an ensemble of atoms, all the atoms in the focal volume are subjected to these effects. Ionisation of a large number of atoms leads to larger electron density and the screening of Coulomb exponentials by the electrons. When the number of electrons in a Debye sphere become larger, plasma properties dominate the laser–matter interaction. Collective phenomena affects the key parameters like the dielectric permittivity which changes the propagation of the electromagnetic fields and in turn the light–matter interaction. Intense laser-produced plasmas are different compared to the conventional low-density plasmas due to the impulsive generation of the plasma [8]. When light fields are so short that the atoms have no time to move, a solid density system would generate plasma of solid density or even higher density. Since the ponderomotive energy of the electrons, the plasmas is at high energy compared to any conventional plasma. High-density, high-temperature laser produced plasma brings about very rich phenomena as every parameter changes vary dramatically both spatially and temporally

[9]. In the relativistic regime, plasma absorption is known to be very different as the basic electron motion is different. Light penetrates far beyond the normal skin depth [6]. The effect of the magnetic field in the Lorentz force leads to J × B forces that drive electrons along laser propagation [10].

Plasmas made of nanoparticles of matter, also called as nanoclusters, are special in their own right [11–13]. A few thousand atoms bound together by van der Waals forces have solid like intra-cluster density. Solid density plasma with a particle size that is much smaller than the light wavelength have much different properties compared to plasma generated from macroscopic targets. In an extended target, when the laser is focused on a few microns, there is always a continuum of cold or unexposed matter connected to the hot plasma region. Ejection of electrons from the focal volume will induce electron mobility from the unexposed region to reduce the charge densities and quickly reduce the local electrostatic potential. A nanocluster suspended in the vacuum is devoid of such cold electron currents. Charge density is higher and remain higher till cluster disintegrates. Plasma temperature remains higher and electron ion recombination rates are lower. A large number of experiments have demonstrated these interesting features [11–13].

Atomic systems in nanoclusters are ionised once the electric field experienced in strong enough to induce multi-photon ionisation. At higher fields tunnel ionisation followed by the over barrier ionisation contributes to free electrons from the atoms. Electrons that leave the atom are considered to be inner ionised. These electrons continue to interact with the electromagnetic fields and gain ponderomotive energy. Electrons pulled away from the ion cloud in each half cycle of the light field are driven back when the field reverses and contribute to the collisional ionisation of the atoms [14]. Electron cloud oscillates with respect of the ion cloud under the coulomb potential and the electric field of the light. Fraction of electrons that gain energy larger than the binding energy due to cluster potential, leave the cluster and are termed outer ionised [15]. Outer ionisation leads to coulomb pressure while the high-energy electrons contributes to the hydrodynamic pressure. At the end of the laser pulse, thus a hot dense nanoplasma is generated and it takes several ps for the cluster to expand and to disintegrate under the coulomb/hydrodynamics forces to expunge highly charged high- energy ions. Hot dense plasma thus is a bright source of short pulses of hot electrons, ions, and X-rays [16,17].

Most of the experiments in clusters, be it to measure ion spectra, electron spectra, X-ray emission, or light absorption, have all been explored mainly in the nonrelativistic regime [13,16,17]. Even at intensities of about 10^{16} Wcm^{-2}, Ar clusters of about 40,000 atoms results in a mean charge of about 8. Ion kinetic energies extend to about 1 MeV, even though the electron energy is only a few keV [18]. Electron cloud oscillates about the ion cloud and these dipole oscillations bring out an asymmetric charge distribution with highly charged high-energy ions preferentially emitted along the direction of polarisation. As the cluster expands, the electron density decreases and when the local plasma frequency matches the laser frequency there is a resonant enhanced light absorption and ion/electron emission. Dipole oscillations of the electron cloud are largely known to be responsible for the efficient nanoplasma heating and the high charge states formed in the cluster [19].

The question we ask in this article is about how the nanoplasma generation changes, if relativistic intensities are used. If the Lorentz force equation drives the electrons differently compared to the dipole oscillations, there would be J × B forces on the electron cloud and electrons could be driven along the direction of the laser propagation. Would the electron ion collisions and the inverse bremsstrahlung heating change the charge propensity in the cluster? We present experiments on the measurement of the ion charge-state propensity distribution (ICSPD) of argon clusters in the relativistic regime for two different cluster sizes. The peak charge state remains unchanged at 8+ in both the cases. The smaller clusters have dominantly low charge states (<8+). For larger clusters, there is a dramatic change and the spectrum is dominated by higher charge states (>8+). In both the cases, intensity is large enough to generate up to 12+ charge states by the over-barrier ionisation. Screening of the field could be responsible for the changes in the spectrum. It appears that the high charge states generated are correlated with the electron impact ionisation. Smaller clusters with lesser number of atoms do not seem to offer sufficient electron–ion collisions to produce the higher charge states. The results point out the need for a detailed simulations of the nanoplasma with the relativistic effects to fully comprehend the observations.

2   Experiments

The experiments presented here use a Thompson parabola spectrometer (TPS) that is described in detail elsewhere [20] and only the salient features are discussed here. A TPS samples ions through a small collimating aperture and disperses positive ions entering like a pencil beam by electric and magnetic fields designed to deflect ions in orthogonal directions. Ions exhibit a parabolic motion and are pushed away from the propagation axis. After the field deflection region, ions are allowed to drift through a small region to increase the spatial dispersion and the ions splat on a position-sensitive ion detector. Microchannel plate (MCP) detector coupled to a phosphor screen and an imaging camera are used to acquire the images of the ion splat position for each laser pulse exposed to the clusters. Ion drift from the undeflected beam path depends on the m/q of the ions and kinetic energy. Splat position of the ions of similar m/q appears as a parabolic trace. Ions closer to the origin (the splat position of the beam when the deflection fields are switched off) have the highest kinetic energy and the lowest energy ions are dispersed far away from the detector. Ions of different m/q give a different parabolic trace. Kinematic equations of the ion motion in the fields and the drift region are used to invert the splat position to m/q and the ion kinetic energy [20]. The technique thus enables charge resolved ion kinetic energy spectrum (CRKES) apart from the ion charge propensity distribution (ICSPD) [21].

Applying the TPS for clusters offers a problem because of the extended nature of the ion source size. Clusters are typically produced by a supersonic jet and sampled in high vacuum system by a skimmer. Even at a cm down the nozzle the cluster beam extends to about a cm. Cluster in the focal volume exceeding an intensity of 10^{13} Wcm^{-2} can contribute to the ionisation signals. In our experimental conditions, the ionisation regime extends to about 3 mm. Ion sampling, therefore, needs an additional pair of slits very close to the focal volume to restrict the ion sampling and only then, well- resolved parabolic traces of ions are measurable. In our experiments, we use a pair movable slits that are motorised and dynamically adjusted to chose the fraction of the focal volume sampled by the TPS. This allows us to sample ions from the relativistic intensity regime and enables the measurements described here [20].

A supersonic jet with a 750 μm wide, 45° conical nozzle operated with different backing pressures of Ar was used to produce different sizes of clusters. Rayleigh scattering measurements [16] in conjunction with the use of the Hagena parameter are used to estimate the cluster sizes. A 500 μm skimmer is used to sample the cluster in a differentially pumped interaction chamber with a base pressure of about 10^{-6} Torr. At these conditions, there is no inter-cluster interactions and the ions fly of the detector without resulting in any charge transfer or recombination reactions. Ion explosion from individual cluster exposed to the intensities is thus measured. 35 fs pulses at a pulse energy of about 600 mJ are focused down to 10μm waist size to achieve peak intensities of $\sim 4 \times 10^{18}$ Wcm^{-2}.

3   Results and Discussion

Figure 1 shows the Thomson parabola image for ion explosion for two cluster sizes Ar$_{1400}$ and Ar$_{36,500}$ at relativistic intensity of 4×10^{18} Wcm^{-2} achieved with 40 fs pulses. The variable slits in the TPS probe are set to image ions generated only in the 300 μm Rayleigh range and so ion emission from the low intense domains are avoided. Since only a small number of clusters are present in the focal area, the ion count rates are very low and necessitated an image acquisition from around 72,000 laser shots to acquire the TPS image. The ion distribution peaks at Ar^{8+} for both the cluster sizes [21]. In Fig. 1a parabolic lines are far away from the origin and a 100 keV ion energy line is indicated. At higher cluster size, the parabola is drawn more closer to the origin and a 250 keV ion energies line is indicated in Fig. 1b. From these images we derive a histogram, by integrating the ion yield in the each parabolic line, to get the relative ion yield for the different charge states. Figure 2 gives the normalised ion charge state propensity distribution (ICSPD). As can be seen with smaller clusters, charge propensity is dominated by the lower charge states while for larger clusters, the high charge states are prominent. For bigger clusters, 8+ is still the dominant ion. But, the lower charge states are depleted considerably and the higher charge propensity has increased considerably. The mean charge here is about 9 while it is about 6.8 with Ar$_{1400}$.

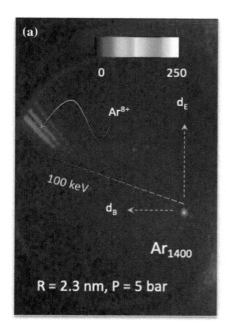
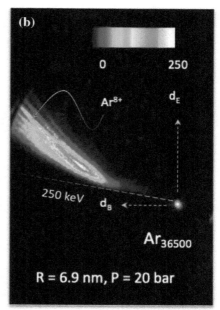

Fig. 1. Thompson parabola images at relativistic intensities **a, b** are ion images obtained from a Thompson parabola spectrometer for two different backing pressures in supersonic jet nozzle at $\sim4 \times 10^{18}$ Wcm^{-2} intensity. The size in the legend indicates the mean cluster size. Line indicates the maximum energy points and indicate that the maximum ion energy observed is 100 keV in case of 2.3 nm (Ar$_{1400}$) and 250 keV in case of 6.9 nm (Ar$_{36,500}$) clusters

To understand these observations, we look into the ionisation mechanism. The ionisation is by (a) multiphoton ionisation, (b) over-the-barrier ionisation, (OBI) (c) collisional ionisation by laser-driven electrons, and (d) collective plasma resonance mechanism. The first three are single particle response mechanisms. Multiphoton initiates the ionisation at the early time of the laser pulse and OBI follows. OBI is a threshold process and the threshold intensity is given to be [3]

$$I_{obi} = 4 \times 10^9 (I_p)^4 / q^2 \tag{4}$$

where I_p is the ionisation potential for atom to a q+ state and I_{obi} gives the intensity threshold in Wcm^{-2}. Table 1 gives the OBI threshold intensity for the different charge states of Ar. At non-relativistic intensities ($<3 \times 10^{16}$ Wcm^{-2}) OBI threshold is such that direct laser field ionisation can generate at the most 8+. There is a big jump in ionisation energy once eight electrons are removed and the 9+ requires electron removal from the L shell of Ar. Correspondingly, there is a big jump in OBI threshold and only relativistic fields can generate 9+ and above directly by field ionisation. At $\sim4 \times 10^{18}$, Ar^{12+} charge state with an ionisation energy of about 618 eV can be achieved by over the barrier ionisation.

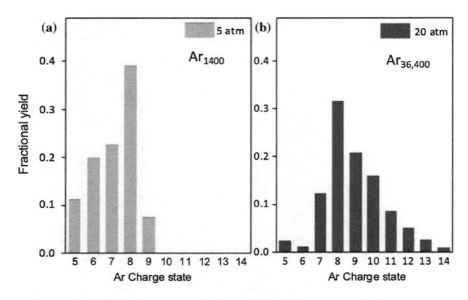

Fig. 2. Relativistic ion emission **a** and **b** are the ICSPD of ion emission from Ar_{1400} and $Ar_{36,000}$ clusters, respectively at relativistic intensity of 4×10^{18} Wcm^{-2}

If all the atoms in the cluster experience the applied relativistic intensity, ideally one would expect only charge states beyond 12+ for measurements done at $\sim 4 \times 10^{18}$. It is clear from the measurements that are presented in Fig. 2 that this is not observed. However, three issues can affect measurements. (i) Focal volume effect: At the focal waist, the beam has a Gaussian intensity profile. Slits used in the experiments restrict the ionisation zone sampled only along the laser propagation direction. In the transverse direction, ionisation in the low intensity focal volume can not be avoided. This can contribute to ionisation signals from clusters exposed to lower non-relativistic intensity. For smaller clusters generated at a lower backing pressure one would expect that the low-intensity focal volume contribution to be less compared to the measurements at high backing pressure with larger clusters. Since low-intensity focal volume contribute to lower charge states, it is expected that the low charge propensity is lower for experiments with Ar_{1400} clusters than at $Ar_{36,500}$ clusters. Experimental results are contrary to these expectations. (ii) Nanoplasma shielding can reduce the intensity experienced by atoms. A bigger cluster would have more atoms shielded by the field than with smaller clusters. So, if the plasma shielding is a dominant factor, then larger cluster should yield a lager fraction of lower charge states. Results presented in Fig. 2 are contrary to these expectations. (iii) Electron–ion recombination after the laser pulse is incident and can affect the charge state distribution. Extensive discussion on recombination in single isolated clusters [22] have concluded that due to the high electron temperature and large electrostatic fields on the clusters, charge reduction is single isolated

Table 1. Over the barrier ionisation threshold intensity for different charge states of Ar along with their ionisation energies

Ar^{q+} (q)	Ionisation energy (eV)	OBI threshold intensity (Wcm^{-2})
1	15.76	2.46×10^{14}
2	27.66	5.82×10^{14}
3	40.74	1.22×10^{15}
4	59.81	3.19×10^{15}
5	75.02	5.06×10^{15}
6	91.00	7.62×10^{15}
7	124.32	1.95×10^{16}
8	143.46	2.65×10^{16}
9	422.45	1.57×10^{18}
10	478.69	2.10×10^{18}
11	538.96	2.79×10^{18}
12	618.26	4.06×10^{18}
13	686.10	5.24×10^{18}
14	755.74	6.66×10^{18}
15	854.77	9.49×10^{18}
16	918.03	1.11×10^{19}
17	4120.88	3.99×10^{21}
18	4426.22	4.73×10^{21}

clusters in negligibly small. For the experimental conditions used in these experiments, inter-cluster recombination processes [23] and charge transfer reaction are avoided. So this feature may not affect the measurements.

Apart from the field ionisation, ionisation by electron collisions is considered to be a dominant feature in the nano-plasmas. In an inverse bremsstrahlung mechanism, electrons gain energy from the field and disperse the energy in electron–ion collisions. In collective mechanism, the dielectric properties of the plasma modify the electro magnetic field interaction. Here again, the imaginary part of the dielectric permitivity that contributes to the field attenuation and energy deposition in the plasma fields depends on the electron-ion collisions. So, it is anticipated that electron-ion collisions control the charge states, especially in the generation of the higher charge states that may not be possible with the field ionisation alone. In Fig. 2b, it is noticeable that the decrease in charge state propensity appears to have exponential dependence of the charge state. This decrease in the charge propensity could be reflective of the decrease in electron impact ionisation cross section. To examine this possibility, we use the well known Lotz formalism [24,25] to evaluate the electron impact ionisation cross section of Ar. Electron collision with Ar^{q+} with q varied from 8 to 14 is

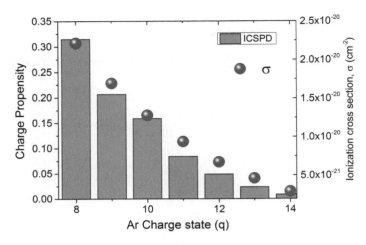

Fig. 3. ICSPD of $Ar_{36,500}$ clusters at relativistic intensity of 4×10^{18} Wcm^{-2} for charge states beyond 8+. Projected in the plot is also the electron impact ionisation cross section evaluated using the Lotz formalism and is given in the right-side axis of the plot

evaluated at an electron energy of 200 keV which represents the ponderomotive energy. At these high energies, electron impact ionization cross section reaches an asymptotic limit and do not vary significantly with the change in electron energy. Figure 3 shows the evaluated electron impact ionisation cross section projected with the ion charge state propensity for charge states beyond 8+. As can be seen the charge propensity very closely follows the decrease in electron impact ionisation cross section for the higher charge states. So is it possible that the higher ionisation state generation is controlled by the electron impact ionisation in the bigger clusters. For the smaller clusters, the electron ion collisions are expected to be less frequent, since the cluster size is smaller and therefore the high charge state propensity may be lower.

At relativistic intensities, pulse profile and pulse contrast are known to affect the plasma generation. In the present experiments, ns contrast is about 10^{-6} and the ps contrast can be as large as 10^{-5} at 10 ps and 10^{-4} at 2–4 ps [26]. This means an intensity of 10^{14} Wcm^{-2} in the 2–4 ps regime prior to the main pulse. Clusters experiencing this intensity would expand under Coulomb pressure. For a Ar_{1400} cluster experiments presented in Fig. 1, the maximum average ion energy is about 50 keV and it raises to about 200 keV for the $Ar_{36,500}$ clusters. Even if we assume that this average energy is achieved 5 ps ahead of the main pulse, a Ar_{1400} cluster would expand to about 2 μm. These atomic ions would still be in the focal volume and should be subjected to the OBI by the main pulse. The very fact there is a dramatic change is ICSPD with the cluster size would mean that the pre-pulse does not destroy the cluster before the main pulse is incident. Pre-pulse expansion can only expose atoms (to the main pulse and should result in field ionisation contributing to the higher charge states. So, the pulse contrast effects cannot explain the charge states distribution measured in the present experiments.

It is possible that the nanoplasma shielding is strong and OBI is not brought out in all the atoms in the cluster. The key question that opens is about the plasma shielding in a 2 nm cluster and how they change at relativistic intensities. With the different arguments presented with the experimental measurements, it is clear that comprehending the ion charge states formed by cluster ionization at relativistic intensities is not straightforward. Detailed calculations of the nanoplasma generation and evolution with the inclusion of relativistic effects is necessary to understand the experimental measurements. At present, there are no PIC models that can evaluate these ionisation dynamics in detail and we hope that the experiments offer inspiration to bring out the nanoplasma dynamics at relativistic intensities.

4   Summary and Conclusions

We present an experimental study of the Ar cluster ionisation at relativistic intensities of about $\sim 4 \times 10^{18}$ Wcm^{-2} by using a Thompson parabola spectrometer. Ion charge propensity distribution presented for different cluster sizes, Ar_{1400} and $Ar_{36,500}$, bring out very important differences in cluster ionisation with the cluster size. The measurements are not reflective of the dominance of over the barrier ionisation. Electron impacting ionisation seem to be important to generate the high charge states and charge propensity correlates with the electron impacting ionisation cross section. Larger clusters appear to be necessary for effective electron impact ionisation and generation of higher charge states.

References

1. Boyd, R.W.: Nonlinear Optics. Academic Press, San Diego (2003)
2. Ammosov, M.V., Delone, N.B., Krainov, V.P.: Sov. Phys. JETP **64**, 1191 (Engl. Transl.) (1986)
3. Protopapas, C.H.K.M., Knight, P.L.: Rep. Prog. Phys. **60**, 389 (1997)
4. Lein, M.: J. Phys. B: At. Mol. Opt. Phys. **40**(16), R135R173 (2007)
5. Lichters, R., Meyer-ter-Vehn, J., Pukhov, A.: Phys. Plasmas **3**, 3425 (1996)
6. Mourou, G.A., Tajima, T., Bulanov, S.V.: Rev. Mod. Phys. **78**, 309 (2006)
7. Gumbrell, E.T., et al.: New J. Phys. **10**, 123011 (2008)
8. Kruer, W.L.: The Physics of Laser Plasma Interactions. Westview Press (Reprint edition, 2001)
9. Gibbon, P.: Short Pulse Laser Interactions with Matter: An Introduction. Imperial College Press (2005)
10. Kruer, W.L., Estabrook, K.G.: Phys. Fluids **28**, 430 (1985)
11. Krainov, V.P., Smirnov, M.B.: Phys. Rep. **370**, 237 (2002)
12. Saalmann, U., Siedschlag, C., Rost, J.M.: J. Phys. B. **39**, R39 (2006)
13. Fennel, T., et al.: Rev. Mod. Phys. **82**, 1793 (2010)
14. Moll, M., Bornath, T., Schlanges, M., Krainov, V.P.: Phys. Plasmas **19**, 033303 (2012)

15. Gets, A.V., Krainov, V.P.: J. Phys. B: At. Mol. Opt. Phys. **39**, 1787–1795 (2006)
16. Jha, J.: Ph.D. thesis, Tata Institute of Fundamental Research (2006)
17. Rajeev, R.: Ph.D. thesis, Tata Institute of Fundamental Research (2012)
18. Kumarappan, V., Krishnamurthy, M., Mathur, D.: Phys. Rev. Lett. **87**, 085005 (2001)
19. Kumarappan, V., Krishnamurthy, M., Mathur, D.: Phys. Rev. A **66**, 033203 (2002)
20. Rajeev, R., et al.: Rev. Sci. Instrum. **82**, 083303 (2011)
21. Rajeev, R., Rishad, K.P.M., Madhu Trivikram, T., Narayanan, V., Brabec, T., Krishnamurthy, M.: Phys. Rev. A **85**, 023201 (2012)
22. Krainov, V.P., Sofronov, A.V.: JETP **103**, 35–38 (2006)
23. Rajeev, R., Madhu Trivikram, T., Rishad, K.P.M., Narayanan, V., Krishnakumar, E., Krishnamurthy, M.: Nat. Phys. **9**, 185 (2013)
24. Lotz, W.: Z. Phys. **216**, 241 (1968)
25. https://www-amdis.iaea.org/
26. Dalui, M., Kundu, M., Tata, S., Lad, A.D., Jha, J., Ray, K., Krishnamurthy, M.: AIP Adv. **7**, 095018 (2017)

Spectroscopic Studies of $^1\Sigma^+$ States of HfH^+ and PtH^+ Molecular Ions

Renu Bala[1(✉)], H. S. Nataraj[1], and Minori Abe[2]

[1] Department of Physics, Indian Institute of Technology Roorkee,
Roorkee 247667, India
rbala@ph.iitr.ac.in; hnrajfph@iitr.ac.in
[2] Department of Chemistry, Tokyo Metropolitan University, 1-1 Minami-Osawa,
Hachioji, Tokyo 192-0397, Japan
abeminoriabe@gmail.com

Abstract. We have studied the potential energy curves (PECs) for $^1\Sigma^+$ electronic states of HfH^+ and PtH^+ molecular ions using the self-consistent field (SCF) method and the coupled-cluster method with single and double excitations (CCSD). The spectroscopic constants derived from these potential energy curves are reported and compared with the available calculations in the literature. Further, the permanent dipole moment (PDM) of these molecular ions are calculated using the finite-field approach by applying a weak external electric field in the perturbative regime. Furthermore, vibrational parameters are obtained by solving the vibrational Schrödinger equation numerically.

1 Introduction

The polar diatomic molecules are of particular interest to experimentalists who are seeking for a non-zero electric dipole moment of an electron (eEDM). The latter plays an important role while exploring the physics beyond the standard model of particle physics. Apart from atoms, the current best limit on eEDM also comes from the cold molecular experiments on ThO [1], HfF^+ [2] and YbF [3]. The theoretical calculations of spectroscopic parameters and properties of such molecules containing heavy atoms is a challenge because of their complex electronic structure. As the eEDM hitherto has evaded detection, new systems are being proposed and explored constantly [4–6]. Two such molecular candidates: PtH^+, and HfH^+ have been proposed by Meyer et al. [7] in which they have obtained the PECs for $^{1,3}\Sigma$, $^{1,3}\Pi$ and $^{1,3}\Delta$ electronic states of these ions, using multi-reference configuration interaction method with single and double excitations (MRCISD), available in MOLPRO quantum chemistry package. Nevertheless, in their work, equilibrium bond lengths (R_e), and rotational constants (B_e) are reported only for $^3\Delta$ states. The spectroscopic constants such as, R_e, electronic transition energy (T_e), and harmonic frequency (ω_e) have been studied by Skripnikov et al. [8] for the four lowest electronic states

© Springer Nature Singapore Pte Ltd. 2019
P. C. Deshmukh et al. (eds.), *Quantum Collisions and Confinement of Atomic and Molecular Species, and Photons*, Springer Proceedings in Physics 230,
https://doi.org/10.1007/978-981-13-9969-5_18

($^3\Delta_3$, $^1\Sigma^+$, $^3\Delta_2$, and $^3\Delta_1$) of PtH$^+$ ion, using MRCISD method. Further, they have used 60-core electron generalized relativistic effective core potential (GRECP) for the Pt atom. The diatomic constants R_e, ω_e, T_e, and dissociation energy (D_e) for $^1\Sigma^+$ state at the non-relativistic (NR) level, for $^1\Sigma^+$ and $^3\Delta$ state using perturbation theory (PT), for $^1\Sigma^+$, $^3\Pi$, and $^3\Delta$ state using spin-free no-pair method corrected to second order in the external potential (PVP) have been reported for PtH$^+$ ion by Dyall [9]. Zurita et al. [10] have performed Hartree–Fock SCF calculations for the low-lying electronic states of PtH$^+$, with the relativistic ten-electron pseudopotential (PP10). Quite recently, several electronic states of PtH and PtH$^+$ have been computed by Shen et al. [11] using MRCI + Q method that includes the single and double excitations plus Davidson's cluster correction. Further, spin–orbit coupling (SOC) effects are included via the state-interaction (SI) approach.

Dyall [9] and Ohanessian et al. [12] have predicted the $^1\Sigma^+$ state as the ground state of PtH$^+$, whereas in the recent studies, [7,8,11] $^3\Delta_3$ state is reported as the true ground state of this ion. Further, two contradictory results for the ordering of first and second excited states of PtH$^+$ are available in the literature. According to [8,11], $^1\Sigma^+$ is lower in energy than $^3\Delta_2$ state, whereas an opposite trend is obtained in [7].

The first study on HfH$^+$ ion has been done by Ohanessian et al. [12] using generalized valence bond plus CISD method together with the effective core potentials. They have reported relative energies and spectroscopic parameters for the lowest four ($^3\Delta$, $^3\Pi$, $^3\Phi$, and $^3\Sigma^-$) electronic states. The Ω dependent PECs of $^1\Sigma$ and $^3\Delta$ state have been studied by Meyer et al. [7]. In their work, ground state of HfH$^+$ ion is a toss-up between $^1\Sigma$ and $^3\Delta_1$. Therefore, more detailed relativistic calculations are required to resolve this issue.

In the current work, we have performed fully relativistic potential energy calculations and the spectroscopic constants, *viz.*, equilibrium bond length (R_e), dissociation energy (D_e), harmonic frequency (ω_e), anharmonic constant ($\omega_e x_e$), and rotational constant (B_e and α_e) of $^1\Sigma^+$ states of HfH$^+$ and PtH$^+$ molecular ions, at SCF and CCSD level of correlation. The permanent dipole moment (PDM), (μ), at equilibrium internuclear distance, R_e, has been calculated using finite-field method for both the ions. Using PECs obtained at the level of CCSD, energies (E_v), rotational constants (B_v), and wavefunctions of vibrational states are obtained by solving the vibrational Schrödinger equation numerically.

In the next section, we briefly describe the computational method, while in Sect. 3, we show results and discussions for the properties calculated and in Sect. 4 we conclude.

2   Computational Method

The relativistic energy calculations for $^1\Sigma^+$ state of PtH$^+$ and HfH$^+$ molecular ions are performed using Dirac–Fock–Coulomb Hamiltonian with DIRAC15 program [13]. The contribution from two-electron (SS|SS) integrals is taken in an approximate manner, as suggested in [14]. The energy calculations at CCSD level of correlation are carried out with the RELCCSD module of DIRAC15 software. The uncontracted correlation consistent polarized valence quadruple zeta (cc-pVQZ) basis set for H [6s 3p 2d 1f] atom [15] and Dyall valence basis set of similar level, dyall.v4z, for Pt [78s 30p 19d 13f 4g 2h] and Hf [72s 30p 19d 13f 4g 2h] [16] are used in conjunction with Gaussian charge distribution and C_{2v} molecular point group symmetry. The nuclear masses of 1.007825, 179.9465457, and 194.964766 atomic mass unit (amu) are used for H, Hf and Pt, respectively. The filled orbitals having energy less than $-5E_h$ are taken as a frozen core. With this constraint, we have correlated 32 electrons for PtH$^+$ and 26 electrons for HfH$^+$ in CCSD calculations. Further, the virtual energy threshold is set at $5E_h$ for both the molecular cations that include 102 and 138 virtual orbitals for PtH$^+$ and HfH$^+$ ion, respectively. The potential energy calculations are performed for a bond distance ranging from, 1 to 12 Å, with a step size of 0.5 Å while around the equilibrium point, a finer step size of about 0.001 Å is adopted. Using VIBROT program available in MOLCAS [17] software package, the calculated potential energies are fitted to an analytical form of potential function using cubic splines. The dissociation energies are calculated as a difference between the energies at equilibrium point and those computed at asymptotic distance by numerical extrapolation. Further, the Numerov method is utilized to solve the ro-vibrational Schrödinger equation numerically to obtain the spectroscopic constants, with $J = 0$ rotational state. The energies of ro-vibrational states ($E_{v,J}$) can be expressed as [18]:

$$E_{v,J} = \omega_e \left(v + \frac{1}{2} \right) - \omega_e x_e \left(v + \frac{1}{2} \right)^2 + \cdots + B_v J(J+1) - \cdots \qquad (1)$$

where v and J are vibrational and rotational quantum numbers, respectively. The rotational constant, B_v, corresponding to vibrational state v can be defined as

$$B_v = B_e - \alpha_e \left(v + \frac{1}{2} \right) + \cdots \qquad (2)$$

where $B_e (= h/8\pi^2 c\mu R_e^2)$ is the rotational constant at the equilibrium point, and α_e is the vibration–rotation interaction constant. Here, c and $\mu [= m_1 m_2/(m_1 + m_2)]$ represents the velocity of light and reduced mass of the molecular system, respectively.

The total energy E of the system in the presence of external electric field of strength ϵ is given by [19]:

$$E(\epsilon) \approx E_0 - \mu\epsilon - \frac{1}{2}\alpha\epsilon^2 \qquad (3)$$

where μ and α represent the dipole moment and the static dipole polarizability, respectively. It is evident from (3) that the dipole moment can be calculated as the first derivative of total energy with respect to applied electric field. First, the field-dependent energies are calculated by adding different values of ϵ in the range; 1.03×10^6 and 2.57×10^6 V/cm, to the Hamiltonian using DIRAC15 software. Thereafter, the PDMs are obtained by fitting the first-order polynomial to these energies calculated at the equilibrium bond length. The range of perturbative electric field is chosen by comparing PDM results obtained using finite-field approach and those calculated directly using the expectation value approach in the DIRAC15 software, at the SCF level.

3   Results and Discussion

It has to be noted that the atomic units for distance (1 au = 0.52917721 Å), dipole moment (1 au = 2.54174691 D), and electric field (1 au = 5.142×10^9 V/cm) are used, unless otherwise mentioned, throughout the paper. Figure 1a, b show the potential energy curves for $^1\Sigma^+$ state of HfH$^+$ and PtH$^+$ molecular ions, respectively, at SCF and CCSD level of correlation. The large difference in the PECs at SCF and CCSD level of theory indicates that the correlation effects are large in both the molecular ions considered. The spectroscopic constants calculated from the PECs, together with the available results in the literature, are tabulated in Table 1. Our value of R_e (=2.868 au) for PtH$^+$ ion, reported at the level of CCSD, differs from other available calculations in the literature [8–12], by a maximum of 3.1%. The computed value of $\omega_e = 2385$ cm^{-1} for PtH$^+$, at CCSD level in the present study, agrees well with the value of 2399 cm^{-1}, reported in [12] at CISD level of correlation, whereas it differs from other calculations [8–11], by a maximum of 11%. The calculated dissociation energy, $D_e = 4.20$ eV, of PtH$^+$ cation at CCSD level of correlation is considerably larger than any of the available results [7,9–12]. The potential depth of HfH$^+$, $D_e = 2.49$ eV reported in [7], is larger by 0.06 eV than our result at CCSD level. The difference between potential depths reported in our work and those reported in the literature could be due to the relativistic effects, correlation method employed, size of the correlation space and the choice of basis sets. There is no other calculation for the

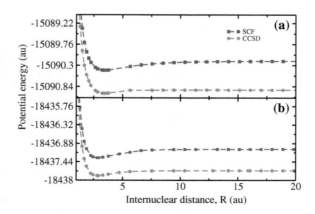

Fig. 1. Potential energy curves obtained at SCF and CCSD level of correlation for $^1\Sigma^+$ state of **a** HfH$^+$, and **b** PtH$^+$ molecular ions

Table 1. Computed spectroscopic constants and molecular properties of HfH$^+$ and PtH$^+$ molecular ions in their $^1\Sigma^+$ state compared with the available results in the literature

Molecule	Method	R_e (au)	D_e (eV)	ω_e (cm^{-1})	$\omega_e x_e$ (cm^{-1})	B_e (cm^{-1})	α_e (cm^{-1})	μ (au)	References
HfH$^+$	SCF	3.473	6.49	1799	19.8	4.97	0.09	1.515	This work
	CCSD	3.432	2.43	1909	18.0	5.11	0.07	1.218	This work
	MRCISD	–	2.49	–	–	–	–	–	[7]
PtH$^+$	SCF	2.864	7.26	2307	21.9	6.94	0.10	0.353	This work
	CCSD	2.868	4.20	2385	42.57	7.28	0.19	0.159	This work
	MRCISD	–	3.14	–	–	–	–	–	[7]
	MRCISD	2.825	–	2428	–	–	–	–	[8]
	NR	2.828	0.42	2464	–	–	–	–	[9]
	PT	2.836	0.89	2546	–	–	–	–	[9]
	PVP	2.830	1.34	2572	–	–	–	–	[9]
	DHF[a]	2.876	1.22	2426	–	–	–	–	[9]
	SCF	2.783	1.31	2650	–	–	–	–	[10]
	MRCISD + Q	2.887	3.03	2320	25.9	7.222	–	–	[11]
	CISD	2.870	2.875	2399	–	–	–	–	[12]

[a]DHF: Dirac–Hartree–Fock

spectroscopic constants, except that of D_e in [7], available in the literature for the $^1\Sigma^+$ electronic state of HfH$^+$ ion.

Those results, for the spectroscopic constants of the molecular ions considered in this work, available in the literature are calculated either at the non-relativistic level or at the scalar-relativistic level. Further, the highest correlation method used among all the available calculations is the MRCISD, which is not size extensive, unlike the CCSD method that we have employed in this work. Further, the size of the basis sets and the active space considered in our work are larger than that in any other published works on these molecular systems. We, therefore, believe that our relativistic calculations performed are more reliable than any other available result.

In order to calculate the contributions from relativistic effects, we have performed both non-relativistic and relativistic energy calculations near the equilibrium point at the SCF level, for both the diatomic molecules. The relativistic contributions for the total energies of HfH$^+$ and PtH$^+$ are found to be 5.3% and 6.3%, respectively. Further, the correlation contributions to the spectroscopic parameters of the molecular ions are calculated by comparing those results calculated at the SCF and the CCSD levels. They are found to be 0.041 au and 0.004 au for the equilibrium bond length R_e, -4.06 eV and -3.06 eV for the dissociation energy D_e, and 110 cm^{-1} and 78 cm^{-1} for the harmonic frequency ω_e, respectively, for HfH$^+$ and PtH$^+$.

The value of PDM, at the equilibrium point, is found to be 1.218 au and 0.159 au for HfH$^+$ and PtH$^+$, respectively, at the CCSD level of theory. The correlation contribution to the PDM is 19.6% and 54.9% for HfH$^+$ and PtH$^+$, respectively for the QZ basis set and at the CCSD level of theory. As there is no experimental

Table 2. Vibrational energies and rotational constants for low-lying vibrational levels of HfH$^+$ and PtH$^+$ at CCSD level of correlation

v	HfH$^+$		PtH$^+$	
	E_v (cm^{-1})	B_v (cm^{-1})	E_v (cm^{-1})	B_v (cm^{-1})
0	915.97	5.033	1170.49	7.183
1	2738.40	4.906	3443.80	6.973
2	4544.27	4.794	5638.28	6.774
3	6294.63	4.698	7753.57	6.580
4	7973.99	4.604	9793.36	6.386
5	9584.91	4.496	11757.69	6.191
6	11126.79	4.377	13646.17	5.993
7	12594.18	4.246	15457.33	5.791
8	13980.91	4.094	17188.82	5.586
9	15277.33	3.914	18837.53	5.374
10	16465.74	3.689	20399.65	5.155

or other theoretical results on the PDM of these ions in the literature, to the best of our knowledge, our result will serve as a benchmark. The calculated vibrational parameters such as the vibrational energies and the vibrationally coupled rotational constants are given in Table 2 for a few low-lying vibrational levels of $^1\Sigma^+$ electronic state of both ions. The relative energy separation between the adjacent vibrational levels and the value of rotational constant decreases, as anticipated, as one goes from lower to higher vibrational states.

4   Conclusions

In summary, theoretical investigation of the lowest $^1\Sigma^+$ state of HfH$^+$ and PtH$^+$ has been carried out using SCF and CCSD levels of correlation taking optimized quadruple zeta basis sets. Our results of the spectroscopic constants are in reasonable agreement with the existing results wherever available. The value of PDM for both the ions is reported for the first time. In addition, both the relativistic and the correlation effects have been calculated for both the ions. Further, the energies, rotational constants, and wavefunctions of different vibrational states are obtained by solving the vibrational Schrödinger equation and the results are reported here. Our results may serve as benchmarks for the future experimental and theoretical investigations of HfH$^+$ and PtH$^+$ molecular ions. Furthermore, the accurate PEC calculations of $^3\Delta$ states are required in order to resolve the issue of the true ground state of these ionic systems and such calculations are being performed.

Acknowledgements. The relevant calculations, whose results are presented in this work, were performed on the computing facility available in the Department of Physics, IIT Roorkee and in the Department of Chemistry, TMU, Japan.

References

1. Baron, J., Campbell, W.C., DeMille, D., Doyle, J.M., Gabrielse, G., Gurevich, Y.V., Hess, P.W., Hutzler, N.R., Kirilov, E., Kozyryev, I., O'Leary, B.R., Panda, C.D., Parsons, M.F., Spaun, B., Vutha, A.C., West, A.D., West, E.P., Collaboration, A.C.M.E.: Methods, analysis, and the treatment of systematic errors for the electron electric dipole moment search in thorium monoxide. New J. Phys. **19**, 073029 (2017). https://doi.org/10.1088/1367-2630/aa708e

2. Cairncross, W.B., Gresh, D.N., Grau, M., Cossel, K.C., Roussy, T.S., Ni, Y., Zhou, Y., Ye, J., Cornell, E.A.: Precision measurement of the electron's electric dipole moment using trapped molecular ions. Phys. Rev. Lett. **119**, 153001 (2017). https://doi.org/10.1103/PhysRevLett.119.153001

3. Hudson, J.J., Kara, D.M., Smallman, I.J., Sauer, B.E., Tarbutt, M.R., Hinds, E.A.: Improved measurement of the shape of the electron. Nature **473**, 493–496 (2011). https://doi.org/10.1038/nature10104

4. Shafer-Ray, Neil E.: Possibility of 0-g-factor paramagnetic molecules for measurement of the electron's electric dipole moment. Phys. Rev. A **73**, 034102 (2006). https://doi.org/10.1103/PhysRevA.73.034102

5. Meyer, E.R., Bohn, J.L.: Prospects for an electron electric-dipole moment search in metastable ThO and ThF$^+$. Phys. Rev. A **78**, 010502(R) (2008). https://doi.org/10.1103/PhysRevA.78.010502

6. Prasannaa, V.S., Vutha, A.C., Abe, M., Das, B.P.: Mercury monohalides: suitability for electron electric dipole moment searches. Phys. Rev. Lett. **114**, 183001 (2015). https://doi.org/10.1103/PhysRevLett.114.183001

7. Meyer, E.R., Bohn, J.L., Deskevich, M.P.: Candidate molecular ions for an electron electric dipole moment experiment. Phys. Rev. A **73**, 062108 (2006). https://doi.org/10.1103/PhysRevA.73.062108

8. Skripnikov, L.V., Petrov, A.N., Titov, A.V., Mosyagin, N.S.: Electron electric dipole moment: Relativistic correlation calculations of the P, T-violation effect in the $^3\Delta_3$ state of PtH$^+$. Phys. Rev. A **80**, 060501(R) (2009). https://doi.org/10.1103/PhysRevA.80.060501

9. Dyall, K.G.: Relativistic effects on the bonding and properties of the hydrides of platinum. J. Chem. Phys. **98**, 9678–9686 (1993). https://doi.org/10.1063/1.464346

10. Zurita, S., Rubio, J., Illas, F., Barthelat, J.C.: Ab initio electronic structure of PtH$^+$, PtH, Pt$_2$, and Pt$_2$H from a one-electron pseudopotential approach. J. Chem. Phys. **104**, 8500–8506 (1996). https://doi.org/10.1063/1.471600

11. Shen, K., Suo, B., Zou, W.: Theoretical study of low-lying Ω electronic states of PtH and PtH$^+$. J. Phys. Chem. A **121**, 3699–3707 (2017). https://doi.org/10.1021/acs.jpca.7b0305

12. Ohanessian, G., Brusich, M.J., Goddard, W.A.: Theoretical study of transition-metal hydrides. 5. HfH$^+$ through HgH$^+$, BaH$^+$, and LaH$^+$. J. Am. Chem. Soc. **112**, 7179–7189 (1990)

13. Bast, R., Saue, T., Visscher, L., Jensen, H.J.A.: In: Bakken, V., Dyall, K.G., Dubillard, S., Ekstroem, U., Eliav, E., Enevoldsen, T., Fasshauer, E., Fleig, T., Fossgaard, O., Gomes, A.S.P., Helgaker, T., Henriksson, J., Ilias, M., Jacob, Ch.R.,

Knecht, S., Komorovsky, S., Kullie, O., Laerdahl, J.K., Larsen, C.V., Lee, Y.S., Nataraj, H.S., Nayak, M.K., Norman, P., Olejniczak, G., Olsen, J., Park, Y.C., Pedersen, J.K., Pernpointner, M., Di Remigio, R., Ruud, K., Salek, P., Schimmelpfennig, B., Sikkema, J., Thorvaldsen, A.J., Thyssen, J., Stralen, J. van, Villaume, S., Visser, O., Winther, T., Yamamoto, S. (Contributors) DIRAC, a relativistic ab initio electronic structure program, Release DIRAC15 (2015). http://www.diracprogram.org

14. Visscher, L.: Approximate molecular relativistic Dirac-Coulomb calculations using a simple Coulombic correction. Theor. Chim. Acta. **98**, 68–70 (1997). https://doi.org/10.1007/s002140050280

15. Dunning, Jr., T.H.: Gaussian basis sets for use in correlated molecular calculations. I. The atoms boron through neon and hydrogen. J. Chem. Phys. **90**, 1007–1023 (1989). https://doi.org/10.1063/1.456153

16. Dyall, K.G.: Relativistic double-zeta, triple-zeta, and quadruple-zeta basis sets for the 5d elements Hf−Hg. Theor. Chem. Acc. **112**, 403–409 (2004). https://doi.org/10.1007/s00214-004-0607-y

17. Karlström, G., Lindh, R., Malmqvist, P.-Å., Roos, B.O., Ryde, U., Veryazov, V., Widmark, P.-O., Cossi, M., Schimmelpfennig, B., Neogrády, P., Seijo, L.: MOLCAS: a program package for computational chemistry. Comput. Mat. Sci. **28**, 222–239 (2003). https://doi.org/10.1016/S0927-0256(03)00109-5

18. Banwell, C.N.: Fundamentals of Molecular Spectroscopy, 3rd edn. Tata McGraw-Hill Publishing Company Limited

19. Maroulis, G.: Hyperpolarizability of H_2O revisited: accurate estimate of the basis set limit and the size of electron correlation effects. Chem. Phys. Lett. **289**, 403–411 (1998). https://doi.org/10.1016/S0009-2614(98)00439-4

Elemental Constitution Detection of Environmental Samples of Delhi Using XRF Spectroscopy

Ruchika Gupta[1,2(✉)], Kajol Chakraborty[1,2], Ch. Vikar Ahmad[1,2], Chirashree Ghosh[3], and Punita Verma[1(✉)]

[1] Department of Physics, Kalindi College, University of Delhi, East Patel Nagar, New Delhi 110008, India
{ruchikagupta.du, drpunitaverma.nature}@gmail.com
[2] Department of Physics and Astrophysics, University of Delhi, New Delhi 110007, India
[3] Department of Environmental Studies, University of Delhi, New Delhi 110007, India

Abstract. Metal toxicity has proven to be a major threat to human life as there are several health risks associated with it. Metal bio-magnification within living systems is a persisting problem. Due to anthropogenic activities, the concentration of various metals in environmental samples (plants, soil, air, water, etc.) is increasing at an alarming rate thus leading to their contamination. Investigations on determining the contamination level of environmental samples have been going on for a substantial amount of time in India as well as all over the world. Electrochemical, chemical or spectroscopic techniques have been popular for the abovementioned purpose. These methods although being easier and compatible, lag behind in sensitivity and accuracy. Thus, there is a need for a better technique which can detect contaminants in trace and ultra-trace amounts and also has multi-element detection capacity. Present work is an effort to identify contamination in different soil and plant samples collected from diverse land use sites in Delhi using X-Ray Fluorescence (XRF) spectroscopy. XRF offers the advantages of higher sensitivity (detection in the range of ppm), multi-element detection capability and easier sample preparation over other techniques. The samples were collected from two sites having different biological toxicity. Concentrations of the elements detected in the samples have been estimated from the X-ray spectra of the samples.

1 Introduction

In densely populated cities like Delhi, industrial as well as domestic waste has inadvertently become a source of toxic elements which has dramatic effects on the ecosystem. There is a continuous discharge of wastes and effluents to the environment, i.e., in soil, water, and air. Most of the waste materials in the vicinity of the industrial area are dumped without any prior treatment which then directly pollutes soil, water systems as well as air. All these wastes from industries and other household sources contain high amounts of heavy elements such as As, Cd, Cr, Cu, Fe, Hg, Mn, Ni, Pb,

© Springer Nature Singapore Pte Ltd. 2019
P. C. Deshmukh et al. (eds.), *Quantum Collisions and Confinement of Atomic and Molecular Species, and Photons*, Springer Proceedings in Physics 230,
https://doi.org/10.1007/978-981-13-9969-5_19

and Zn [1, 2]. Contamination of soil and plants by the significant amount of metals constitutes biological toxicity. This toxicity has proven to be a major threat and there are several health risks associated with it. The repercussions become more severe with the reduction in dimension of contaminating particles. As the particle size decreases, its ability to seep through soil and vegetation increases. Consumption of food grown on such soil by humans or animals paves way for the pollutants to enter the body and cause diseases like cancer [3]. A large number of investigations all over the world have discussed the effects of pollution on human health [4–6]. Consequently, there is a need to determine and evaluate the contamination of environmental samples and make efforts to develop a methodology which could reduce or better overcome this environmental deterioration.

Investigations on determining the contamination level of environmental samples have been going on for a substantial amount of time in India as well as all over the world. Electrochemical, chemical, or spectroscopic techniques have been popular for the abovementioned detection purpose. These methods, although being easier and compatible, lack sensitivity for multi-element detection and accuracy. Thus, there is a need for a better technique which can detect contaminants in trace and ultra-trace amounts. Present work is an effort to identify contamination in different soil and plant samples collected from diverse land use sites in Delhi using XRF spectroscopy. X-Ray Fluorescence (XRF) spectroscopy, specifically a versatile method for composition analysis, is nondestructive and has high sensitivity with multi-element detection capability. This method has high sensitivity in detecting elements with the atomic number in the range of $Z > 18$ in trace and ultra-trace amounts in both thick as well as in thin samples.

The underlying principle in X-Ray Fluorescence (XRF) procedure is based on measuring the energy and intensity of the characteristic X-rays emitted from the sample under investigation. In this technique, the sample is irradiated by a collimated beam of photons, thus creating vacancies in the atomic shells of elements present in the sample. These vacancies are subsequently occupied by the transition of higher shell electrons leading to the emission of X-rays characteristic to the elemental composition of the sample. The emitted X-rays are then detected by a suitable X-ray detector, visible as a spectrum in the computer after appropriate data acquisition system and software analysis. The intensity measurement of the X-rays gives a concentration of the individual elements present in the sample imparting quantitative estimation of the elemental composition of the sample. Previous investigations have used X-ray radioactive sources such as ^{55}Fe, X-ray tubes or synchrotron radiation to investigate various samples using K, L, and M shell/subshell X-ray emission to detect trace and ultra-trace elements like chromium, arsenic, selenium, silver, mercury, lead, etc., in all types of environmental samples. This technique has been extensively used by environmentalists as well as physicists to determine impurities in a sample [3].

1.1 Previous Investigations on Determination of Elemental Constitution of Environmental Samples Using XRF

The implementation of XRF as a pollutant determinant tool in environmental samples has gained momentum due to its various merits. One of them is the absence of exhaustive procedures involved in sample preparation, analysis, storage, etc., as in the case of other established electrochemical techniques like Atomic Absorption Spectroscopy (AAS). From the literature, it was evident that environmental samples including different types of soils, various medicinal plants, water samples as well as aerosols have been analyzed using XRF to determine their elemental constitution and quantify the pollutants present in them. Mittal et al. [7] in 1993 determined the calcium and potassium concentrations in roots and leaves of spinach and radish using XRF taking care of inter-element effects of calcium and potassium. The enhancement of potassium K X-rays due to the presence of calcium K X-rays in the sample was avoided by selective excitation of the sample. Deep et al. [8] studied the potassium variation in (*Vigna radiate*) sprouts with lunar phases following the method of selective excitation in thick samples. Bandhu et al. [9] analyzed the pollutant concentration in aerosol samples of Chandigarh, India. Similarly, Alrakabi et al. [10] and Singh et al. [11] reported the metal contamination found in groundwater samples of Chandigarh. Deep et al. [12] analyzed different water samples using XRF. The results indicated the presence of Ca in water in the range 0.1–100 ppm which attributes to its hardness. In Spain, Queralt et al. [13] investigated the presence of macro, micro, and nonessential elements in medicinal plants. Antoaneta et al. [14] reported the decrease of heavy metal concentration in soil with increasing distance from metallurgical work sites in Romania. Cruz et al. [15] from Barcelona and Maiana et al. [16] from Argentina examined aerosol samples under different conditions. McCumber et al. [17] from USA measured concentration of lead in soil.

It is inferred that XRF finds wide ranging applications in detecting pollutants in various environmental samples which include aerosols, plants, soil, and water. The multi-element detection capacity of XRF technique is evident from [9] where sixteen elements were detected in aerosol samples. XRF is an efficient technique to detect elements at ultra-trace levels which is clear from the fact that calcium was found in the range of 0.1–100 ppm in water samples as reported by Deep et al. [12] and lead was detected in a range of 10–170 ppm in soil samples as reported by McCumber et al. [17]. Thus, in the present work, XRF spectroscopy has been employed for analysis of environmental plant and soil samples as detailed here.

2   Methodology and Experiment

The methodology consisted of three parts, viz., identification of sites and collection of samples, turning powdered samples into pellets, and finally irradiation of pellets and data acquisition.

2.1   Sample Collection

Air pollution shows a seasonal variation. Moreover, during winter in Delhi, on average air pollution rises by more than twice the amount in comparison to other seasons (Spring and Summer). Among all pollutants, Particulate matter (PM), PM_{10} and $PM_{2.5}$, are of greater concern for public health because it is a carrier of all airborne metal elements. According to Central Pollution Control Board (CPCB) portal, on an average day of a January month (last winter phase), the concentration of $PM_{2.5}$ exceeded 200 mg/m^2, in the range of "Very Poor" rating of AQI (Air Quality Index) value compared to other seasons. Macro and microclimatic reasons for rising pollution levels in Delhi during winter month include biomass burning from inside and outside Delhi or brick kiln's operation in Delhi as major causes of pollution. These seasonal increments collude with meteorological factors, especially during winter months. Natural "**Thermal Inversion Process**" (cool air stagnates over the city keeping pollution close to the ground) creates a persistent winter fog which further worsens the problem.

For the abovementioned reasons, plant leaves and soil samples were collected during peak winter month (January) in Delhi from two sites; one a heavy traffic junction and crowded market place referred here as Site 1 and another, a ridge area with greenery and less traffic referred here as Site 2. These two sites were chosen to obtain a comparison between metal concentrations in environmental samples obtained from both the sites. The collected plant and soil samples from two sites in the month of January 2017 are summarized in Table 1.

2.2   Powder to Pellet/Target Preparation

The collected plant samples were first washed to remove surface impurities and then oven-dried thoroughly to evaporate all moisture content. The dried leaves were ground and sieved through a sieve mesh with an aperture width of 53 μm yielding a very fine powder. The soil samples from both sites were also grinded and then sieved to obtain a fine powder. These fine powdered samples were subsequently converted to thick pellets [7] using a pellet making die setup along with a 15 ton hydraulic press.

2.3   Irradiation of Pellets and Data Acquisition

The experiment was performed at XRF Lab, Punjabi University, Patiala, India. Figure 1 is a schematic representation of the geometrical arrangement experimental set-up. The experimental setup is described elsewhere in detail [18]. The sample pellets were irradiated with X-ray photons from a 50 kV X-ray tube equipped with Rh anode. An Amptek Si-PIN X-ray detector having a resolution of 145 eV at 5.9 keV was used to detect the X-rays emitted from the sample pellets.

Table 1. Sample information

Sample	Species	Sample scientific name	Sites	
			Site 1 Heavy traffic junction	Site 2 Ridge area
Plant	Banyan	*Ficus benghalensis*	✓	✓
	Ashok	*Polyalthia longifolia*	✓	✓
	Putranjiva	*Putranjiva roxburghii*	✓	–
	Mulberry	*Morus alba*	✓	–
Soil type	Sandy loam to clay loam		✓	✓

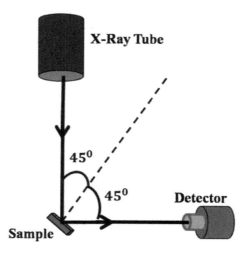

Fig. 1. 90° single reflection geometrical arrangement of 50 kV X-ray tube with Rh anode, sample S and an Amptek Si-PIN X-ray detector. Distance between X-ray tube window and sample = 3.5 cm; distance between sample and X-ray detector = 0.5 cm. Sample diameter = 2.5 cm [18]

2.4   Analysis

For irradiation, the tube was operated at a constant voltage and filament current i.e. 17 kV and 0.01 mA, respectively. The peaks in the recorded sample spectrum were identified with the help of X-ray energy table [19]. The energy calibration of the detector was done by irradiating pellets of different elements and taking their Kα peak as reference. The fractional concentrations were estimated for all the identified elements according to the following formula:

$$Conc.\ of\ X\ in\ sample = \left[\frac{Area\ under\ peak\ of\ X\ in\ sample}{Area\ under\ peak\ of\ X\ in\ standard}\right] Conc.\ of\ X\ in\ standard$$

(1)

where X = Element detected whose concentration has to be calculated (in mg/g) and standard implies any known compound of element X (for example CaO can be used as a standard for measuring the concentration of Ca in the sample). The sample and standard were irradiated under same tube voltage and filament current. The experimental setup (tube–sample–detector geometry) was also the same for both the irradiations. The flux variation of the tube while irradiating sample and standard was taken into account by normalizing both spectra with respect to the area of Rh Lα peak. Both the irradiations were done within a time interval of 30 min to minimize drift in tube and detector conditions as well as the electronic setup.

Since the experimental setup was inside a glass chamber, a background spectrum was recorded by replacing the sample with borax and subtracted from each recorded sample spectrum as shown Fig. 2.

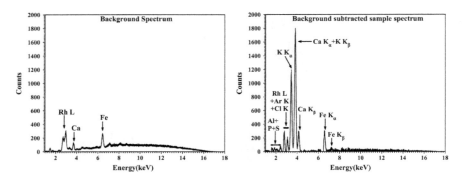

Fig. 2. Left: background spectrum recorded with the borax target. Right: sample X-ray spectrum with subtracted background. Both spectra have been recorded for 1000 s

The characteristic peaks in the spectrum indicate the elemental composition of the sample. As mentioned in Sect. 2.1 plant and soil samples from two sites having different toxicities were collected which are summarized in Table 1. The background subtracted spectra of different samples are shown in the figures below. Figure 3 shows X-ray spectra of plant samples from Site 1 in January 2017. Figure 4 shows X-ray spectra of plant samples collected from Site 2 in January 2017.

Figure 5 shows the background-subtracted X-ray spectra of soil samples from site 1 and site 2.

Standards are basically compounds of the elements detected in the environmental samples. Thick pellets of these compounds were also irradiated under the same tube conditions to estimate the elemental concentration in the sample by using equation (1).

3   Results and Discussions

In plant samples, K, Ca, and Fe were detected whereas, in case of soil samples apart from these, Ti and Mn were also found. In the case of *Putranjiva roxburghii*, a small amount of Mn was detected. Table 2 and Table 3 show the calculated concentrations of different elements detected in plant and soil samples, using equation 1, respectively.

The minimum detection limit of the experimental setup was also determined for different elements by exciting known samples at their edge energies as shown in Table 4.

It has been observed that concentrations of K, Ca in plants and K, Ca, Ti and Fe in soils are much higher than their MDL's while Fe in plants and Mn in soils are just indications with estimated amounts less than their MDLs. On average, it was found that the X-ray tube setup had a minimum detection limit of around 1 mg/g for high Z elements. The qualitative investigation, peak intensities, of plant and soil samples

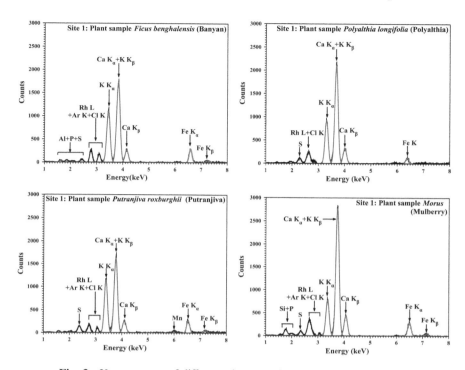

Fig. 3. X-ray spectra of different plant samples collected in from site 1

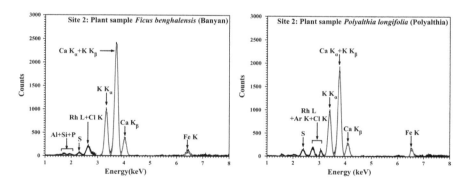

Fig. 4. X-ray spectra of plant samples collected from site 2

showed that the present XRF setup can efficiently detect middle Z (atomic number) elements, i.e., from K to Fe as can be seen from Figs. 6 and 7.

Preliminary calculations showed the presence of Ca in substantial amount in plant and soil samples of both sites. In the case of soil samples, high amounts of Fe were also detected as shown in Fig. 7. Future investigations would comprise the determination of

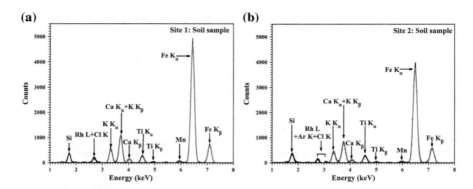

Fig. 5. X-ray spectra of soil samples from **a** site 1 and **b** site 2

Table 2. Fractional concentrations of various elements found in plant samples collected from site 1 and site 2

Plant sample	Element	Concentration (mg/g)	
		Site 1	Site 2
Ficus benghalensis	K	39.0	31.6
	Ca	60.0	79.4
	Fe	02.0	01.0
Polyalthia longifolia	K	30.7	31.3
	Ca	70.3	59.3
	Fe	00.9	01.1
Putranjiva roxburghii	K	37.6	–
	Ca	55.5	–
	Mn	00.7	–
	Fe	02.2	–
Morus alba	K	26.9	–
	Ca	93.1	–
	Fe	03.5	–

Table 3. Fractional concentrations of various elements found in soil

Element	Concentration (mg/g)	
	Site 1	Site 2
K	36.1	31.7
Ca	71.7	60.4
Ti	09.9	11.0
Mn	02.1	01.7
Fe	82.2	74.3

Table 4. Minimum detection limit (MDL) of the experimental setup for different elements found in the samples

Element	MDL (mg/g)
Ca	3.4
Ti	3.8
Fe	2.6

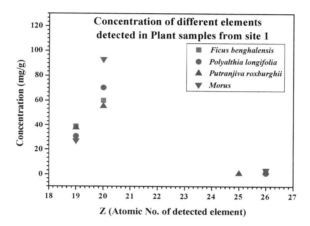

Fig. 6. Concentration of different elements detected in plant samples from site 1

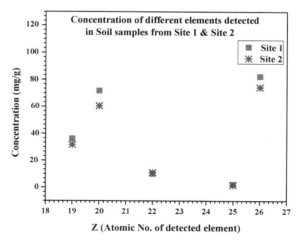

Fig. 7. Concentration of different elements detected in soil samples from site 1 and site 2

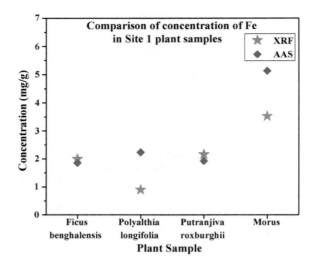

Fig. 8. Comparison between XRF and AAS concentrations of Fe in plant samples of site 1

Ca and Fe levels in soil samples, as vegetation is grown on these soils, and study their effect on the human body as well as agriculture.

The plant samples of Site 1 were also analyzed using Atomic Absorption Spectroscopy (AAS) method. A comparison was drawn between the AAS and XRF calculated concentrations of Fe in plant samples taken from Site 1 as shown in Fig. 8.

From Fig. 8, it can be seen that in case of *Polyalthia longifolia* and *Morus alba*, AAS shows higher concentrations of Fe whereas both XRF and AAS calculated concentrations are nearly same for *Ficus benghalensis* and *Putranjiva roxburghii*. This anomaly will be studied further in detail to isolate the possible reasons for such deviation, in the future.

4   Conclusion

Various plant and soil samples from two different regions of Delhi were analyzed for their elemental constitution using XRF spectroscopy. Details of the preliminary measurements performed to quantify the efficiency of the setup have been presented. Investigations will be continued in the future to assess metal contamination level in environmental samples on a large scale basis to identify and mark randomly highly metal accumulated vegetation species and soil types. After knowing the accurate value of spatio-temporal contamination level of samples, various suitable control methodologies can be devised to stop further toxicity followed by implementation of better scientific ways to reduce the level of pollution. This will ultimately lead to sustainable management of environment.

Acknowledgements. Professor Raj Mittal, Nuclear Science Laboratories, Physics Department, Punjabi University, Patiala, India is gratefully acknowledged for her constant guidance. We are also grateful to her for making the XRF facility of Nuclear Science Laboratories, Physics Department available for conducting the experiment. The authors also acknowledge the help and support of Ms. Preeti, Research Scholar, Nuclear Science Laboratories, Physics Department in performing the experiment successfully. One of the authors R. G. thanks IUAC, New Delhi, UFUP project 58324 for the scholarship, K. C. and Ch. V. A. acknowledge University of Delhi for the Ph.D. scholarship.

References

1. Arora, B.R.: Pollution potential of municipal waste water in Punjab. Indian J. Agric. Sci. **45**, 80–85 (1985)
2. Su, C.: A review on heavy metal contamination in the soil worldwide: situation, impact and remediation techniques. Environ. Skept. Crit. **3**(2), 24 (2014)
3. Verma, P., Ahmad, Ch.V., Gupta, R., Chakraborty, K., Sharma, T., Gupta, A., Gupta, S.: X-Ray fluorescence and detection of contamination in environmental samples: trace and ultra-trace levels. Yrly. Acad. J. 49–57 (2017). Kalindi College, University of Delhi. ISSN: 2348-9014
4. Beckett, K.P., Freer-Smith, P.H., Taylor, G.: Urban woodlands: their role in reducing the effects of particulate pollution. Environ. Pollut. **99**(3), 347–360 (1998)
5. Kumar, A., Scott Clark, C.: Lead loadings in household dust in Delhi, India. Indoor Air **19** (5), 414–420 (2009)
6. Rizwan, S.A., Nongkynrih, B., Gupta, S.K.: Air pollution in Delhi: its magnitude and effects on health. Indian J. Commun. Med. **38**(1), 4 (2013). Official publication of Indian Association of Preventive & Social Medicine
7. Mittal, R., Allawadhi, K.L., Sood, B.S., Singh, N., Kumar, A., Kumar, P.: Determination of potassium and calcium in vegetables by X-Ray fluorescence spectrometry. X-Ray Spectrom. **22**, 413–417 (1993)
8. Deep, K., Mittal, R.: Macronutrient K variation in mung bean sprouts with lunar phases. Eur. Sci. J. **10**(9), 295–306 (2014)
9. Bandhu, H.K., Puri, S., Garg, M.L., Singh, B., Shahi, J.S., Mehta, D., Swietlicki, E.S., Dhawan, D.K., Mangal, P.C.: Elemental composition and sources of air pollution in the city of Chandigarh, India, using EDXRF and PIXE techniques. Nucl. Instrum. Methods Phys. Res. Sect. B **160**(1), 126–138 (2000)
10. Alrakabi, M., Singh, G., Bhalla, A., Kumar, S., Kumar, S., Srivastava, A., Rai, B., Singh, N., Shahi, J.S., Mehta, D.: Study of uranium contamination of ground water in Punjab state in India using X-Ray fluorescence technique. J. Radioanal. Nucl. Chem. **294**(2), 221–227 (2012)
11. Singh, G., Bhalla, A., Kumar, S., Alrakabi, M., Srivastava, A., Rai, B., Singh, N., Shahi, J. S., Mehta, D.: Reply to query related to Study of uranium contamination of ground water in Punjab state in India using X-Ray fluorescence technique. J. Radioanal. Nucl. Chem. **298**(1), 731–733 (2013)
12. Deep, K., Mittal, R.: Calcium hardness analysis of water samples using EDXRF technique. J. Nucl. Phys. Mater. Sci. Radiat. Appl. **2**(1), 105–113 (2014). Chitkara University
13. Queralt, I., Ovejero, M., Carvalho, M.L., Marques, A.F., Llabrés, J.M.: Quantitative determination of essential and trace element content of medicinal plants and their infusions by XRF and ICP techniques. X-Ray Spectrom. **34**(3), 213–217 (2005)

14. Ene, A., Bosneagă, A., Georgescu, L.: Determination of heavy metals in soils using XRF technique. Rom. J. Phys. **55**(7–8), 815–820 (2010)
15. Minguillón, M.C., Cirach, M., Hoek, G., Brunekreef, B., Tsai, M., Hoogh, K. de., Jedynska, A., Kooter, I.M., Nieuwenhuijsen, M., Querol, X.: Spatial variability of trace elements and sources for improved exposure assessment in Barcelona. Atmos. Environ. **89**, 268–281 (2014)
16. Achad, M., López, M.L., Ceppi, S., Palancar, G.G., Tirao, G., Toselli, B.M.: Assessment of fine and sub-micrometer aerosols at an urban environment of Argentina. Atmos. Environ. **92**, 522–532 (2014)
17. McCumber, A., Strevett, K.A.: A geospatial analysis of soil leads concentrations around regional Oklahoma airports. Chemosphere **167**, 62–70 (2017)
18. Gupta, S., Deep, K., Jain, L., Ansari, M.A., Mittal, V.K., Mittal, R.: X-Ray fluorescence (XRF) set-up with a low power X-Ray tube. Appl. Radiat. Isot. **68**(10), 1922–1927 (2010)
19. Bearden, J.A.: X-Ray wavelengths. Rev. Modern Phys. **39**(1), 78 (1967)

Quantum Hall States for $\alpha = 1/3$ in Optical Lattices

Rukmani Bai[1,2], Soumik Bandyopadhyay[1,2], Sukla Pal[1], K. Suthar[1], and D. Angom[1(✉)]

[1] Physical Research Laboratory, Ahmedabad 380009, Gujarat, India
angom@prl.res.in
[2] Indian Institute of Technology Gandhinagar, Gandhinagar 382355, Gujarat, India

Abstract. We examine the quantum Hall (QH) states of the optical lattices with square geometry using Bose–Hubbard model (BHM) in presence of artificial gauge field. In particular, we focus on the QH states for the flux value of $\alpha = 1/3$. For this, we use cluster Gutzwiller mean field (CGMF) theory with cluster sizes of 3×2 and 3×3. We obtain QH states at fillings $\nu = 1/2, 1, 3/2, 2, 5/2$ with the cluster size 3×2 and $\nu = 1/3, 2/3, 1, 4/3, 5/3, 2, 7/3, 8/3$ with 3×3 cluster. Our results show that the geometry of the QH states is sensitive to the cluster sizes. For all the values of ν, the competing superfluid (SF) state is the ground state and QH state is the metastable state.

1 Introduction

Ultracold Bosons in optical lattices (OLs) [6] have been the subject of intense research since the experimental realization of Bose– Einstein condensate (BEC) in OLs [3,15,16,26]. The rapid developments in the control of BECs in OLs have made these systems an elegant experimental tool to explore the foundations of strongly correlated quantum many-body systems. Till recently, many of these were limited to the realm of theoretical studies. The near ideal, defect-free, experimental realizations of OLs make these excellent proxies to explore quantum many-body effects in condensed matter systems. One remarkable recent experimental development is the introduction of synthetic magnetic fields in OLs [1,30]. This makes the physics of quantum Hall (QH) effect accessible to OLs. An important outcome is the study of Harper–Hofstadter model [18,20] and observation of fractal spectrum [8] for interacting Bosons in OL with synthetic magnetic field [23]. The scope to investigate the interplay of lattice geometry, synthetic magnetic fields, and strong interactions have made these systems an excellent platform to explore exotic quantum many-body phases.

In the quantum description, the energies of electrons in an external magnetic field are quantized into Landau levels. These have large degeneracy and in a lattice these correspond to the Bloch bands [18] and is sensitive to applied

P. C. Deshmukh et al. (eds.), *Quantum Collisions and Confinement of Atomic and Molecular Species, and Photons*, Springer Proceedings in Physics 230,
https://doi.org/10.1007/978-981-13-9969-5_20

magnetic field. The key point is the geometrical phase an electron acquires when completing a loop in the cyclotron motion. Thus, neutral atoms in OLs can mimic the physics of electrons in magnetic fields if a geometrical phase can be induced to the atoms. This is achieved through the generation of a synthetic magnetic field in OLs through an artificial gauge potential [7,24,27,28] using lasers. Then, an atom hopping around a single unit cell in the OL, also called a plaquette, acquires Peierls' phase [36] of $\Phi = 2\pi\alpha$. Where α is the flux quanta per plaquette and it is related to the strength of the synthetic magnetic field. In the condensed matter systems realizing high α require magnetic fields $\approx 10^3$ Tesla. In this respect, the OLs have the advantage that by suitable choice of external as well as internal parameters, various topological states such as fractional QH (FQH) states are obtainable with the current experimental realizations [34,39] and a variety of FQH-like states can be expected to emerge from these systems.

A paradigmatic model which describes BECs in OLs is the Bose–Hubbard model (BHM) [11,22]. In this model, the kinetic energy of the bosons competes with the onsite interaction and drives a quantum phase transition (QPT) from superfluid (SF) to bosonic Mott insulator (MI) phase [16,40]. Various theoretical methods, such as mean field theory [11], strong coupling expansion [12,13,32,43], quantum Monte Carlo [44], density matrix renormalization group [37] have been used to study the role of quantum fluctuation and short range on site interaction on QPT. The SF phase is compressible with finite SF order parameter and phase coherent; MI phase, on the other hand, is incompressible with zero-order parameter and shows integer commensurate filling per lattice site. In contrast to these two phases, the QH states are incompressible states with zero-order parameter and have incommensurate filling. Several previous works [4,10,14,17,21,25,31,34,35,39,42] have theoretically explored the existence of FQH states in OLs using BHM with synthetic magnetic fields, the bosonic counterpart of the Harper–Hofstadter model [18,20]. These theoretical works have also examined the possible signatures of the FQH states. One of the possibilities is the measurement of two-point correlation function in the bulk and in the edge of the lattice [19]. Such measurements may be experimentally possible using the concepts from quantum information theory [9,41].

In the present work, we use cluster Gutzwiller mean field (CGMF) theory to provide a better description of the atom–atom correlations. It is proven to be more accurate than single site Gutzwiller theory and previous works have also used CGMF theory to study both the integer quantum Hall (IQH) and FQH states. In [31] the appearance of incompressible QH ground state in the hardcore limit with stripe order for $\alpha = 1/5$ and $\nu = 1/2$ is reported. Similarly, using reciprocal cluster mean field (RCMF) analysis, a competing FQH state is reported for $\alpha = 1/4$ in the recent study by Hügel et al. [21]. Motivated by the above observation of QH states, we explore the possible QH states for $\alpha = 1/3$ with different cluster size in the hardcore limit and demonstrate the dependence of the QH state geometry on the cluster size. For the present studies, we have considered 3×2 and 3×3 clusters and report the improvement in the description of the QH states with the increase of cluster sizes. This stems from more accurate accounting of the correlation effect with larger cluster sizes. We have also performed a comparative study between the obtained QH and SF states with both the cluster sizes and detect the emergence of various patterns.

2   Theory

We study a system of spinless bosonic atoms at $T = 0\,K$, confined in a two-dimensional (2D) optical lattice of square geometry under the influence of artificial gauge field [1,2,23,30]. In the Landau gauge $\mathbf{A} = (A_x, 0, 0)$ with $A_x = 2\pi\alpha q$, the system is described by the following Hamiltonian [22,23,34,35,39], BHM Hamiltonian where Peierls substitution is incorporated in the nearest neighbor (NN) hopping [18,20,36],

$$\hat{H} = -\sum_{\langle jk \rangle} \left(e^{i2\pi\alpha q} J_x + J_y\right) \hat{b}_j^\dagger \hat{b}_k + \sum_j \hat{n}_j \left[\frac{U}{2}(\hat{n}_j - 1) - \mu\right], \tag{1}$$

$j \equiv (p, q)$ corresponds to the lattice site index where q is the index of the lattice site along y-axis, $\hat{b}_j$ ($\hat{b}_j^\dagger$) are the bosonic annihilation (creation) operators, $\hat{n}_j$ is the occupation number operator at jth lattice site, J_x (J_y) are the hopping strengths between two neighboring sites along x (y) direction, U corresponds to the onsite interaction and μ is the chemical potential. Based on the experimental realizations, we consider isotropic hopping $J_x = J_y = J$, and repulsive onsite interaction energy ($U > 0$). In the presence of synthetic magnetic field, the atoms acquire a phase $2\pi\alpha$ upon hopping around a plaquette, where, α is the number of flux quanta per plaquette, and it has values $0 \leq \alpha \leq 1/2$. In the absence of synthetic magnetic field ($\alpha = 0$), the Hamiltonian (1) reduces to the familiar BHM Hamiltonian which admits two possible phases—MI and SF [11,16,22]. The MI phase appears in the strongly interacting regime ($J/U \ll 1$), while SF phase occurs in the limit ($J/U \gg 1$). In homogeneous system, where the OL does not include any background potential, the phase-boundary between MI and SF forms lobes of different fillings and in presence of magnetic field the MI-lobes are enhanced [33]. We employ the single site Gutzwiller mean field method (SGMF) and CGMF to analyse the system in presence of synthetic magnetic field and obtain the QH states.

2.1   Mean Field Theory and Gutzwiller Approximation

Following the mean field [38] calculations of the BHM, we decompose the creation and annihilation operators in (1) into mean field and fluctuation around the mean field, that is, $\hat{b}_j = \phi_j + \delta\hat{b}_j$, with $\phi_j = \langle \hat{b}_j \rangle$, and similarly, $\hat{b}_j^\dagger = \phi_j^* + \delta\hat{b}_j^\dagger$, with $\phi_j^* = \langle \hat{b}_j^\dagger \rangle$. Neglecting the term quadratic in the fluctuations, the mean field Hamiltonian is

$$\hat{H}^{\mathrm{MF}} = -\sum_{\langle jk \rangle} \left[\left(J_x e^{i2\pi\alpha q} + J_y\right)\left(\hat{b}_j^\dagger \phi_k + \phi_j^* \hat{b}_k - \phi_j^* \phi_k\right) + \mathrm{H.c.}\right]$$
$$+ \sum_j \left[\frac{U}{2}\hat{n}_j(\hat{n}_j - 1) - \mu\hat{n}_j\right]. \tag{2}$$

We can, therefore, express the total Hamiltonian as the sum of single site mean field Hamiltonians

$$\hat{h}_j = -\left[\left(J_x e^{i2\pi\alpha q} + J_y\right)\left(\phi_k^* \hat{b}_j - \phi_k^* \phi_j\right) + \text{H.c.}\right] \tag{3}$$
$$+ \frac{U}{2}\hat{n}_j(\hat{n}_j - 1) - \mu\hat{n}_j.$$

The next step is to diagonalize the Hamiltonian (2) for each site separately. For this, we consider the Gutzwiller ansatz, that is the ground state of the entire lattice is the direct product of the ground states of all the individual sites, and can be written in the Fock basis as

$$|\Psi_{\text{GW}}\rangle = \prod_j |\psi\rangle_j = \prod_j \sum_{n=0}^{N_b} c_n^{(j)} |n\rangle_j, \tag{4}$$

with the normalization condition $\sum_n |c_n^{(j)}|^2 = 1$. Here, N_b is the occupation number state maximum number of particles at a site and $c_n^{(j)}$ corresponds the complex coefficients for the ground state $|\psi\rangle_j$ at the jth site. Gutzwiller ansatz is the exact solution of the system in the strongly interacting regime ($J \ll U$).

For the numerical computations, we consider $N_b = 10$ and choose an initial guess of ϕ. Then, we diagonalize the Hamiltonian for each site and retain the ground state as the state $|\psi\rangle_j$ in $|\Psi_{\text{GW}}\rangle$. Then, using this $|\psi\rangle_j$, we calculate new ϕ for the next iteration and this cycle is continued till convergence is reached. To distinguish the different phases, we compute the SF order parameter at each site, and for the jth lattice site SF order parameter is

$$\phi_j = \langle\Psi_{\text{GW}}|\hat{b}_j|\Psi_{\text{GW}}\rangle = \sum_{n=1}^{N_b} \sqrt{n} c_{n-1}^{*(j)} c_n^{(j)}. \tag{5}$$

From the above expression it is evident that ϕ_j is zero in the MI phase of the system since only one of the co-efficients in (4) is nonzero, and it is finite for the SF phase due to the different c_ns contribution. We also compute the average lattice occupancy or density at each of the lattice site as

$$\rho_j = \langle\Psi_{\text{GW}}|\hat{n}_j|\Psi_{\text{GW}}\rangle = \sum_{n=0}^{N_b} n|c_n^{(j)}|^2. \tag{6}$$

These are the essence of SGMF theory.

2.2   Theory of CGMF

In the SGMF Hamiltonian (4), we decouple the hopping terms between two neighbouring sites by considering the mean field or SF order parameter ϕ. Thus, we could write the Hamiltonian of the entire system as the sum of Hamiltonians of individual sites and implement it as a site wise computations. However, this approximation is inadequate to incorporate the correlation effects arising from the NN hopping. To remedy this short coming, which assumes great importance to describe strongly correlated states like QH states, previous works have relied on CGMF [31]. To derive the CGMF Hamiltonian, we consider the entire lattice size as $K \times L$, which we divide into W clusters (C) of size $M \times N$, i.e., $W = (K \times L)/(M \times N)$. The case of $M = N = 1$ corresponds to the SGMF theory. In CGMF Hamiltonian, the hopping term is decomposed into two parts. First is the actual hopping term in the internal link of the cluster (δC) and the second term takes care of the boundary via mean fields. Our recent study [4] describes the decomposition of hopping term and CGMF method more clearly. After decomposition, the Hamiltonian for a single cluster (C) is expressed in the following way

$$
\begin{aligned}
\hat{H}_C = & - \sum_{p,q \in C} \left[\left(e^{i2\pi\alpha q} J_x \hat{b}^\dagger_{p+1,q} \hat{b}_{p,q} + \text{H.c.} \right) + \left(J_y \hat{b}^\dagger_{p,q-1} \hat{b}_{p,q} + \text{H.c.} \right) \right] \\
& - \sum_{p,q \in \delta C} \left[\left(e^{i2\pi\alpha q} J_x \langle a^*_{p,q} \rangle \hat{b}_{p,q} + \text{H.c.} \right) + \left(J_y \langle a^*_{p,q} \rangle \hat{b}_{p,q} + \text{H.c.} \right) \right] \\
& + \sum_{p,q \in C} \left[\frac{U}{2} \hat{n}_{p,q} (\hat{n}_{p,q} - 1) - \mu \hat{n}_{p,q} \right],
\end{aligned}
\tag{7}
$$

where $\langle a_{p,q} \rangle = \sum_{p',q' \in \mathscr{C}} \langle b_{p',q'} \rangle$. Then, we use Gutzwiller ansatz and the local cluster wavefunction in a Fock basis can be expressed as

$$
|\psi_c\rangle = \sum_{n_1, n_2, \ldots, n_{MN}} C_{n_1, n_2, \ldots, n_{MN}} |n_1, n_2, \ldots, n_{MN}\rangle,
\tag{8}
$$

with n_i being the index of the occupation number state of ith lattice site within the cluster, and $C_{n_1, n_2, \ldots, n_{MN}}$ is the amplitude of the cluster Fock state $|n_1, n_2, \ldots, n_{MN}\rangle$. Here also, the total Hamiltonian of the system can be written as the sum of all the individual cluster Hamiltonians [29]. The SF order parameter ϕ is computed for each cluster in the similar way as discussed in SGMF method. The next step is to find the ground state and we adopt the similar process as is described in SGMF theory. We take the initial solution for ϕ, construct the Hamiltonian matrix elements for a single cluster and diagonalize it. After diagonalization, we consider the lowest ground state $|\psi_c\rangle$ for the cluster and calculate the new ϕ and repeat the cycle until we get the converged solution. Here, it is worth to be mention that in case of CGMF, the convergence is very sensitive to the initial conditions [25, 31] and we use the method of successive over-relaxation for the better convergence [5].

3   Results and Discussions

We start our computations by considering a single cluster for $\alpha = 1/3$, with 3×2 and 3×3 clusters. We choose these clusters as Hamiltonian becomes periodic over a 3×1 magnetic unit cell in the Landau gauge for this flux value of α. We obtain

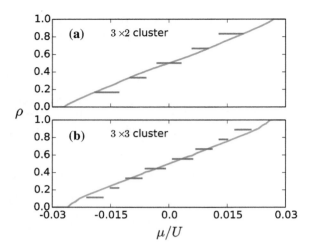

Fig. 1. The variation in the number density ρ for $\alpha = 1/3$ as function of μ. The states in SF phase are compressible and have nonzero superfluid order parameter ϕ. As a result, ρ varies linearly with μ and the green curve represents the SF states. For specific values of filling factor ν there are states with constant ρ, represented by the blue lines, and these correspond to the QH states. **a** Results from 3×2 cluster, the plateaus or the constant ρ values correspond to $\nu = n/2, n = 1, 2, \ldots, 5$ and the corresponding ρ values are $n/6$. **b** Results from 3×3, the platueas correspond to $\nu = n/3, n = 1, 2, \ldots, 8$ and the corresponding ρ values are $n/9$

QH states from the CGMF and characterize them based on the compressibility $\kappa = \partial\rho/\partial\mu$, where the density for the cluster is $\rho = \sum_j \langle\psi_c| \hat{n}_j |\psi_c\rangle /(K \times L)$. As the QH states are incompressible $\kappa = 0$ for these states, and κ is finite for the compressible SF states. Therefore, $\rho(\mu)$ of QH states has plateaus for different fillings ν and $\rho(\mu)$ is linear for the SF phase. In the Fig. 1, the plateaus corresponding to constant ρ indicate the existence for the QH states. Our computations, as mentioned earlier, are in the hardcore boson limit where $\rho < 1$. We obtain the QH states at $\nu = n/2$, with $n = 1, 2, \ldots, 5$ by taking the 3×2 cluster, and at $\nu = n/3$, with $n = 1, 2, \ldots, 8$ by taking the 3×3 cluster. The QH states are enhanced with the larger cluster size as mentioned above. Here, 3×3 cluster is close to exact diagonalization (ED) as the central lattice site has exact hopping contributions from the nearest neighbor sites. And, indeed, the diagonalization of the cluster can be transformed into ED with minor modifications in the computations of the Hamiltonian matrix elements. One main reason for enhancement in the QH states with 3×3 cluster is that it describes correlations effects more accurately compared to the 3×2 cluster and hence, the results are more accurate. Further, we show the density plots for the QH and SF states for the larger lattice system. For this, we take 12×12 lattice sites and $J/U = 0.01$. This system size, and hopping energies are kept the same for all the QH and SF states discussed in the rest of the manuscript. The IQH state at filling $\nu = 1$, $\mu/U = -0.008$ with density $\rho = 1/3$ is shown in Fig. 2. The IQH state has stripe

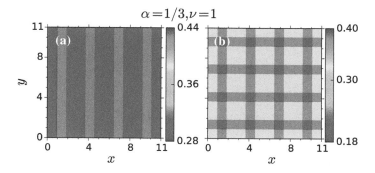

Fig. 2. The IQH state for $\alpha = 1/3$ and $\nu = 1$ with **a** 3×2 cluster and **b** with 3×3 cluster. The IQH state switches from stripe to checkerboard geometry with the mentioned cluster sizes

pattern with 3×2 cluster and transforms into checkerboard pattern with 3×3 cluster. The transformation from the stripe to checkerboard pattern is observed for the other IQH state of $\nu = 2$ as well. We observe that all the IQH and FQH states have stripe pattern except the $\nu = 3/2$ state, which has homogeneous density with $\rho = 0.5$ with 3×2 cluster. And the corresponding SF states have a zigzag pattern in the density ρ and in the SF order parameter ϕ. One of the FQH state for $\nu = 5/2$ and corresponding SF state is shown in the Fig. 3. As discussed earlier, we do not observe the half-integer FQH states with 3×3 cluster, but do observe the FQH states at the one third fillings. One of the FQH and SF state with 3×3 cluster for $\nu = 1/3$ is shown in the Fig. 4. Here, with 3×3 cluster, we observe all the FQH states have checkerboard pattern and all the SF states have the diagonal stripe pattern. We find that in all the cases SF states is the ground state and QH state is metastable state.

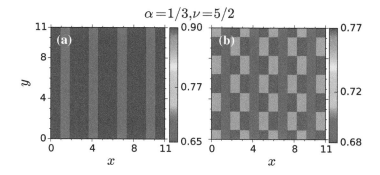

Fig. 3. a The FQH state with $\alpha = 1/3$ and $\nu = 5/2$ with 3×2 cluster, it has a stripe pattern in the density with vanishing SF order parameter. **b** The analogous SF state with zigzag pattern in the density as well as in the SF order parameter

$$\alpha = 1/3, \nu = 1/3$$

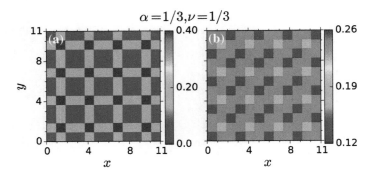

Fig. 4. a The FQH state with $\alpha = 1/3$ and $\nu = 1/3$ with 3×3 cluster with a checkerboard pattern in the density and vanishing SF order parameter. **b** The analogous SF state with diagonal stripe pattern in the density as well as in the SF order parameter

One pertinent question that needs to be addressed is the experimental detection of the QH states we have discussed. In this regard, the two-point correlation function, say along the y-axis for a fixed x coordinate $C_x(y)$, is an observable based on which we can identify the QH states. It may be measured through quantum probes [9,41], and is defined as

$$C_x(y) = \langle \hat{b}_x^\dagger(y) \hat{b}_0(y) \rangle, \tag{9}$$

where the expectation is taken with cluster wavefunction. In a recent work [19], it is reported that for QH states, $C_x(y)$ decays as inverse power law at the edge. While in the bulk, it initially decays exponentially as QH state is gapped and has a power-law tail associated with the edge. We also observe these trends of $C_x(y)$ for the QH states discussed in this work.

4   Conclusion

We obtain QH states by considering the two cluster sizes as 3×2 and 3×3 in the CGMF theory. With the larger cluster 3×3, we approach ED as the CGMF provides an exact description of the hopping term for the central lattice site. We obtain QH states with fillings $\nu = n/2$, $n = 1, 2, \ldots, 5$ and corresponding density $\rho = n/6$ with 3×2 cluster size. However, with 3×3 cluster, we obtain a larger set of QH states with fillings $\nu = n/3$, $n = 1, 2, \ldots, 8$ and corresponding density $\rho = n/9$. We also observe the competing SF states corresponding to all the QH states. We find that the SF state is the ground state and QH state is metastable state in all the cases. We have demonstrated that the QH states change geometry from stripe to checkerboard by switching 3×2 to 3×3 cluster size. On the other hand, SF state has zigzag pattern with both the cluster sizes, specially with 3×3 the zigzag pattern is equivalent to diagonal stripe pattern. Thus, we have established that to obtain correct density pattern of the QH states, it is essential to consider larger cluster sizes in the CGMF theory.

References

1. Aidelsburger, M., Atala, M., Lohse, M., Barreiro, J.T., Paredes, B., Bloch, I.: Realization of the Hofstadter Hamiltonian with ultracold atoms in optical lattices. Phys. Rev. Lett. **111**, 185,301 (2013). https://doi.org/10.1103/PhysRevLett.111. 185301

2. Aidelsburger, M., Atala, M., Nascimbène, S., Trotzky, S., Chen, Y.A., Bloch, I.: Experimental realization of strong effective magnetic fields in an optical lattice. Phys. Rev. Lett. **107**, 255,301 (2011). https://doi.org/10.1103/PhysRevLett.107. 255301

3. Anderson, B.P., Kasevich, M.A.: Macroscopic quantum interference from atomic tunnel arrays. Science **282**, 1686 (1998). https://doi.org/10.1126/science.282.5394. 1686. http://science.sciencemag.org/content/282/5394/1686

4. Bai, R., Bandyopadhyay, S., Pal, S., Suthar, K., Angom, D.: Bosonic quantum Hall states in single-layer two-dimensional optical lattices (2018). Phy. Rev. A **98**, 023606 (2018). https://doi.org/10.1103/PhysRevA.98.023606

5. Barrett, R., Berry, M., Chan, T., Demmel, J., Donato, J., Dongarra, J., Eijkhout, V., Pozo, R., Romine, C., van der Vorst, H.: Templates for the Solution of Linear Systems: Building Blocks for Iterative Methods. SIAM (1994). https://doi.org/10. 1137/1.9781611971538

6. Bloch, I., Dalibard, J., Zwerger, W.: Many-body physics with ultracold gases. Rev. Mod. Phys. **80**, 885 (2008). https://doi.org/10.1103/RevModPhys.80.885

7. Dalibard, J., Gerbier, F., Juzeliūnas, G., Öhberg, P.: Colloquium: artificial gauge potentials for neutral atoms. Rev. Mod. Phys. **83**, 1523 (2011). https://doi.org/ 10.1103/RevModPhys.83.1523

8. Dean, C.R., Wang, L., Maher, P., Forsythe, C., Ghahari, F., Gao, Y., Katoch, J., Ishigami, M., Moon, P., Koshino, M., Taniguchi, T., Watanabe, K., Shepard, K.L., Hone, J., Kim, P.: Hofstadters butterfly and the fractal quantum hall effect in moiré superlattices. Nature **497**, 598 (2013). https://doi.org/10.1038/nature12186

9. Elliott, T.J., Johnson, T.H.: Nondestructive probing of means, variances, and correlations of ultracold-atomic-system densities via qubit impurities. Phys. Rev. A **93**, 043,612 (2016). https://doi.org/10.1103/PhysRevA.93.043612

10. Fisher, M.P.A., Weichman, P.B., Grinstein, G., Fisher, D.S.: Boson localization and the superfluid-insulator transition. Phys. Rev. B **40**, 546 (1989). https://doi. org/10.1103/PhysRevB.40.546

11. Freericks, J.K., Krishnamurthy, H.R., Kato, Y., Kawashima, N., Trivedi, N.: Strong-coupling expansion for the momentum distribution of the Bose-Hubbard model with benchmarking against exact numerical results. Phys. Rev. A **79**, 053,631 (2009). https://doi.org/10.1103/PhysRevA.79.053631

12. Freericks, J.K., Monien, H.: Strong-coupling expansions for the pure and disordered Bose-Hubbard model. Phys. Rev. B **53**, 2691–2700 (1996). https://doi.org/ 10.1103/PhysRevB.53.2691

13. Gerster, M., Rizzi, M., Silvi, P., Dalmonte, M., Montangero, S.: Fractional quantum hall effect in the interacting hofstadter model via tensor networks. Phys. Rev. B **96**, 195,123 (2017). https://doi.org/10.1103/PhysRevB.96.195123

14. Greiner, M., Bloch, I., Mandel, O., Hänsch, T.W., Esslinger, T.: Exploring phase coherence in a 2D lattice of Bose-Einstein condensates. Phys. Rev. Lett. **87**, 160,405 (2001). https://doi.org/10.1103/PhysRevLett.87.160405

15. Greiner, M., Mandel, O., Esslinger, T., Hänsch, T.W., Bloch, I.: Quantum phase transition from a superfluid to a Mott insulator in a gas of ultracold atoms. Nature (London) **415**, 39 (2002). https://doi.org/10.1038/415039a

16. Hafezi, M., Sørensen, A.S., Demler, E., Lukin, M.D.: Fractional quantum Hall effect in optical lattices. Phys. Rev. A **76**, 023,613 (2007). https://doi.org/10.1103/PhysRevA.76.023613

17. Harper, P.G.: Single band motion of conduction electrons in a uniform magnetic field. Proc. Phys. Soc. A **68**, 874 (1955). http://iopscience.iop.org/0370-1298/68/10/304

18. He, Y.C., Grusdt, F., Kaufman, A., Greiner, M., Vishwanath, A.: Realizing and adiabatically preparing bosonic integer and fractional quantum Hall states in optical lattices. Phys. Rev. B **96**, 201,103 (2017). https://doi.org/10.1103/PhysRevB.96.201103

19. Hofstadter, D.R.: Energy levels and wave functions of Bloch electrons in rational and irrational magnetic fields. Phys. Rev. B **14**, 2239 (1976). https://doi.org/10.1103/PhysRevB.14.2239

20. Hügel, D., Strand, H.U.R., Werner, P., Pollet, L.: Anisotropic Harper-Hofstadter-Mott model: competition between condensation and magnetic fields. Phys. Rev. B **96**, 054,431 (2017). https://doi.org/10.1103/PhysRevB.96.054431

21. Jaksch, D., Bruder, C., Cirac, J.I., Gardiner, C.W., Zoller, P.: Cold bosonic atoms in optical lattices. Phys. Rev. Lett. **81**, 3108 (1998). https://doi.org/10.1103/PhysRevLett.81.3108

22. Jaksch, D., Zoller, P.: Creation of effective magnetic fields in optical lattices: the Hofstadter butterfly for cold neutral atoms. New J. Phys. **5**, 56 (2003). https://doi.org/10.1088/1367-2630/5/1/356

23. Jiménez-García, K., LeBlanc, L.J., Williams, R.A., Beeler, M.C., Perry, A.R., Spielman, I.B.: Peierls substitution in an engineered lattice potential. Phys. Rev. Lett. **108**, 225,303 (2012). https://doi.org/10.1103/PhysRevLett.108.225303

24. Kuno, Y., Shimizu, K., Ichinose, I.: Bosonic analogs of the fractional quantum Hall state in the vicinity of Mott states. Phys. Rev. A **95**, 013,607 (2017). https://doi.org/10.1103/PhysRevA.95.013607

25. Lewenstein, M., Sanpera, A., Ahufinger, V., Damski, B., Sen(De), A., Sen, U.: Ultracold atomic gases in optical lattices: mimicking condensed matter physics and beyond. Adv. Phys. **56**, 243 (2007). https://doi.org/10.1080/00018730701223200

26. Lin, Y.J., Compton, R.L., Jimenez-Garcia, K., Phillips, W.D., Porto, J.V., Spielman, I.B.: A synthetic electric force acting on neutral atoms. Nat. Phys. **7**, 531 (2011). https://doi.org/10.1038/nphys1954

27. Lin, Y.J., Compton, R.L., Perry, A.R., Phillips, W.D., Porto, J.V., Spielman, I.B.: Bose-Einstein condensate in a uniform light-induced vector potential. Phys. Rev. Lett. **102**, 130,401 (2009). https://doi.org/10.1103/PhysRevLett.102.130401

28. Lühmann, D.S.: Cluster Gutzwiller method for bosonic lattice systems. Phys. Rev. A **87**, 043,619 (2013). https://doi.org/10.1103/PhysRevA.87.043619

29. Miyake, H., Siviloglou, G.A., Kennedy, C.J., Burton, W.C., Ketterle, W.: Realizing the Harper Hamiltonian with laser-assisted tunneling in optical lattices. Phys. Rev. Lett. **111**, 185,302 (2013). https://doi.org/10.1103/PhysRevLett.111.185302

30. Natu, S.S., Mueller, E.J., Das Sarma, S.: Competing ground states of strongly correlated bosons in the Harper-Hofstadter-Mott model. Phys. Rev. A **93**, 063,610 (2016). https://doi.org/10.1103/PhysRevA.93.063610

31. Niemeyer, M., Freericks, J.K., Monien, H.: Strong-coupling perturbation theory for the two-dimensional Bose-Hubbard model in a magnetic field. Phys. Rev. B **60**, 2357 (1999). https://doi.org/10.1103/PhysRevB.60.2357

32. Oktel, M.O., Niţă, M., Tanatar, B.: Mean-field theory for Bose-Hubbard model under a magnetic field. Phys. Rev. B **75**, 045,133 (2007). https://doi.org/10.1103/PhysRevB.75.045133

33. Palmer, R.N., Jaksch, D.: High-field fractional quantum Hall effect in optical lattices. Phys. Rev. Lett. **96**, 180,407 (2006). https://doi.org/10.1103/PhysRevLett. 96.180407

34. Palmer, R.N., Klein, A., Jaksch, D.: Optical lattice quantum Hall effect. Phys. Rev. A **78**, 013,609 (2008). https://doi.org/10.1103/PhysRevA.78.013609

35. Peierls, R.E.: On the theory of diamagnetism of conduction electrons. Z. Phys. **80**, 763 (1933). https://doi.org/10.1007/BF01342591

36. Peotta, S., Chien, C.C., Di Ventra, M.: Phase-induced transport in atomic gases: from superfluid to Mott insulator. Phys. Rev. A **90**, 053,615 (2014). https://doi. org/10.1103/PhysRevA.90.053615

37. Sheshadri, K., Krishnamurthy, H.R., Pandit, R., Ramakrishnan, T.V.: Superfluid and insulating phases in an interacting-boson model: mean-field theory and the RPA. EPL **22**, 257 (1993). https://doi.org/10.1209/0295-5075/22/4/004. http:// stacks.iop.org/0295-5075/22/i=4/a=004

38. Sørensen, A.S., Demler, E., Lukin, M.D.: Fractional quantum Hall states of atoms in optical lattices. Phys. Rev. Lett. **94**, 086,803 (2005). https://doi.org/10.1103/ PhysRevLett.94.086803

39. Stöferle, T., Moritz, H., Schori, C., Köhl, M., Esslinger, T.: Transition from a strongly interacting 1D superfluid to a Mott insulator. Phys. Rev. Lett. **92**, 130,403 (2004). https://doi.org/10.1103/PhysRevLett.92.130403

40. Streif, M., Buchleitner, A., Jaksch, D., Mur-Petit, J.: Measuring correlations of cold-atom systems using multiple quantum probes. Phys. Rev. A **94**, 053,634 (2016). https://doi.org/10.1103/PhysRevA.94.053634

41. Umucalılar, R.O., Oktel, M.O.: Phase boundary of the boson Mott insulator in a rotating optical lattice. Phys. Rev. A **76**, 055,601 (2007). https://doi.org/10.1103/ PhysRevA.76.055601

42. Umucalilar, R.O., Mueller, E.J.: Fractional quantum Hall states in the vicinity of Mott plateaus. Phys. Rev. A **81**, 053,628 (2010). https://doi.org/10.1103/ PhysRevA.81.053628

43. Wang, T., Zhang, X.F., Hou, C.F., Eggert, S., Pelster, A.: High-order strong-coupling expansion for the Bose-Hubbard model (2018). arXiv:1801.01862

44. Wessel, S., Alet, F., Troyer, M., Batrouni, G.: Quantum Monte Carlo simulations of confined bosonic atoms in optical lattices. Phys. Rev. A **70**, 053,615 (2004). https://doi.org/10.1103/PhysRevA.70.053615

Laser-Induced Fluorescence Spectroscopy of the $\tilde{C}^2\Pi_r - \tilde{X}^2\Sigma^+$ Transition in LaNH Molecule in Supersonic Free-Jet

Soumen Bhattacharyya[1,2], Sheo Mukund[1,2], and S. G. Nakhate[1,2(✉)]

[1] Atomic and Molecular Physics Division, Bhabha Atomic Research Centre, Mumbai 400 085, India
nakhate@barc.gov.in
[2] Homi Bhabha National Institute, Anushaktinagar, Mumbai 400094, India

Abstract. The $\tilde{C}^2\Pi_r - \tilde{X}^2\Sigma^+$ system of the linear LaNH molecule in the ultraviolet spectral region has been studied by the laser-induced fluorescence (LIF) spectroscopy in free-jet. Recently, we reported the rotational analysis of the origin band (000 - 000) of the $\tilde{C} - \tilde{X}$ transition, and the detection of the low-lying $\tilde{A}'^2\Delta_r$ electronic state (Bhattacharyya et al. in Chem Phys Lett 692, 1–6 (2018), [9]). Along with the origin band, we detected three weaker bands, two of them originate from the fundamental and second bend vibrations ν_2(π, bend) of the ground state. The experimental rotational spectrum of one of the band observed at 23,158 cm^{-1} along with the rotational analysis is presented. Rotational analysis confirms the band as $\tilde{C}^2\Pi_{3/2}(001,{}^2\Pi) - \tilde{X}^2\Sigma^+(02^00,{}^2\Sigma)$. A true rotational analysis for the band at 23,252 cm^{-1} was not possible, however, on the basis of contour fit to the experimental spectrum, we could tentatively assign the band as $\tilde{C}^2\Pi_{3/2}(01^10,{}^2\Delta) - \tilde{X}^2\Sigma^+(01^10,{}^2\Pi)$.

1 Introduction

The lanthanide sulphides have potentials to be used as solar energy conversion materials, pigments, infrared window materials, and phosphor host media [1]. The lanthanide oxides and imides have attracted considerable interest for their use in cat-alytic applications. Imides based organometallic lanthanide complexes have been synthesized recently [2]. Studies on the possibility of treating ammonia nitrogen wastewater using doped La/Fe/TiO$_2$ photocatalyst by oxidizing ammonia nitrogen to nitrogen has been reported [3]. The process of actual development of the industrial catalyst for automotive, petroleum industry, nitrogen fixation, water gas shift reaction and so on is certainly complex. However, at the heart of these complex reactions is a common reaction step—the oxidative addition of gaseous reagents with a transition metal. Elucidating the mechanism of this common reaction step plays a vital role in the development of these homogeneous and heterogeneous catalysts. The reaction steps in a complex heterogeneous catalysis reaction include activations of H–H, C–H, C–C and N–H bonds to a transition metal centre to form a precursor complex. Understanding the electronic structure of transition metal-containing small imides like ScNH, YNH and LaNH plays an important role in this cause. The electronic structure provides

© Springer Nature Singapore Pte Ltd. 2019
P. C. Deshmukh et al. (eds.), *Quantum Collisions and Confinement of Atomic and Molecular Species, and Photons*, Springer Proceedings in Physics 230,
https://doi.org/10.1007/978-981-13-9969-5_21

information on the nature of bonding of transition metals with smaller molecules and radicals.

The monoxides, imides and sulphides of Sc, Y and La form an isovalent family of molecules whose optical spectra in the yellow-red region is dominated by the $\tilde{A}^2\Pi_r - \tilde{X}^2\Sigma^+$ and $\tilde{B}^2\Sigma^+ - \tilde{X}^2\Sigma^+$ electronic transitions. The only theory available for the ground and low-lying electronic states of these molecules is for YNH by Das and Balasubramanian [4]. The authors had shown that the YNH molecule in linear form is more stable than the NYH bent molecule, and the symmetry of the linear YNH ground state is $^2\Sigma^+$. The singly occupied σ molecular orbital has dominant electron density from the yttrium metal's $5s$ orbital. The electron promotion from the metal $ns\sigma$ to the $(n-1)d\pi$ and $(n-1)d\sigma$ orbitals is accountable for the $\tilde{A} - \tilde{X}$ and $\tilde{B} - \tilde{X}$ transitions, respectively. The experimental data for the ScNH [5] and YNH [6] molecules are limited to the origin bands of the $\tilde{A} - \tilde{X}$ and $\tilde{B} - \tilde{X}$ transitions. For LaNH molecule, substantial spectroscopic work related to the vibrational structure of the ground state, and the rotational spectra for the $\tilde{A} - \tilde{X}$ and $\tilde{B} - \tilde{X}$ transitions were presented by Rixon [7]. Along with the origin bands of the $\tilde{A} - \tilde{X}$ and $\tilde{B} - \tilde{X}$ transitions, rotational transitions between the bending vibrations of the ground and excited states were also presented in the thesis. The Renner–Teller coupling between the excited states $\tilde{A}^2\Pi_r$ and $\tilde{B}^2\Sigma^+$ was evident from the 33% higher vibrational frequency of the $\tilde{B}$ state compared to the $\tilde{X}$ state. The effect of Renner–Teller coupling was also manifested in the rotational spectrum by the fact that the Λ-doubling constant in the $\tilde{A}^2\Pi_r$ state has similar value as the spin-rotation constant, γ in the $\tilde{B}^2\Sigma^+$ state.

In our recent work in LaNH molecule, we presented the rotational spectrum of the origin band of the $\tilde{C}^2\Pi_r - \tilde{X}^2\Sigma^+$ transition and determined rotational constants for the $\tilde{C}$ state. We also experimentally detected the low-lying $\tilde{A}'^2\Delta_r$ electronic state for the first time in a group3 imide molecule. Along with the origin band, we observed three weaker bands lying between the spin–orbit components of the origin band. To make the work uncluttered, we reported the origin band only in the earlier communication. In the present paper, we focus on the rotational spectrum of two of these weak bands observed, respectively, at 23,252 and 23,158 cm^{-1}. Based on the energy consideration, they are assigned, respectively, as the $\tilde{C}^2\Pi_{3/2}(01^10) - \tilde{X}^2\Sigma^+(01^10)$ and $\tilde{C}^2\Pi_{3/2}(001) - \tilde{X}^2\Sigma^+(02^00)$ transitions.

2 Experimental

The experimental details could be found in the earlier work from our laboratory [8]. In brief, using Smalley-type laser vaporization supersonic expansion apparatus, we produced LaNH molecules in free-jet. Two percent NH_3 diluted in helium was used in the gas pulsed valve, which facilitated the reaction between the La-plasma and NH_3. LaNH molecules in the free-jet were excited by a XeCl excimer laser pumped tunable dye laser at right angles to the beam expansion axis, about 50 mm downstream the nozzle. The resulting laser-induced fluorescence (LIF) was collected orthogonal to the free-jet and dye laser beam axes by a lens system. The LIF was dispersed by a monochromator and detected by a Peltier cooled photomultiplier tube. The signal was amplified, integrated and stored on a computer. The rotational spectrum was recorded by scanning the dye laser wavelength. The spectral linewidth of the dye laser was 0.1 cm^{-1}. The wavelength calibration was done by a wavelength metre (HighFinesse WS5) with an absolute accuracy of 2 pm.

3   Results and Discussion

The LIF search scan, in the 15,000–23,750 cm^{-1} spectral range, for the excitation bands of the reaction products of the laser produced lanthanum plasma and ammonia, revealed two intense bands centred at 23,366 and 23,093 cm^{-1} and three weak bands at 23,322, 23,252 and 23,158 cm^{-1} belonging to LaNH molecule. The low-resolution excitation spectrum was presented in Fig. 1 of our recent publication [9]. The carrier of these bands was confirmed from the known values of the ground state vibrations; $v_3(\sigma^+$, La-N stretch$) = 752$ cm^{-1} and $v_2(\pi$, bend$) = 464$ cm^{-1} [7]. For the two intense bands, the lower state, $\tilde{X}^2\Sigma^+$ (000) was confirmed from the lower state combination differences. The dispersed fluorescence spectrum also confirmed the lower state as the ground state. Two weak bands, respectively, at 23,252 and 23,158 cm^{-1} probably belong to the $\tilde{C}^2\Pi_{3/2}(01^10) - \tilde{X}^2\Sigma^+ (01^10)$ and $\tilde{C}^2\Pi_{3/2}(001) - \tilde{X}^2\Sigma^+ (02^00)$ transitions. A crude approximation can be made to locate these two bands energetically by

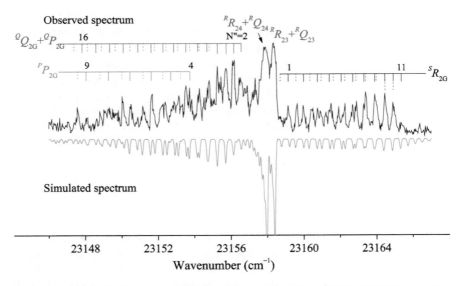

Fig. 1. Rotationally resolved spectrum of the $\tilde{C}^2\Pi_{3/2}(001,^2\Pi) - \tilde{X}^2\Sigma^+ (02^00,^2\Sigma)$ transition in LaNH. The branches are marked in the spectrum by solid and dashed vertical lines which, respectively, represent transitions from G = 3 and G = 4 hyperfine components of the lower state. The simulated spectrum has been generated using the constants in Table 1 and for rotational temperature of 40 K and Gaussian linewidth of 0.1 cm^{-1}

assuming that the fundamental $v_2(\pi,$ bend), (01^10) and $v_3(\sigma^+,$ La-N stretch) frequency, (001) of the $\tilde{C}$ state is same as that of the $\tilde{A}^2\Pi$ state of 380 and 693 cm^{-1}, respectively [7]. This place the former and later bands at 23,282 and 23,162 cm^{-1}, respectively. The energy of the $\tilde{X}^2\Sigma^+ (02^00,^2\Sigma)$ state was measured 896 cm^{-1} in [7].

3.1 Rotational Analysis of the $\tilde{C}^2\Pi_{3/2}(001) - \tilde{X}^2\Sigma^+ (02^00)$ Transition

The rotationally resolved spectrum of the band centred at 23,158 cm^{-1} is presented in Fig. 1. The $\tilde{C}^2\Pi_{3/2}$ electronic state belongs to Hund's case (a). The $\tilde{C}^2\Pi_{3/2}(001,^2\Pi) - \tilde{X}^2\Sigma^+ (02^00,^2\Sigma)$ system, where the notation indicates electronic and vibronic symmetries of the upper and lower states, is expected to have four apparent branches [10]. These four branches are ^PP, $^QP + {}^QQ$, $^RQ + {}^RR$ and ^SR. In terms of the pattern-forming quantum number J' for the upper $^2\Pi$ state and N'' for the lower $^2\Sigma^+$ state, these four branches correspond to values of $J'-N''$ of 3/2, 1/2, $-1/2$ and $-3/2$. The $\tilde{X}^2\Sigma^+ (02^00,^2\Sigma)$ lower state is described by $b_{\beta S}$ coupling; La has magnetically active nuclei with nuclear spin, I = 7/2, and since the unpaired electron in σ molecular orbital of LaNH is mainly 6s electron of La atom, spin (S) of the electron couples to I more strongly than the rotation (N). I and S couple to form G which in turn couples to N to form the total angular momentum F. For LaNH $\tilde{X}^2\Sigma^+$ state, I = 7/2 and S = 1/2, which gives G = 3 and 4, with the two G levels split by $4b_F$. The lower state has been identified from the combination differences of the isolated rotational lines. We used PGOPHER [11] to assign rotational lines of all the branches and to fit the experimental line positions. We followed a modified branch labelling scheme [12] over the commonly occurring $^{\Delta N}\Delta J_{F_i'F_i''}(N'')$, in which the F_i'' label is replaced by the value of G''. The four branches for a $\tilde{C}^2\Pi_{3/2}(001,^2\Pi) - \tilde{X}^2\Sigma^+ (02^00,^2\Sigma)$ transition are $^PP_{2G''}$, $^QP_{2G''} + {}^QQ_{2G''}$, $^RQ_{2G''} + {}^RR_{2G''}$ and $^SR_{2G''}$, where G'' = 3, 4.

Usually, the Ω quantum number of the upper electronic state is determined from the first rotational lines. We could not observe first rotational lines isolated in the rotational branches at the employed resolution of 0.1 cm^{-1} of the dye laser. The intensity in the band is approximately one-fourth of the origin band, and this hindered us to record the rotational spectrum at higher resolution of the dye laser by introducing intra-cavity etalon. Once the intra-cavity etalon is used it reduces the dye laser pulse energy and produces a spectrum for the weak band with unacceptable signal to noise ratio. In our earlier work on the origin band, the first ^SR branch G''-pairs, $^SR_{23}(0)$ and $^SR_{24}(0)$, could not be observed isolated at lower resolution of ~ 0.1 cm^{-1} and were merged with the broad $^RQ_{23} + {}^RR_{23}$ branch lines. The $^SR_{23}(0)$ line was observed isolated only when the spectrum was recorded at a higher resolution of 0.05 cm^{-1}. For the present spectrum as well, the first isolated line in this branch is $^SR_{24}(1)$. It was straightforward to identify the $^SR_{2G''}$ branch lines from $N'' = 1$–11 for both the G'' = 3 and 4 components. These $^SR_{2G''}$ branch lines were observed mostly isolated at the employed experimental resolution of 0.1 cm^{-1}. The $^QP_{2G''} + {}^QQ_{2G''}(N'')$ branch lines from $N'' = 2$–16, which combine with the corresponding $^SR_{2G''}(N'')$ lines were identified in the spectrum from the lower state combination differences. The $^RQ_{2G''} + {}^RR_{2G''}$ branch is intense, however, observed unresolved. This branch has a clear double band head structure

Table 1. Molecular constants (in cm^{-1}) for the $\tilde{C}^2\Pi_{3/2}(001,^2\Pi)$ and $\tilde{X}^2\Sigma^+(02^00,^2\Sigma)$ states of LaNH molecule. The symmetry notations shown inside the bracket indicate vibronic symmetry for the corresponding electronic state

Constants	$\tilde{C}^2\Pi_{3/2}(001,^2\Pi)$	$\tilde{X}^2\Sigma^+(02^00,^2\Sigma)$
T	$x + 23057.547(4)$	x^a
A	200^b	–
B	$0.285730(38)$	$0.30649(25)$
$D \times 10^6$	–	$8.82(59)$
$q \times 10^4$	–	4.788^c
b_F	–	0.11095^c

[a]$x = 896$ cm^{-1} from DF measurements [7]

[b]The spectrum was fitted using spin–orbit parameter, $A = 200$ cm^{-1}, in the upper state. Therefore, the fitted band origins are shifted 100 cm^{-1} to lower energies from the sub-band origins

[c]The constants are held fixed to the values for the $\tilde{X}^2\Sigma^+(01^10)$ state from [7]

separated by ~ 0.44 cm^{-1}. Many rotational lines of this branch were observed blended with the $^Q P_{2G''} + {}^Q Q_{2G''}$ branch lines, which were identified in the spectrum from $N'' = 2$–16. The $^P P_{2G''}$ branch lines were identified from $N'' = 4$–9. The rotational spectrum was observed mostly free from local rotational perturbations at the employed resolution.

Rotational constants for both the upper and lower states were determined from the least squares fitting of the rotational wavenumbers using PGOPHER. The lower state was split into two states where the $G'' = 4$ component was placed $4b_F$ (=0.4438 cm^{-1}) higher in energy than the $G'' = 3$ component. The constants for the lower state was initially held fixed to the known values for the $v_2(\pi, \text{bend}) = 1$ state of the $\tilde{X}^2\Sigma$ state, $\tilde{X}^2\Sigma^+(01^10,^2\Pi)$ [7]. The molecular constants are listed in Table 1, and the rotational line list is provided in Table 2. The rotational temperature ~ 40 K and Gaussian linewidth ~ 0.1 cm^{-1} were determined from the intensity contour fit.

3.2   23,252 cm^{-1} Band

The band observed at 23,252 cm^{-1} perhaps belongs to the $\tilde{C}^2\Pi_{3/2}(01^10,^2\Delta)$ – $\tilde{X}^2\Sigma^+(01^10,^2\Pi)$ transition. This consists of two overlapping bands corresponding to the two-spin components, $\Omega = 5/2$ and $3/2$ of the $^2\Delta$ vibronic state of the $\tilde{C}^2\Pi$ electronic state. For the present spectrum, which was recorded at a moderate resolution of 0.1 cm^{-1}, it was not possible to assign the rotational lines unambiguously and carry out a rotational analysis. However, our simulated spectrum, with two energetically close-lying upper states of $^2\Delta$ vibronic symmetry, and the lower state as $\tilde{X}^2\Sigma^+(01^10,^2\Pi)$ reproduces the experimental spectrum well (Fig. 2). The simulated spectrum was obtained with the following rotational constants (in cm^{-1}) for the upper state $-T(\tilde{C}^2\Pi_{3/2}(01^10,^2\Delta_{5/2}) = 23618.4$, $T(\tilde{C}^2\Pi_{3/2}(01^10,^2\Delta_{3/2}) = 23617.0$, $B = 0.2727$, $A = 200$, $q = 0.008$, $D = 1 \times 10^{-6}$. The rotational constants other than T were

Table 2. The observed and calculated transition energies (in cm^{-1}) for the $\tilde{A}^2\Pi_{3/2}(001,^2\Pi_{3/2}) - \tilde{X}^2\Sigma^+(02^00,^2\Sigma)$ band of LaNH molecule

Branch	N''	Observed	Obs. − calc.	Branch	N''	Observed	Obs. − calc.
$^PP_{23}$	4	23153.73	0.02	$^QP_{24} + {}^QQ_{24}$	2	23156.16	0.03
	5	23152.70	0.05		3	23155.74	0.02
	6	23151.62	0.06		4	23155.26	−0.01
	7	23150.46	0.03		5	23154.81	0.02
	8	23149.27	0.01		6	23154.26	0.00
	9	23148.04	−0.02		7	23153.73	0.03
					8	23153.08	−0.03
$^PP_{24}$	4	23153.26	−0.01		9	23152.44	−0.04
	5	23152.27	0.06		10	23151.81	−0.01
	6	23151.18	0.07		11	23151.18	0.05
	7	23150.02	0.04		12	23150.46	0.03
	8	23148.81	−0.01		13	23149.64	−0.05
	9	23147.61	−0.01		14	23148.90	−0.03
					15	23148.12	−0.03
$^QP_{23} + {}^QQ_{23}$	2	23156.57	0.00		16	23147.39	0.03
	3	23156.16	0.00				
	4	23155.74	0.02	$^SR_{23}$	1	23159.15	−0.07
	5	23155.26	0.04		2	23159.99	−0.01
	6	23154.69	−0.01		3	23160.75	0.01
	7	23154.19	0.04		4	23161.41	−0.03
	8	23153.52	−0.03		5	23162.06	−0.03
	9	23152.86	−0.07		6	23162.68	−0.04
	10	23152.27	0.00		7	23163.30	0.00
	11	23151.62	0.04		8	23163.89	0.03
	12	23150.89	0.02		9	23164.45	0.08
	13	23150.08	−0.05		10	23164.90	0.04
	14	23149.33	−0.04		11	23165.35	0.03
	15	23148.60	0.01				
	16	23147.83	0.03	$^SR_{24}$	1	23158.70	−0.08
					2	23159.57	0.01
$^QP_{24} + {}^QQ_{24}$	2	23156.16	0.03		3	23160.33	0.04
	3	23155.74	0.02		4	23160.95	−0.04
	4	23155.26	−0.01		5	23161.62	−0.04
	5	23154.81	0.02		6	23162.26	−0.01
	6	23154.26	0.00		7	23162.87	0.01
	7	23153.73	0.03		8	23163.40	−0.01
	8	23153.08	−0.03		9	23163.98	0.05
	9	23152.44	−0.04		10	23164.45	0.04
	10	23151.81	−0.01		11	23164.90	0.03
	11	23151.18	0.05				

(continued)

Table 2. (continued)

Branch	N″	Observed	Obs. − calc.	Branch	N″	Observed	Obs. − calc.
$^QP_{24} + {}^QQ_{24}$	12	23150.46	0.03				
	13	23149.64	−0.05				
	14	23148.90	−0.03				
	15	23148.12	−0.03				
	16	23147.39	0.03				

assigned same for both the vibronic states. The rotational constants for the lower state, $\tilde{X}^2\Sigma^+(01^10)$, were held fixed to the following values from [7]; $T = 464$, $B = 0.306263$, $D = 2.8 \times 10^{-7}$, $q = 0.000478$ and $4b_F = 0.4438$. For an unambiguous rotational analysis, the rotational spectrum needs to be investigated at a higher resolution.

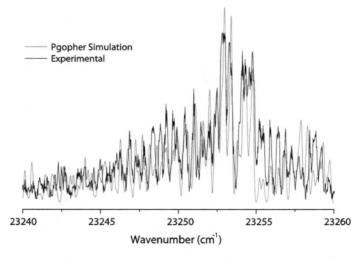

Fig. 2. Experimental and simulated spectrum of the $\tilde{C}^2\Pi_{3/2}(01^10,^2\Delta) - \tilde{X}^2\Sigma^+(01^10,^2\Pi)$ transition

4   Conclusion

The laser-induced fluorescence spectrum from the second bend vibration of the ground state, $\nu_2(\pi, \text{bend}) = 2$ to the fundamental stretch vibration $\nu_3(\sigma^+, \text{La-N stretch}) = 1$ of the $\tilde{C}^2\Pi_{3/2}$ state has been studied in the free-jet of LaNH molecules. The rotational spectrum has been recorded at a resolution of 0.1 cm^{-1} by scanning the wavelength of a tunable dye laser. The rotationally resolved LIF spectrum of the $\tilde{C}^2\Pi_{3/2}(001,^2\Pi)$ – $\tilde{X}^2\Sigma^+(02^00,^2\Sigma)$ transition has been analyzed to obtain the molecular constants for the upper and lower states. The rotational constants for the $\tilde{C}^2\Pi_{3/2}(01^10,^2\Delta)$ – $\tilde{X}^2\Sigma^+(01^10,^2\Pi)$ transition could be derived from a contour fit to the experimental spectrum. To derive true rotational constants, the spectrum is to be obtained at a higher resolution.

References

1. Marin, C.M.: Synthesis and applications of lanthanide sulfides and oxides. Ph.D. thesis, University of Nebraska (2016)
2. Schädle, D., Maichle-Mössmer, C., Törnroos, K., Anwander, R.: Holmium(III) supermesityl-imide complexes bearing methylaluminato/gallato ligands. Inorganics **3**, 500–510 (2015)
3. Luo, X., Chen, C., Yang, J., Wang, J., Yan, Q., Shi, H., Wang, C.: Characterization of La/Fe/TiO$_2$ and its photocatalytic performance in ammonia nitrogen wastewater. Int. J. Environ. Res. Public Health **12**, 14626–14639 (2015)
4. Das, K., Balasubramanian, K.: Geometries and energy separations of low-lying states of YNH and NYH. J. Chem. Phys. **93**, 6671–6675 (1990)
5. Steimle, T.C., Xin, J., Marr, A.J., Beaton, S.: Detection and characterization of scandium imide, ScNH. J. Chem. Phys. **106**, 9084–9094 (1997)
6. Simard, B., Jakubek, Z., Niki, H., Balfour, W.J.: High resolution molecular beam study of the origin band of the $\tilde{B}^2\Sigma^+ - \tilde{X}^2\Sigma^+$ system of yttrium imide. J. Chem. Phys. **111**, 1483–1493 (1999)
7. Rixon, S.J.: High resolution electronic spectra of some new transition metal-bearing molecules. Ph.D. thesis, University of British Columbia (2004)
8. Nakhate, S.G., Mukund, S., Bhattacharyya, S.: Radiative lifetime measurements in neutral zirconium using time-resolved laser induced fluorescence in supersonic free-jet. J. Quant. Spectrosc. Radiat. Transf. **111**, 394–398 (2010)
9. Bhattacharyya, S., Mukund, S., Nakhate, S.G.: Observation and rotational analysis of the (0,0) $\tilde{C}^2\Pi - \tilde{X}^2\Sigma^+$ transition in LaNH molecule and detection of $\tilde{A}'^2\Delta_r$ state. Chem. Phys. Letts. **692**, 1–6 (2018)
10. Herzberg, G.: Molecular Spectra and Molecular Structure, I. Spectra of Diatomic Molecules. Van Nostrand Reinhold, New York (1950)
11. Western, C.M.: PGOPHER: a program for simulating rotational structure, University of Bristol, Bristol (2016). http://pgopher.chm.bris.ac.uk
12. Krechkivska, O., Morse, M.D.: Resonant two-photon ionization spectroscopy of jet-cooled tantalum carbide, TaC. J. Chem. Phys. **133**, 054309-1-8 (2010)

Photoionization of Acetylene Doped in Helium Nanodroplets by EUV Synchrotron Radiation

Suddhasattwa Mandal[1]([✉]), Ram Gopal[2], S. R. Krishnan[3], Robert Richter[4], Marcello Coreno[5], Marcel Mudrich[6], Hemkumar Srinivas[7], Alessandro D'Elia[8], Bhas Bapat[9], and Vandana Sharma[10]

[1] Indian Institute of Science Education and Research Pune, Pune 411008, Maharashtra, India
suddhasattwa.mandal@students.iiserpune.ac.in
[2] TIFR Centre for Interdisciplinary Sciences, Hyderabad 500107, Telangana, India
ramgopal@tifrh.res.in
[3] Indian Institute of Technology Madras, Chennai 600036, Tamil Nadu, India
srkrishnan@iitm.ac.in
[4] Elettra-Sincrotrone Trieste, Strada Statale 14-km 163.5, 34149 Basovizza Trieste, Italy
robert.richter@elettra.eu
[5] Consiglio Nazionale delle Ricerche – Istituto di Struttura della Materia, 34149 Trieste, Italy
marcello.coreno@cnr.it
[6] Department of Physics and Astronomy, Aarhus University, 8000 Aarhus C, Denmark
mudrich@phys.au.dk
[7] Max-Planck-Institut für Kernphysik, 69117 Heidelberg, Germany
hemkumar.srinivas@mpi-hd.mpg.de
[8] Department of Physics, University of Trieste, 34127 Trieste, Italy
delia@iom.cnr.it
[9] Indian Institute of Science Education and Research Pune, Pune 411008, Maharashtra, India
bhas.bapat@iiserpune.ac.in
[10] Indian Institute of Technology Hyderabad, Sangareddy 502285, Telangana, India
vsharma@iith.ac.in

Abstract. Photoionization process of acetylene doped in helium nanodroplets is studied with EUV synchrotron radiation with photon energies between 20 and 26 eV by Photoelectron-Photoion Coincidence (PEPICO) experiment by detecting photoelectrons in coincidence with the photoions using electron velocity map imaging (VMI) spectrometer and ion time of flight (TOF) spectrometer. Acetylene is ionized in the droplet via Penning ionization at 21.6 eV photon energy. For photon energy of 23.9 eV and above the photoionization threshold of He, charge transfer ionization occurs in acetylene following autoionization and direct ionization in the droplet respectively.

© Springer Nature Singapore Pte Ltd. 2019
P. C. Deshmukh et al. (eds.), *Quantum Collisions and Confinement of Atomic and Molecular Species, and Photons*, Springer Proceedings in Physics 230, https://doi.org/10.1007/978-981-13-9969-5_22

1 Introduction

Helium nanodroplet is used as an ideal host matrix for spectroscopic study of embedded molecules because of its non-reactiveness with the embedded molecules, transparency to infrared, visible and ultraviolet light and ability to ro-vibronically cool the embedded molecule. However, upon photo-excitation of the He nanodroplets with extreme ultraviolet radiation, typically for photon energies between 20 and 26 eV, He droplet strongly absorbs the incident light and excites to electronically excited states or ionizes to the continuum. Following this excitation or ionization, He nanodroplet exhibits a very complex photodynamic between the droplet and the dopant [4,5]. Due to the transfer of excitation energy from the droplet to the dopant, ionization may occur in the dopant via Penning ionization process or charge transfer between the neutral dopant and the ionized droplet may lead to dopant ionization via charge transfer ionization process.

The effect of He environment in ionization process of acetylene with EUV synchrotron radiation has been studied at the GasPhase beamline of Elettra Synchrotron Facility, Italy. The photoion yields for different ionic fragments at different photon energies enabled us to identify different ionization processes in acetylene. We observed $C_2H_2^+$ ion yield peaks at 21.6 and 23.9 eV photon energies which occur through Penning ionization process following the de-excitation of excited He droplet and charge transfer ionization process following the autoionization of He droplet respectively. Charge transfer ionization process is noticed between the directly ionized He atoms in the droplet and the dopant C_2H_2 at photon energies above the ionization threshold of He.

We have recorded the photoelectrons with a Velocity Map Imaging (VMI) spectrometer in coincidence with the photoions detected with a Time of Flight (TOF) mass spectrometer for photon energies between 20 and 26 eV. From this detection procedure, the photoelectron spectra (PES) correlated to different ionic fragments from the droplet are obtained and these mass correlated PES helped us to gain insight into the ionization processes of acetylene in the droplet.

We have also looked into the isomerization process in acetylene under the He droplet environment. Acetylene cation $[HC = CH]^+$ is a well-studied molecular ion in which isomerization occurs through H atom migration from one C atom to other C atom upon absorption of EUV radiation. The isomerization of gas phase isolated acetylene occurs at an energy of 16.7 eV [9] which is well below the photoexcitation energy of He nanodroplets, therefore He nanodroplet environment is expected to play an important role in the isomerization process. The signature of isomerization in acetylene is the formation of CH_2^+ ion, therefore the CH_2^+ and $C_2H_2^+$ ion yields at different photon energies have been monitored. We have not observed any substantial change in the ion yield ratio of the isomeric ion (CH_2^+) to its parent ion $(C_2H_2^+)$ for the case of doped droplet ionization, when compared with the same ratio due to effusive acetylene ionization at these energies. This suggests that there is no enhancement or inhibition of isomerization process by the host He matrix.

2   Experimental Details

We have performed our experiment at the GasPhase beamline [11] of Elettra Synchrotron facility, Trieste, Italy. The schematic of the experimental setup has been shown in Fig. 1. The detailed information about the experimental setup could be found elsewhere [7,15]. The He droplets are generated by supersonically expanding pressurized high purity helium gas ($He6.0$) through a cold nozzle of 5μm diameter into the vacuum. The size of the droplet depends on the backing pressure and the nozzle temperature. For our experiment, we have kept the backing pressure to be constant at 50 bar and varied the nozzle temperature (T_{noz}) from 14 to 22 K. From the previous studies, the mean droplet size could be estimated to be 2000–15000 atoms per droplet at these expansion conditions [16,17].

After passing through the skimmer of 0.4 mm diameter orifice, the droplet beam is doped with acetylene gas via an effusive gas jet of acetylene in the doping chamber. The amount of doping is controlled by the pressure of the effusive acetylene gas in the doping chamber. We have varied the doping chamber pressure (P_{DC}) from 6.5×10^{-7} mbar to 4.5×10^{-6} mbar, where single to multiple doping condition of acetylene molecules into the droplet has been achieved. A mechanical chopper is introduced between pure He droplet source and the doping chamber to eliminate the background signal coming from the effusive gas from the doped droplet signal. Chopper open signal corresponds to ions and electrons

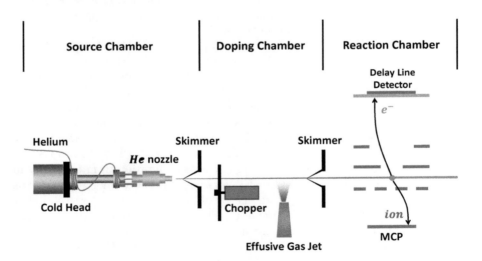

Fig. 1. Schematic of the experimental setup. He droplets are created in the source chamber by supersonic expansion of the helium gas through the cold nozzle. In the doping chamber, the droplets are doped with the effusive gas. At last the doped droplet beam reaches the reaction chamber, where it is ionized by XUV synchrotron radiation, and creates photoelectrons and ions which are then detected in coincidence via VMI and TOF spectrometer respectively

forming in the droplet and the effusive gas that is accompanied by the droplet. On the other hand, chopper close signal denotes ions and electrons forming in the effusive gas only. Therefore, to get the ionization signal only from the droplet, chopper close signal has been subtracted from the chopper open signal.

The acetylene doped helium droplet beam is then passed through the second skimmer to the reaction chamber, where the droplet beam interacts with the EUV synchrotron radiation of the GasPhase beamline of Elettra. The photon energy of the beamline can be varied between 13 and 900 eV with a good energy resolution of $E/\triangle E \geq 10^4$. However, we have used photon energy between 20 and 26 eV, which is required for electron excitation in the He droplet and ionization to a few electonvolts above the first ionization threshold of He in the droplets. The peak intensity of the radiation was kept around 15 Wm^{-2} with repetition rate of 500 MHz.

We used a Velocity Map Imaging (VMI) spectrometer to detect the photoelectrons in coincidence with the photoions which are being detected with a Time of Flight (TOF) mass spectrometer. This experimental detection scheme enabled us to perform Photoelectron Photoion Coincidence (PEPICO) experiment, where we could obtain the photoelectron energy spectrum (PES) correlated to different ionic fragments. To get the full three-dimensional velocity distribution of the photoelectrons from the two dimensional VMI images, MEVELER method, introduced by Bernhard Dick [6], has been used. For calibrating the VMI spectrometer, effusive helium gas has been ionized with different photon energies from 25 to 40 eV. Since the ionization energy of atomic helium (24.58 eV) is well known, the photoelectrons with well-known kinetic energies have been imaged on the VMI spectrometer at different photon energies. Since in the VMI spectrometer, same energy electrons are mapped on the 2D position sensitive detector, photoelectrons having specific constant kinetic energies form concentric circles in the MEVELER inverted images. From these images, the relationship between the kinetic energies of the photoelectrons and the radii of the circles has been obtained. With this method, we have calibrated the VMI spectrometer.

3   Results and Discussion

Upon irradiation of EUV radiation, both pure and doped He droplets exhibit very interesting excitation and ionization behaviour [4,5]. For photon energy between 20.5 and 23.0 eV, He droplets are excited with very high cross section into the energy band derived from $1s2s$ and $1s2p$ atomic He states. Following this excitation, inter-band and intra-band relaxation happen in the droplet which lead to He^* and He_2^* excimers formation inside the droplet [1,10,18,19]. These He^* and He_2^* excimers then migrate to the surface of the droplet due to repulsive interaction between the excimers and the droplet either by resonant hopping process or by fast nuclear motion to the surface [1,3,14]. Depending on size of the droplet, these He^* are either released from the droplet or trapped at the surface and eventually relaxed into the long lived $1s2s^{1,3}S$ He states or vibrationally excited He_2^* states. The latter are eventually evaporated out of the droplet by vibrational relaxation through coupling with the droplet [3]. In case of doped He droplet, when the droplet is relaxed to its ground state, the relaxation energy can be transferred to the dopant atom or molecule and excite or ionize the dopant by Penning ionization process [5].

When the droplet is excited by 23–24.6 eV photons, higher lying Rydberg states of the He_2^* [1,2,19] are accessed which can autoionize by very complex mechanism. This leads to emission of very low kinetic energy (<1 meV) electron [12,13]. Now for dopant present in the droplet, the vacancy created in the He droplet, can be transferred to the dopant atom or molecule and produce dopant ion via charge transfer process following the autoionization. For photon energy above the ionization threshold of He ($IP_{He} = 24.6$ eV), He^+ ions are created in the droplet by direct ionization of He. For doped droplet, the positive hole created in He^+ can eventually migrate to the dopant and can ionize the droplet by charge transfer process. In the autoionization and direct ionization regime, ionization in the droplet leads to fragmentation in the droplet, producing small He cluster ions (He_N^+) [8].

In this work, we have studied photoionization process of acetylene doped in the helium nanodroplet by measuring the dependence of photoion yield of He_2^+, CH_2^+ and $C_2H_2^+$ ion at different photon energies (see Fig. 2). The nozzle temperature and doping chamber pressure are maintained at 16 K and 6.6×10^{-7} mbar respectively. The green dashed lines correspond to the atomic energy levels of He. Below the ionization threshold of He atom, we can see that, He_2^+ ion yield peaks around 23.1 and 23.8 eV photon energies which correspond to $1s3p^1P$ and $1s4p^1P$ excited levels of atomic He respectively. For photon energy

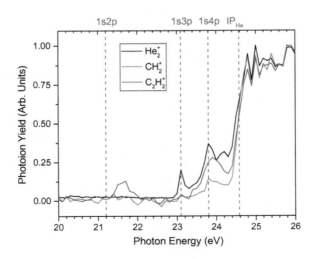

Fig. 2. Photoion yield at different photon energies for He_2^+, CH_2^+, and $C_2H_2^+$ ions coming from droplet at 16 K nozzle temperature and doping chamber pressure of 6.6×10^{-7} mbar. The dashed lines are corresponding to atomic excitation energies and photoionization threshold energy of He atom

above ionization potential of He (IP_{He}), direct ionization of the droplet can be seen. Above 23.0 eV photon energy, photoion yield spectrum for $C_2H_2^+$ and CH_2^+ ions are almost similar to that of He_2^+, which implies that there is initial autoionization ($< IP_{He}$) or direct ionization ($\geq IP_{He}$) of He in the droplet which is followed by charge transfer ionization of acetylene. Around 21.6 eV photon energy which corresponds to $1s2p^1P$ excitation band of He droplet, we can see a peak in the photoion yield of $C_2H_2^+$. Ionization in acetylene happens at this energy through the Penning ionization process.

There is negligible $C_2H_2^+$ ion yield coming from the droplet, before the Penning ionization and $C_2H_2^+$ ions are detected only when there are excitation and autoionization or direct ionization of the He in the droplet even though ionization threshold of C_2H_2 (11.3 eV) is much below the Penning ionization energy. One can argue that, C_2H_2 is only excited or ionized inside the droplet via the initial excitation or ionization of He droplet environment.

There are two major states of $[HCCH]^+$, $X^2\Pi_u$ (11.3 eV) and $A^2\Sigma_g^+$ (16.7 eV), in which the C_2H_2 can be populated due to the energy transfer to the C_2H_2, following the relaxation of excited He droplet to its ground state in the Penning ionization regime. Now the $A^2\Sigma_g^+$ state of $[HCCH]^+$ is unstable towards isomerization since there is no potential barrier for isomerization to $[H_2CC]^+$ in the $\tilde{X}^2B_1$ (13.2 eV) and $\tilde{X}^2B_2$ (16.7 eV) states. On the other hand, $X^2\Pi_u$ does not lead to isomerization because of the finite potential differences [9]. For different photon energies from 20 to 24.5 eV, the ratio of CH_2^+ ion yield to the parent $C_2H_2^+$ ion yield in case of the doped droplet has been observed to be the same as the ratio when measured with effusive C_2H_2. If there are any changes in the relative populations of the isomeric state ($A^2\Sigma_g^+$) and non-isomeric state ($X^2\Pi_u$) of $C_2H_2^+$ then we should expect to see otherwise. Therefore, we argue that there is no enhancement or suppression of the isomerization process due to the He droplet environment.

The photoelectron spectra correlated to $C_2H_2^+$ ions and He_2^+ ions at 26 eV photon energy for $T_{noz} = 16K$ and $P_{DC} = 6.6 \times 10^{-7}$ mbar are shown in Fig. 3. We can see that PES correlated to both the ions are exactly the same and there is a sharp peak in the PES at the photoelectron kinetic energy of 1.5 eV which is approximately equal to the excess photon energy above the ionization threshold (24.58 eV) of He atom. Therefore we can say that the photoelectrons detected in coincidence with the $C_2H_2^+$ ions are actually coming from the direct ionization of He atom inside the droplet which is followed by the charge transfer ionization of C_2H_2. The PES correlated to $C_2H_2^+$ at the Penning ionization and autoionization regimes will be discussed in other papers.

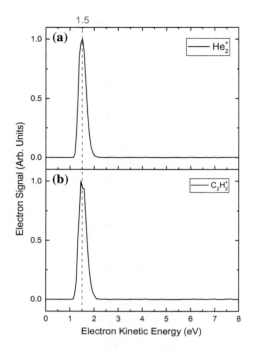

Fig. 3. Photoelectron Spectra correlated to **a** He_2^+ and **b** $C_2H_2^+$ ions at 26 eV photon energy for T_{noz} =16 K and $P_{DC} = 6.6 \times 10^{-7}$ mbar. The blue dashed line represents the peak kinetic energy of the PES

4   Conclusion

Acetylene is ionized through Penning ionization process and charge transfer ionization process following the autoionization and direct ionization of He droplet in the He nanodroplet environment. In the Penning ionization regime, there is no evidence to the role of droplet environment leading to either enhancement or suppression of isomerization process in acetylene.

References

1. Bnermann, O., Kornilov, O., Leone, S.R., Neumark, D.M., Gessner, O.: Femtosecond extreme ultraviolet ion imaging of ultrafast dynamics in electronically excited helium nanodroplets. IEEE J. Sel. Topics Quantum Electron. **18**(1), 308–317 (2012)
2. Bnermann, O., Kornilov, O., Haxton, D.J., Leone, S.R., Neumark, D.M., Gessner, O.: Ultrafast probing of ejection dynamics of rydberg atoms and molecular fragments from electronically excited helium nanodroplets. J. Chem. Phys. **137**(21), 214302 (2012)

3. Buchenau, H., Toennies, J.P., Northby, J.A.: Excitation and ionization of 4he clusters by electrons. J. Chem. Phys. **95**(11), 8134–8148 (1991)
4. Buchta, D., Krishnan, S.R., Brauer, N.B., Drabbels, M., OKeeffe, P., Devetta, M., Di Fraia, M., Callegari, C., Richter, R., Coreno, M., Prince, K.C., Stienkemeier, F., Ullrich, J., Moshammer, R., Mudrich, M.: Extreme ultraviolet ionization of pure he nanodroplets: mass-correlated photoelectron imaging, penning ionization, and electron energy-loss spectra. J. Chem. Phys. **139**(8), 084301 (2013)
5. Buchta, D., Krishnan, S.R., Brauer, N.B., Drabbels, M., O'Keeffe, P., Devetta, M., Di Fraia, M., Callegari, C., Richter, R., Coreno, M., Prince, K.C., Stienkemeier, F., Moshammer, R., Mudrich, M.: Charge transfer and penning ionization of dopants in or on helium nanodroplets exposed to EUV radiation. J. Phys. Chem. A **117**(21), 4394–4403 (2013). PMID: 23638683
6. Dick, B.: Inverting ion images without abel inversion: maximum entropy reconstruction of velocity maps. Phys. Chem. Chem. Phys. **16**, 570–580 (2014)
7. Fechner, L., Grüner, B., Sieg, A., Callegari, C., Ancilotto, F., Stienkemeier, F., Mudrich, M.: Photoionization and imaging spectroscopy of rubidium atoms attached to helium nanodroplets. Phys. Chem. Chem. Phys. **14**, 3843–3851 (2012)
8. Fröchtenicht, R., Henne, U., Toennies, J.P., Ding, A., Fieber-Erdmann, M., Drewello, T.: The photoionization of large pure and doped helium droplets. J. Chem. Phys. **104**(7), 2548–2556 (1996)
9. Jiang, Y.H., Rudenko, A., Herrwerth, O., Foucar, L., Kurka, M., Kühnel, K.U., Lezius, M., Kling, M.F., van Tilborg, J., Belkacem, A., Ueda, K., Düsterer, S., Treusch, R., Schröter, C.D., Moshammer, R., Ullrich, J.: Ultrafast extreme ultraviolet induced isomerization of acetylene cations. Phys. Rev. Lett. **105**, 263002 (2010)
10. Kornilov, O., Bnermann, O., Haxton, D.J., Leone, S.R., Neumark, D.M., Gessner, Oliver: Femtosecond photoelectron imaging of transient electronic states and Rydberg atom emission from electronically excited he droplets. J. Phys. Chem. A **115**(27), 7891–7900 (2011). PMID: 21688802
11. O'Keeffe, P., Bolognesi, P., Coreno, M., Moise, A., Richter, R., Cautero, G., Stebel, L., Sergo, R., Pravica, L., Ovcharenko, Y., Avaldi, L.: A photoelectron velocity map imaging spectrometer for experiments combining synchrotron and laser radiations. Rev. Sci. Instrum. **82**(3), 033109 (2011)
12. Peterka, D.S., Kim, J.H., Wang, C.C., Poisson, L., Neumark, D.M.: Photoionization dynamics in pure helium droplets. J. Phys. Chem. A **111**(31), 7449–7459 (2007). PMID: 17571863
13. Peterka, D.S., Lindinger, A., Poisson, L., Ahmed, M., Neumark, Daniel M.: Photoelectron imaging of helium droplets. Phys. Rev. Lett. **91**, 043401 (2003)
14. Scheidemann, A., Schilling, B., Peter Toennies, J.: Anomalies in the reactions of helium(1+) with sulfur hexafluoride embedded in large helium-4 clusters. J. Phys. Chem. **97**(10), 2128–2138 (1993)
15. Shcherbinin, M., LaForge, A.C., Sharma, V., Devetta, M., Richter, R., Moshammer, R., Pfeifer, T., Mudrich, M.: Interatomic coulombic decay in helium nanodroplets. Phys. Rev. A **96**, 013407 (2017)
16. Stienkemeier, F., Lehmann, K.K.: Spectroscopy and dynamics in helium nanodroplets. J. Phys. B At. Mol. Opt. Phys. **39**(8), R127 (2006)
17. Toennies, J.P., Vilesov, A.F.: Superfluid helium droplets: a uniquely cold nanomatrix for molecules and molecular complexes. Angew. Chem. Int. Ed. **43**(20), 2622–2648 (2004)

18. von Haeften, K., de Castro, A.R.B., Joppien, M., Moussavizadeh, L., von Pietrowski, R., Möller, T.: Discrete visible luminescence of helium atoms and molecules desorbing from helium clusters: the role of electronic, vibrational, and rotational energy transfer. Phys. Rev. Lett. **78**, 4371–4374 (1997)
19. von Haeften, K., Laarmann, T., Wabnitz, H., Mller, T.: The electronically excited states of helium clusters: an unusual example for the presence of Rydberg states in condensed matter. J. Phys. B At. Mol. Opt. Phys. **38**(2), S373 (2005)

Positron Collision Dynamics for C_2–C_3 Hydrocarbons

Suvam Singh$^{(\boxtimes)}$, Pankaj Verma, Nafees Uddin, Paresh Modak,
Nidhi Sinha, Himani Tomer, Vishwanath Singh, and Bobby Antony

519, Atomic and Molecular Physics Lab, Department of Physics, Indian Institute
of Technology (Indian School of Mines), Dhanbad 826004, Jharkhand, India
{suvamsingh18sep,pvermaism,nafeesuddin27,
paresh.modak301,sinhanidhi00,himanitomer22,
singhvishwanath42}@gmail.com, bobby@iitism.ac.in

Abstract. This article presents positron scattering cross sections for simple hydrocarbons, viz., ethane, ethene, ethyne, propane, and propyne. The present work focuses on the calculation of different elastic and inelastic interactions due to positron impact with simple hydrocarbons in terms of inelastic cross sections (both with and without positronium formation channel), electronic excitation cross section (Q_{exc}) and momentum transfer cross section (Q_{mtcs}). Knowing that positron-plasma study is one of the trending fields, the calculated data have diverse plasma and astrophysical modeling applications. Due to the lack of previous studies, all the cross sections presented have been reported for the first time.

1 Introduction

Introduction of sophisticated trapping methods and storage [1] has now enabled researchers around the world to perform a number of positron scattering experiments. Similarly, many theories have come up to back these measurements. Even though these technological and computational developments have helped in finding the cross sections for many systems, there is still a dearth of data in molecules such as the present ones. The present work is a small step in improving the scenario of cross-section data for positron scattering from simple hydrocarbons. Being an antiparticle of electron, positron annihilates electrons. In addition to that, they can combine with electrons to form neutral plasmas or pair plasmas. These plasmas have significance in the growth of extreme astrophysical objects such as pulsars and black holes [2]. However, laboratory experiments on pair plasmas such as positron accumulation experiment (PAX) [3] and a positron-electron experiment (APEX) [3] have proved that positron collisions have various applications in the field of plasma science as well.

Study of various cross sections for C_2–C_3 hydrocarbons (viz., ethane, ethene, ethyne, propane, and propyne) via positron scattering is undertaken in the present endeavor. This work is believed to be important due to numerous applications of the chosen targets in the field of astrophysics and plasma physics. For example, the plasma-facing material of almost all modern operating fusion devices contains carbon as one of the major element. Chemical erosion caused by plasma-wall interactions is an

© Springer Nature Singapore Pte Ltd. 2019
P. C. Deshmukh et al. (eds.), *Quantum Collisions and Confinement of Atomic and Molecular Species, and Photons*, Springer Proceedings in Physics 230,
https://doi.org/10.1007/978-981-13-9969-5_23

abundant source of simple hydrocarbons which contaminate the hydrogenic plasma [4]. The composition of the hydrocarbon fluxes flowing inside the plasma covers a wide spectrum of molecules from methane to propane [5]. These hydrocarbons play an important role in plasma diagnostics in the diverter region of magnetically confined high-temperature hydrogen plasma [6]. As the energy of ions striking the walls of plasma surface decreases more complex C_2–C_3 hydrocarbons are discharged [4] and, hence the importance of the present study becomes more vital. Moreover, the presence of these hydrocarbons in the planetary and cometary atmosphere makes these targets important in the field of astrophysics [6]. Various cross-section data such as different excitation and momentum transfer cross sections are required to comprehend the behavior of these molecules in various astrophysical and plasma environment. Furthermore, these data are very essential in various modeling raging from RF plasma processing to astrophysical media [7]. In addition to that, lighter linear hydrocarbons such as ethyne are interesting targets due to their role in chemical vapor deposition reactions [8].

The modified form of spherical complex optical potential (SCOP) formalism [9, 10] and complex scattering potential-ionization contribution (CSP-ic) [11, 12] are used in this work to calculate the positron scattering cross sections over a wide energy range for the targets ethane, ethyne, propane, and propyne. The cross sections reported in the present work are inelastic (with and without including the positronium (Ps) formation channel), electronic excitation, and momentum transfer cross section. In our recent article [13], positron scattering total cross section, elastic cross section, direct ionization cross section positronium formation cross section, and total ionization cross section have been reported for the same set of targets by employing an identical method. However, for the present set of cross sections, no previous studies have been reported in the literature. The following section outlines the theoretical methodology followed by results and discussion and finally, conclusions are drawn from the calculated results.

2 Theory

The detail of the theoretical methods employed in this work is already given in the articles [9–12]. However, an overview is provided in this section. The modified form of SCOP [9, 10] and CSP-ic methods [11, 12] are used to calculate various cross sections reported in this work. In this model, the potential experienced by the incident positrons due to the target is represented by the following equation:

$$V_{opt}(r, E_i) = V_{st}(r) + V_{pol}(r) + iV_{abs}(r, E_i) \tag{1}$$

where, V_{st} represents the static potential, which is an interaction potential experienced by an incident positron due to the static electronic charge cloud of the target. V_{pol} is the polarization potential, which occurs because of the distortion of the charge distribution of the target due to the field created by the incident projectile. The first two terms form the real part in (1), which gives rise to the elastic processes occurring during the collision. The complex part of (1) is the absorption potential. This accounts for all the inelastic processes occurring during the scattering process. V_{abs} represents the total loss of scattered flux into all allowed inelastic channels, such as positronium formation, electronic excitation, and ionization channels. Since the present model treats the scattering problem in the form of spherical potential, hence rotational and vibrational channels cannot be included in the calculation.

The charge density and static potential have been formulated by the potential field parameters of Cox and Bonham [14], whereas the correlation polarization potential given by Zhang et al. [15] is used for the formulation of polarization potential. The present work uses the absorption model of Reid and Wadehra's [16] to calculate various inelastic processes occurring during the collision.

Accurate evaluation of charge density of the targets is very essential in the present model since all the interaction potentials depend on this. Since the parameters given by Cox and Bonham [14] are only for atoms, the multi-scattering center (MSC) approach is employed to derive the charge density for molecules. In this method, instead of considering the collision between the incident positron and the whole molecule, independent collision between the incident positron and a part of the molecule is visualized. To develop the charge density of each center, the charge density of lighter hydrogen atom is expanded from the center of the carbon atom. However, if the bond length of C–C is smaller than the diameter of the carbon atom, then the charge density of carbon atom is expanded from the center of mass of the system. For the present set of targets, such condition arises in case of ethyne and propyne, where the triple bond between two carbon atoms is smaller than the sum of the diameter of the respective carbon atoms. The charge density of these two carbon atoms is expanded from their center of mass. However, for the rest of the molecules, i.e., ethane, ethene, and propane, a simple addition of the charge density of the constituent atoms is done due to large C–C bond length as compared to the diameter of the carbon atom. This is done on the basis of the assumption that if the bond length is large, then the charge density of one atom does not affect the charge density of the other atom of the same molecule. Figure 1 gives the structure of the molecules, whereas Table 1 provides the respective bond lengths.

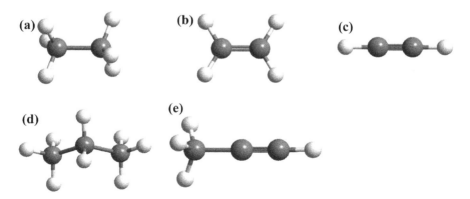

Fig. 1. Molecular structure. **a** Ethane, **b** Ethene, **c** Ethyne, **d** Propane, **e** Propyne

Table 1. Bond lengths [20]

Target	Bond length (Å)
Ethane	C–C = 1.536, C–H = 1.091
Ethene	C=C = 1.339, C–H = 1.086
Ethyne	C≡C = 1.203, C–H = 1.063
Propane	C–C = 1.526, C–H = 1.089, 1.094, 1.094
Propyne	C–C = 1.460, C≡C = 1.207, C–H = 1.060, 1.096

3   Calculation of Different Inelastic Cross Sections

The interaction potential obtained in (1) was used in the Schrödinger equation, which was solved by partial wave method under suitable boundary conditions [17]. The inelastic cross section contains many absorption channels such as positronium formation, electronic excitations, and ionizations. Due to the two center problem, incorporating positronium formation channel in the absorption phenomena is one the most difficult task that one has to deal with while studying positron scattering. A phenomenological method is adopted to overcome this situation. The details of the method can be obtained from our previous articles [9–12]. However, the essential parts are discussed here.

For low incident energies, the convergence of the solution is obtained with less partial waves. However, for positrons with high energies, more than 50 partial waves were required for the convergence. The solutions were obtained in the form of complex phase shifts (δ_l) which were used to calculate the absorption factor (η_l), using the relation $\eta_l = \exp(-2\mathrm{Im}\delta_l)$. This absorption factor forms the main parameter which is essential to calculate the inelastic cross sections, which is given as follows:

$$Q_{inel}(E_i) = \frac{\pi}{k^2} \sum_{l=0}^{\infty} (2l+1)\left(1 - \eta_l^2\right) \tag{2}$$

where l is the number of partial waves and k is the wave vector associated with the scattered wave.

The phenomenological formula proposed by Chiari et al. [18] is employed in the present work to determine the absorption threshold parameter. Since rotational and vibrational excitations are not included in the present calculation, all the inelastic processes are assumed to be prohibited below the absorption threshold. However, beyond this value, the inelastic channels like Ps formation, electronic excitations, and ionizations are included. Although this method is far from being accurate, however, proves to be an effective process to include Ps formation channel in the calculation for large and complex molecules which is not possible to do using any stringent theories due to the complexity of the problem. The threshold parameter employed in this work is given as

$$\Delta(E_i) = \Delta_e - \left(\Delta_e - \Delta_p\right)e^{-(E_i-\Delta_p)/E_m} \tag{3}$$

where E_i is the energy of the incident projectile (positron), Δ_e and Δ_p are the first electronic excitation energy and Ps formation threshold, respectively. E_m is the energy at which the inelastic cross section has the maximum magnitude when Δ_e is used as the inelastic threshold. In the present calculation, ionization energy (IE) is used as the absorption threshold to calculate the absorption potential given by Reid and Wadehra [16] which is then used to calculate the inelastic cross section without Ps formation (Q_{in}) using (2). Equation 3 is used to evaluate the dynamic inelastic threshold for the calculation of inelastic cross section including the Ps formation channel (Q_{inel}).

4 Calculation of Electronic Excitation Cross Section (Q_{exc})

As already mentioned, Q_{in} only contains electronic excitation and ionization channels. Thus, mathematically it can be given as

$$Q_{in} = Q_{exc} + Q_{ion} \tag{4}$$

where, Q_{exc} represents the sum over all the discrete excitation cross sections for all the accessible electronic states and Q_{ion} is the total cross sections of all allowed direct ionization processes. The well-known CSP-ic method [11, 12] is used to estimate the Q_{ion}, which is already reported in [13]. Since Q_{in} is evaluated using the above-mentioned method and Q_{ion} is already known, Q_{exc} can be easily estimated by

$$Q_{exc} = Q_{in} - Q_{ion}. \tag{5}$$

From [11, 12, 19], it is understood that the uncertainty in the estimation of the direct ionization cross section is around 7%. Eventually, this results in an uncertainty of 7% in the determination of Q_{exc} as well.

5 Calculation of Momentum Transfer Cross Section (Q_{mtcs})

Q_{mtcs} provides information about the average momentum transferred by a projectile to the target scattering process. The Q_{mtcs} calculated here is averaged over all the scattering angles. The expression for finding Q_{mtcs} is given as

$$Q_{mtcs}(k) = \frac{4\pi}{k^2} \sum_{l=0}^{\infty} (l+1)\sin^2[\delta_{l+1}(k) - \delta_l(k)] \tag{6}$$

Table 2 enumerates all the parameters used in the calculation, which includes target properties. No other parameter is used in the present calculation, which renders the present data to be easily reproducible.

Table 2. Target properties

Target	Polarizability (Å^3) [20]	IE (eV) [20]	Δ_e (eV)	Δ_p (eV)	E_m (eV)
Ethane	4.226	11.52	10.00	4.72	40
Ethene	4.188	10.51	9.60	3.71	35
Ethyne	3.487	11.40	9.20	4.60	35
Propane	5.921	10.94	9.49	4.14	40
Propyne	5.550	10.36	9.07	3.56	35

6   Results and Discussion

6.1   Electronic Excitation Cross Section (Q_{exc})

Figure 2 presents the total electronic excitation cross section for simple hydrocarbons up to 500 eV. This cross section provides the absorption of flux due to the electronic excitation of the target due to positron collision. However, since the threshold is taken to be the ionization potential of the target, hence, the present cross section only provides information for the electronic states which lie beyond the ionization threshold. The variation in the magnitude of cross section is dependent on the size of the target. Hence, it is observed that the magnitude of Q_{exc} is highest for propane and lowest for ethene. Ethyne being the smallest target is studied in this work, hence, as expected it has the lowest Q_{exc}. Besides, it is noted that in the rising end, the curves are pretty steep whereas it is rather flat in the trailing end of the curve. This signifies that the rate of electronic excitation is high in the low energies which quickly attains a maximum and then gradually decreases as the energy of the positron increases. Due to less interaction time, the cross sections at higher energies decrease. Since the present Q_{exc} is calculated by subtracting the ionization cross section from the inelastic cross section (Q_{in}), hence, it has an uncertainty of 7% in its calculation [12].

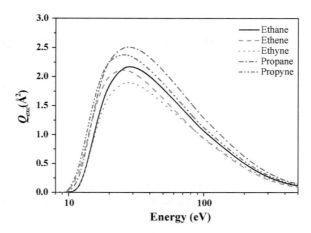

Fig. 2. Positron scattering electronic excitation cross section. Solid line: Ethane, dashed line: Ethene, dotted line: Ethyne, dashed dot line: Propane, dashed dot dot line: Propyne

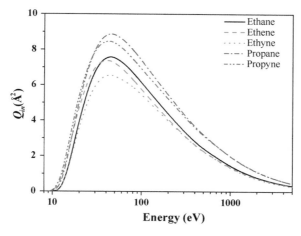

Fig. 3. Positron scattering inelastic cross section without positronium formation channel. Solid line: Ethane, dashed line: Ethene, dotted line: Ethyne, dashed dot line: Propane, dashed dot dot line: Propyne

6.2   Inelastic Cross Section Without Positronium Formation (Q_{in}) (i.e., Electronic Excitation Plus Ionization Cross Section)

Positron scattering inelastic cross section without the positronium formation channel is pictorially represented in Fig. 3. The inelastic cross section reported here consists of all the accessible electronic excited states and the allowed ionization channels. The general trend of the graph is that the Q_{in} increases as the energy increases then it reaches a maximum and after which the curve starts decreasing. The magnitude of cross section depends on the size of the molecule, thus the number of electron plays a role in determining the cross section. Since the number of electrons is highest in propane, hence, it has the largest magnitude of cross section. The magnitude of the cross section is in the order propane > propyne > ethane > ethene > ethyne same as the order of the number of electrons in the targets. On the left end of the peak the curves get steeper as the size increases which is obvious due to the fact that as the size of the target increases the magnitude of cross section increase, however, the peak position remains almost at the same energy, which results in a steeper rising curve. It is worth mentioning that the threshold of Q_{in} is the ionization energy of that particular target.

6.3   Inelastic Cross Section Including the Positronium Formation Channel (Q_{inel}) (i.e., Electronic Excitation Plus Ionization Plus Positronium Formation Cross Section)

Figure 4 provides the positron scattering inelastic cross section which includes the positronium formation channel. Unlike the previous case, here the inelastic threshold is varied with energy according to (3). The magnitude of cross section for each molecule is greater than the cross sections observed in the previous case. This is due to the fact that Q_{inel} includes an additional absorption channel of Ps formation. Hence, it is evident

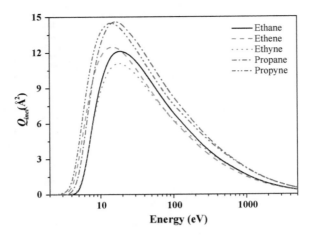

Fig. 4. Positron scattering inelastic cross section including positronium formation channel. Solid line: Ethane, dashed line: Ethene, dotted line: Ethyne, dashed dot line: Propane, dashed dot dot line: Propyne

that the difference between Q_{inel} and Q_{in} gives the Ps formation cross section. It is well-known that Ps formation has major influence at low energies, which adds to the cross sections calculated in Fig. 3. Hence, the peaks in the present curves show a left shift when compared to Fig. 3. Unlike the previous case, the magnitude of cross section not only depends on the number of target electrons but also on Δ_e. From Table 2, it is seen that for molecules containing two carbon atoms, Δ_e decreases as the size of the molecule decrease. Same is seen for molecules containing three carbon atoms as well. It is known that low threshold value increases the cross section, whereas from the previous case it is seen that the cross section decreases as the size of the molecule decrease. These two contrasting variations make the magnitude of the Q_{inel} non-uniform with respect to the size of the target. This could be the reason that the peak positions are different for each of the targets. The general trend of the graph is somewhat like a Gaussian function similar to all the previous results as seen in Figs. 2 and 3. It is noticeable that in the trailing end of the curve, all the values tend to coincide. This is in accordance with the first Born approximation, where the cross section depends majorly on the incident kinetic energy of the projectile.

7   Correlation Plot

The total cross section reported in our recent work [13] agrees well with the experimental studies and was found to lie within the experimental uncertainties. Since the present work uses the same method, the cross sections reported is expected to be reliable and consistent. However, to further ascertain the consistency of the present data a correlation plot has been shown in Fig. 5. Figure 5a shows a correlation plot between Q_{exc} and number of electrons in the molecule at 45 eV. It is observed that there is a linear relationship between the Q_{exc} and number of electrons in the target. Likewise,

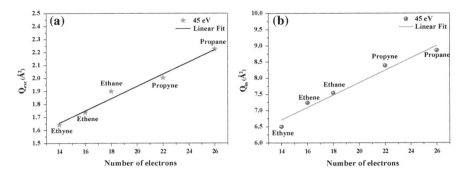

Fig. 5. Correlation plot. **a** Q_{exc} versus the number of electrons, **b** Q_{in} versus the number of electrons

Fig. 5b shows a correlation plot between Q_{in} and number of electrons at 45 eV. This figure also reflects a linear relationship between Q_{in} and number of electrons in the target. The energy is chosen to be 45 eV because of the fact that Q_{exc} and Q_{in} have the maximum magnitude around this energy for all the targets. The intercept and slope of the linear fit of Fig. 5a are 0.995 and 0.047, respectively, whereas for Fig. 5b they are 4.037 and 0.191, respectively. This linear relationship in the two figures demonstrates the reliability of the present data.

8   Momentum Transfer Cross Section

Figure 6 illustrates the positron scattering momentum transfer cross section for simple hydrocarbons from 5 to 300 eV. As evident from the size of the molecule, Q_{mtcs} for propane has the highest magnitude whereas Q_{mtcs} for ethyne is the lowest. It is

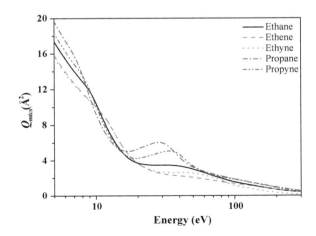

Fig. 6. Positron scattering momentum transfer cross section. Solid line: Ethane, dashed line: Ethene, dotted line: Ethyne, dashed dot line: Propane, dashed dot dot line: Propyne

observed that as the size of the molecule decreases Q_{mtcs} is seen to decrease as well. However, for ethene and ethyne, the nature is slightly different from others. Moreover, a hat like structure is observed for some molecules between the energy range of 15–40 eV. Since Q_{mtcs} is derived from the elastic cross section, which is calculated in presence of absorption processes, the hat like structure could be due to the inelastic channel. Since, in this region the inelastic cross sections have a higher magnitude of cross section. This effect is maximum for propane and propene as they have significantly higher inelastic cross sections than the rest of the targets in the concerned energy range. These $e^{\pm} Q_{mtcs}$ data are valuable in modeling positron transport in gases.

9 Conclusion

This work reports different cross sections due to positron scattering from simple molecules. No such previous studies have been found in the literature. Hence, no comparisons were made for the present calculations. Different thresholds have been employed to calculate various cross sections, which could be helpful to study several threshold effects. The targets undertaken in this work have a variety of applications in the field of plasma and astrophysics. Hence, the calculated cross sections are believed to be useful to the modeling community. The calculations done are free from any adjustable parameters thus making the calculation reproducible. Moreover, the SCOP formalism employed in this work has a distinct advantage over other methods that compute cross sections in the intermediate and high energy range, since it takes less computing. Besides, it can handle complex and large targets with much ease as compared to any other model. Hence, the present work is expected to be appreciated by the atomic and molecular physics community.

References

1. Greaves, R.G., Tinkle, M.D., Surko, C.M.: Creation and uses of positron plasmas. Phys. Plasmas **1**, 1439–1446 (1994). https://doi.org/10.1063/1.870693
2. Sarri, G., Poder, K., Cole, J.M., Schumaker, W., Piazza, A.D., Reville, B., Dzelzainis, T., Doria, D., Gizzi, L.A., Grittani, G., Kar, S., Keitel, C., Krushelnick, K., Kuschel, S., Mangles, S., Najmudin, Z., Shukla, N., Silva, L., Symes, D., Thomas, A., Vargas, M., Vieira, J., Zepf, M.: Generation of neutral and high-density electron–positron pair plasmas in the laboratory. Nat. Commun. **6**(1–8), 6747 (2015). https://doi.org/10.1038/ncomms7747
3. Pedersen, T.S., Danielson, J.R., Hugenschmidt, C., Marx, G., Sarasola, X., Schauer, F., Schweikhard, L., Surko, C.M., Winkler, E.: Plans for the creation and studies of electron–positron plasmas in a stellarator. New J. Phys. **14**, 035010 (2012). https://doi.org/10.1088/1367-2630/14/3/035010
4. Deutsch, H., Becker, K., Janev, R.K., Probst, M., Mark, T.D.: Isomer effect in the total electron impact ionization cross section of cyclopropane and propene (C_3H_6). J. Phys. B: At. Mol. Opt. Phys. **33**, L865 (2000). https://doi.org/10.1088/0953-4075/33/24/102
5. Eckstein, W., Hofer, W., Philipps, V., Roth, J.: Physical Processes of the Interaction of Fusion Plasmas with Solids. Academic press (1996)

6. Jiang, Y., Sun, J., Wan, L.: Total cross sections for electron scattering by polyatomic molecules at 10–1000 eV: C_2H_2, C_2H_4, C_2H_6, C_3H_6, C_3H_8 and C_4H_8. Z. Phys. D: At. Mol. Clust. **34**, 29–33 (1995). https://doi.org/10.1007/BF01443734

7. Charlton, M., Humberston, J.W.: Positron Physics. Cambridge University Press (2001)

8. Makabe, T.: In: Kimura, M., Itikawa, Y. (eds.) Advances in Atomic, Molecular and Optical Physics, vol. 44, pp. 127–154. Academic Press (2001)

9. Singh, S., Dutta, S., Naghma, R., Antony, B.: Theoretical formalism to estimate the positron scattering cross section. J. Phys. Chem. A **120**, 5685–5692 (2016). https://doi.org/10.1021/acs.jpca.6b04150

10. Singh, S., Dutta, S., Naghma, R., Antony, B.: Positron scattering from simple molecules. J. Phys. B: At. Mol. Opt. Phys. **50**, 135202 (2017). https://doi.org/10.1088/1361-6455/aa7550

11. Singh, S., Antony, B.: Positronium formation and ionization of atoms and diatomic molecules by positron impact. EPL **119**, 50006 (2017). https://doi.org/10.1209/0295-5075/119/50006

12. Singh, S., Antony, B.: Study of inelastic channels by positron impact on simple molecules. J. Appl. Phys. **121**, 244903 (2017). https://doi.org/10.1063/1.4989850

13. Singh, S., Antony, B.: Positron induced scattering cross sections for hydrocarbons relevant to plasma. Phys. Plasmas **25**, 053503 (2018). https://doi.org/10.1063/1.5024581

14. Cox, H.L., Bonham, R.A.: Elastic electron scattering amplitudes for neutral atoms calculated using the partial wave method at 10, 40, 70, and 100 kV for Z = 1 to Z = 54. J. Chem. Phys. **47**, 2599–2608 (1967). https://doi.org/10.1063/1.1712276

15. Zhang, X., Sun, J., Liu, Y.: A new approach to the correlation polarization potential-low-energy electron elastic scattering by He atoms. J. Phys. B: At. Mol. Opt. Phys. **25**, 1893–1897 (1992). https://doi.org/10.1088/0953-4075/25/8/021

16. Reid, D.D., Wadehra, J.M.: A quasifree model for the absorption effects in positron scattering by atoms. J. Phys. B: At. Mol. Opt. Phys. **29**, L127–L133 (1996). https://doi.org/10.1088/0953-4075/29/4/002

17. Bransden, B.H., Joachain, C.J.: Physics of Atoms and Molecules. Pearson Education Limited (2003)

18. Chiari, L., Zecca, A., Girardi, S., Trainotti, E., Garcia, G., Blanco, F., McEachran, R.P., Brunger, M.J.: Positron scattering from O_2. J. Phys. B: At. Mol. Opt. Phys. **45**, 215206 (2012). https://doi.org/10.1088/0953-4075/45/21/215206

19. Naghma, R., Mahato, B.N., Vinodkumar, M., Antony, B.: Electron impact total ionization cross sections for atoms with Z = 49–54. J. Phys. B: At. Mol. Opt. Phys. **44**, 105204 (2011). https://doi.org/10.1088/0953-4075/44/10/105204

20. CCCBDB. www.nist.gov/pml/handbook-basic-atomic-spectroscopicdata

Electron-Impact Excitation of Pb$^+$

Swati Bharti, Lalita Sharma$^{(\boxtimes)}$, and Rajesh Srivastava

Indian Institute of Technology Roorkee, Roorkee 247667, India
lalitfph@iitr.ac.in

Abstract. We have used relativistic distorted-wave method to study electron-impact excitation of the fine-structure transitions from the $6s^26p\ ^2P_{1/2,3/2}$ states to the excited $6s^26d\ ^2D_{3/2,5/2}$, $6s6p^2\ ^2D_{3/2,5/2}$ and $6s^27\ s\ ^2S_{1/2}$ states of Pb$^+$ ion. The bound state wavefunctions of the target ion are obtained within the multi-configuration Dirac-Fock approach. The reliability of the calculated wavefunctions is determined by comparing our calculated oscillator strengths for the different transitions with the available measurements and other theoretical calculations. Finally, the cross sections for all the considered fine-structure transitions are reported in the incident electron energy range from the excitation threshold to 100 eV. Cross section results for the resonance transition $6s^26p\ ^2P_{1/2} \rightarrow 6s^26d\ ^2D_{3/2}$ are compared with the only available recent measurements [Gomonai et al. Eur. Phys. J. D **71** 31 (2017)] and good agreement is found at the high incident electron energies.

1 Introduction

Lead has been discovered in interstellar gas by the Goddard High-Resolution Spectrograph (GHRS) onboard the Hubble Space Telescope (HST) [1]. It has also been observed that the cosmic abundance of lead is the highest among the elements heavier than barium (Z > 56). Pb$^+$ is expected to be the dominant ionization stage of gaseous lead in cold, neutral interstellar clouds [2]. The resonance line $6s^26p\ ^2P_{1/2} \rightarrow 6s^26d\ ^2D_{3/2}$ of the Pb$^+$ at 1433.906 Å are reported in the emission spectra of several stars as well as in archival Space Telescope Imaging Spectrograph (STIS) data [1, 2]. Recently, Heidarian et al. [3] have reported the first detection of the $6s^26p\ ^2P_{1/2} \rightarrow 6s6p^2\ ^2D_{3/2}$ transition of Pb$^+$ at 1203.616 Å in the interstellar medium. Thus it is important to have accurate atomic data for a singly charged lead ion to enhance our understanding of the production of elements heavier than iron as well as depletion of heavy elements from the interstellar gas into the dust grains.

Electron-impact excitation processes play an important role in estimating the abundances of species in the plasma as well as temperature and density of the plasma. Most of the earlier theoretical and experimental work have reported wavelengths, transition probabilities and oscillator strengths of the transitions and lifetime of the states of Pb$^+$ ion. Alonso-Medina [4] measured transition probabilities of 30 lines of Pb$^+$ by measuring the intensities of the emission lines of a laser-produced plasma of Pb. Colón and Alonso-Medina [5] calculated transition probabilities for 190 lines, with wavelengths lower than 15,000 Å using the relativistic Hartree–Fock method and configuration interaction in an intermediate coupling scheme. These involved

P. C. Deshmukh et al. (eds.), *Quantum Collisions and Confinement of Atomic and Molecular Species, and Photons*, Springer Proceedings in Physics 230, https://doi.org/10.1007/978-981-13-9969-5_24

transitions from the $6s^2ns$ ($n = 7$–14) $^2S_{1/2}$, $6s^2np$ ($n = 7$–9) $^2P_{1/2;3/2}$, $6s^2nd$ ($n = 6$–15) $^2D_{3/2;5/2}$, $6s^2nf$ ($n = 5$–7) $^2F_{5/2;7/2}$, and $6s6p^2$ ($^4P_{1/2;3/2;5/2}$, $^2D_{3/2;5/2}$, $^2P_{1/2;3/2}$, and $^2S_{1/2}$) levels of Pb⁺. Alonso-Medina et al. [6] also reported first experimental data of the transition probability for the 2203.5 Å line of Pb⁺ arising from the transition $6s^26p$ $^2P_{3/2} \rightarrow 6s^27s$ $^2S_{1/2}$. Recently, Heidarian et al. [3] performed lifetime measurements using beam-foil techniques on the levels of Pb⁺ producing lines at 1203.6 Å ($6s6p^2$ $^2D_{3/2}$) and 1433.9 Å ($6s^26d$ $^2D_{3/2}$). They reported the first detection of the 1203.6 Å line in composite HST/STIS spectra. Quinet et al. [7], determined experimentally as well as theoretically radiative decay rates from the $6s^27s$ $^2S_{1/2}$ level. Safronova et al. [8] performed calculations using relativistic many-body perturbation theory (MBPT) and obtained energies of $6s^2np_j$ ($n = 6$–9), $6s^2ns_{1/2}$ ($n = 7$–9), $6s^2nd_j$ ($n = 6$–8), and $6s^2nf_{5/2}$ ($n = 5$–6) states. They also calculated reduced matrix elements, oscillator strengths, and transition rates for the 72 possible $6s^2nlj$-$6s^2n'l'j$ electric–dipole transitions.

So far, there is only one experimental work reported recently by Gomonai et al. [9] for electron-impact excitation of Pb⁺ ion. They used crossed-beam technique to measure effective cross section of the electron-impact excitation for the resonance line $6s^26p$ $^2P_{1/2} \rightarrow 6s^26d$ $^2D_{3/2}$ of the Pb⁺ ion in the (6–100) eV energy range. Except for this experiment, there is no other theoretical or experimental investigation to study electron-impact excitation of singly charged lead. Therefore, in this paper, we have focused on selected transitions due to electron-impact excitation of Pb⁺ ions. We have considered only important resonance lines detected in the HST/STIS spectra of Pb⁺ ion from the viewpoint of astrophysical applications. We have considered excitations of the $6s^26d$ $^2D_{3/2,5/2}$, $6s6p^2$ $^2D_{3/2,5/2}$ and $6s^27s$ $^2S_{1/2}$ states from the fine-structure levels $^2P_{1/2,3/2}$ of the ground state configuration $6s^26p$. Since lead (Z = 82) is a heavy element with dominant relativistic effects, we have used a fully relativistic distorted-wave (RDW) method to evaluate the scattering amplitude for the excitation processes. We have already applied this method successfully to study electron-impact excitation of neutral lead atom [10]. In this method, the projectile electron wavefunctions are obtained by solving Dirac equation while the bound states of the Pb⁺ ion are calculated within multi-configuration Dirac-Fock (MCDF) framework using GRASP2k code [11]. Since the accuracy of the atomic wavefunctions directly influence the quality of the projectile electron wavefunction and consequently, the scattering parameters, we have compared oscillator strengths of the transitions with previous measurements and theoretical results as well as NIST database. Finally, results are reported for excitation cross sections from threshold to 100 eV electron energy. Cross sections for the resonance line $6s^26p$ $^2P_{1/2} \rightarrow 6s^26d$ $^2D_{3/2}$ are compared with the only available measurements [9].

The paper is arranged as follows: the next section describes briefly the RDW method employed to obtain the cross section for the 8 dipole allowed transitions as mentioned above. Subsequently, results and discussion are presented in Sect. 3 followed by concluding remarks in Sect. 4.

2   Theoretical Method

Using RDW approximation the scattering amplitude for excitation from the state $a \to b$ can be expressed as

$$
\begin{aligned}
&f(J_b, M_b, \mu_b; J_a, M_a, \mu_a; \theta) \\
&= (2\pi)^2 \sqrt{\frac{k_b}{k_a}} \langle \Psi_b^-(1,2,\ldots,N+1) | V_C - U_b(N+1) | \Psi_a^+(1,2,\ldots,N+1) \rangle
\end{aligned} \tag{1}
$$

where J, M refer to the total angular momentum quantum number and its associated magnetic quantum numbers. $\mu_{a/b}$ denote the spin projections of the projectile electron. θ is the angle between the momentum vectors k_a and k_b of the incident and scattered electrons. V_C is the target-projectile Coulomb interaction as given by equation

$$
V_C = -\frac{Z}{r_{N+1}} + \sum_{j=1}^{N} \frac{1}{|r_j - r_{N+1}|} \tag{2}
$$

Here position coordinates of the atomic and projectile electrons with respect to the nucleus are given by, respectively, $r_j(1,\ldots,N)$ and r_{N+1}, Z is the nuclear charge and N is the number of bound electrons in the target. U_b is the distortion potential which depends only on the radial coordinates of the projectile electron. We have chosen U_b to be a spherically averaged static potential of the excited state of the ion. Assuming that the projectile electron does not alter the wavefunctions of the target, the total wave-functions $\Psi_{a/b}^{+/-}$ in the initial/final channels a/b can be represented as

$$
\Psi_{a/b}^{+/-} = A\Phi_{a/b}(1,2,\ldots,N)F_{a/b}^{+/-}(k_{a/b}, N+1) \tag{3}
$$

where $\Phi_{a/b}$ denotes the N-electron target wave functions and $F_{a/b}^{+/-}$ represent the projectile electron distorted-wave function with $\pm$ sign referring to the incoming and outgoing waves. The procedure to obtain the distorted wavefunctions with the help of partial wave expansion method and subsequently, solving coupled Dirac equations is described in our earlier work [12]. The integrated cross section for excitation of atom from J_a to J_b state can be obtained by using the following expression,

$$
\sigma(J_a \to J_b) = \frac{1}{2(2J_a+1)} \sum_{\substack{M_b, M_a \\ \mu_b, \mu_a}} \int |f(J_b, M_b, \mu_b; J_a, M_a, \mu_a; \theta)|^2 d\Omega \tag{4}
$$

We have performed these calculations in the *Collision frame of reference* with quantization axis (z-axis) to be along the incident electron beam direction while the scattering plane is chosen to be the xz-plane.

3   Results and Discussion

In the present work, we have calculated cross sections for electron-impact excitation of the $6s^2 6d$ $^2D_{3/2,5/2}$, $6s6p^2$ $^2D_{3/2,5/2}$ and $6s^2 7s$ $^2S_{1/2}$ states from the fine-structure levels $6s^2 6p$ $^2P_{1/2,3/2}$ of the ground state. These are 8 dipole transitions in total. The reliability of the cross sections depends on the accuracy of the target wavefunctions used in the calculation. Therefore, we first discuss our method of obtaining wavefunctions of the target states and their quality assessment. These wavefunctions are obtained within multi-configuration Dirac-Fock approximation by using GRASP2k code. Pb⁺ is a heavy element with ground state electronic configuration $[Xe]4f^{14}5d^{10}6s^2 6p$ having fully filled $5d$ and $6s$ inner shells and partially filled $6p$ outer shell. The previous theoretical investigations on spectroscopic properties of Pb⁺ [7] have revealed the importance of the correlation between valence and sub valence shells. Therefore, to observe the effect of core-valence interaction we considered here single substitution from inner $5d$ and $6s$ subshells.

We optimized various sets of wavefunctions of the atomic states by including a number of configuration state functions. To ascertain the accuracy of these sets we compared oscillator strengths with other theoretical and experimental values. We found that the set-1, having configurations as shown in Table 1, shows overall best agreement with results from NIST database [13], experimental measurements and other theoretical methods. Adding more configuration to set-1, for example in set-2 which has additional $5d^{10}6s6d^2$ configuration, has very little effects on the oscillator strengths of the transitions as can be seen from Table 2. We observe from Table 2 that the beam-foil measurements from the Toledo Heavy Ion Accelerator (THIA) [3] differ significantly from the NIST values for $6s^2 6d$ $^2D_{3/2} \rightarrow 6s^2 6p$ $^2P_{1/2}$ and $6s^2 6d$ $^2D_{5/2} \rightarrow 6s^2 6p$ $^2P_{3/2}$ transitions. Similarly, there seems to be a wide discrepancy among theoretical results for these two transitions, however, our value is closer to the THIA measurements for both the transitions. For $6s^2 6d$ $^2D_{3/2} \rightarrow 6s^2 6p$ $^2P_{3/2}$ transition, all the theoretical results are in good agreement with the measurements. Our calculations for $6s6p^2$ $^2D_{3/2,5/2} \rightarrow 6s^2 6p$ $^2P_{1/2,3/2}$ transition agree reasonably with the calculation of Heidarian et al. [3]. All theoretical and experimental results are in good agreement for $6s^2 7s$ $^2S_{1/2} \rightarrow 6s^2 6p$ $^2P_{3/2}$.

The bound state wavefunctions of the target ion in the initial and final states are used to calculate the distortion potential [12]. The projectile electron travels under the influence of the distortion potential and its wavefunction is obtained by solving the coupled Dirac equations with appropriate boundary conditions. Thus collecting all necessary constituents required for the transition matrix (1), we finally obtain cross sections.

Table 1. Configuration sets in atomic wavefunction calculation

Set - 1	$5d^{10}6s^2 6p, 5d^{10}6s^2 6d, 5d^{10}6s^2 7s, 5d^9 6s^2 6p^2, 5d^9 6s^2 6d7s, 5d^{10}6s6p^2$ $5d^{10}6s7s^2, 5d^{10}6s^2 7d, 5d^{10}6s^2 8d, 5d^9 6s^2 7s^2, 5d^9 6s^2 6d^2$
Set - 2	$5d^{10}6s^2 6p, 5d^{10}6s^2 6d, 5d^{10}6s^2 7s, 5d^9 6s^2 6p^2, 5d^9 6s^2 6d7s, 5d^{10}6s6p^2$ $5d^{10}6s6d^2, 5d^{10}6s7s^2, 5d^{10}6s^2 7d, 5d^{10}6s^2 8d, 5d^9 6s^2 7s^2, 5d^9 6s^2 6d^2$

Table 2. Comparison of our calculated oscillator strengths of the transitions in Pb^+ with corresponding values from the NIST database [13], other theoretical methods and experiments

Transition levels		f-value				
Upper	Lower	Set-1	Set-2	Other calculations	Measurements	NIST
$6s^26d\,^2D_{3/2} \rightarrow 6s^26p\,^2P^\circ_{1/2}$		0.3079	0.3070	0.86^a, 0.268^b, 0.372^c, 0.4518^d, 0.869^e	$0.321(34)^b$	0.869
$6s^26d\,^2D_{3/2} \rightarrow 6s^26p\,^2P^\circ_{3/2}$		0.0673	0.0674	0.063^b, 0.074^c, 0.06255^d, 0.100^e	$0.064(7)^b$	
$6s^26d\,^2D_{5/2} \rightarrow 6s^26p\,^2P^\circ_{3/2}$		0.1963	0.2004	0.123^b, 0.054^c, 0.5024^d, 0.880^e,	$0.179(14)^b$	0.881
$6s6p^2\,^2D_{3/2} \rightarrow 6s^26p\,^2P^\circ_{1/2}$		1.2532	1.250	1.124^b, 2.02^c	$0.75(3)^b$	
$6s6p^2\,^2D_{3/2} \rightarrow 6s^26p\,^2P^\circ_{3/2}$		0.0004	0.0004	0.0011^b, 0.0003^c	$0.0001(0)^b$	
$6s6p^2\,^2D_{5/2} \rightarrow 6s^26p\,^2P^\circ_{3/2}$		0.4578	0.4476	0.573^b, 0.897^c	$0.204(14)^b$	
$6s^27s\,^2S_{1/2} \rightarrow 6s^26p\,^2P^\circ_{1/2}$		0.1697	0.1696	0.2316^c, 0.139^e, 0.1900^f	–	
$6s^27s\,^2S_{1/2} \rightarrow 6s^26p\,^2P^\circ_{3/2}$		0.1608	0.1616	0.201^e, 0.1820^f	0.1795^c	0.180

References **a**: Cardelli et al. (1993) [1]; **b**: Heidarian et al. (2015) [3]; **c**: Colon & Alonso-Medina (2001) [5]; **d**: Safronova et al. (2005) [8]; **e**: Migdalek (1976a) [14]; **f**: Kunisz and Migdalek (1974) [15];

We have calculated integrated cross sections (ICS) for 8 dipole allowed fine-structure transitions as listed in Table 2 using atomic wavefunctions of set-1 and set-2. These cross sections as a function of incident electron energy are shown in Fig. 1. It can be seen that the two sets of ICSs are in good agreement with each other. All these transitions show the characteristic feature of dipole—allowed transition, i.e., the ICSs fall off as $\ln(E)/E$ with increasing incident electron energy. For excitation of $6s^26p\,^2P_{1/2}$ to $6s^26d\,^2D_{3/2}$ state, we have compared our results with the recent measurements reported by Gomonai et al. [9]. The discrepancy in our results with the experimental data for energies below 80 eV is due to the fact that the measurements include the effect of cascade transitions and the resonance contribution of autoionizing states. Our calculations, being two-state approximation in the first-order perturbation theory,

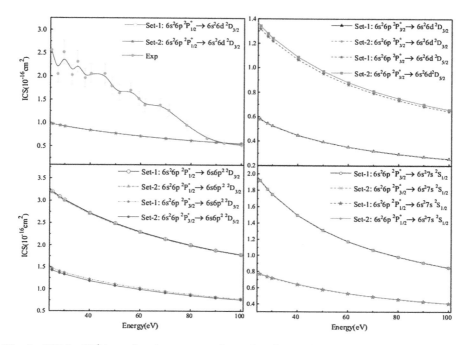

Fig. 1. ICS for Pb$^+$ ion using the two sets of wavefunctions; solid circles with error bar present the experimental results by Gomonai et al. [9]

do not incorporate resonance effects as well as cascading from higher states. However, at 100 eV electron energy value of our cross section is 0.55×10^{-16} cm^2 which is in excellent agreement with the measured value of $(0.5 \pm 0.3) \times 10^{-16}$ cm^2.

We find that the ICS for $6s^2 6p$ $^2P_{3/2}$ to $6s6p^2$ $^2D_{3/2}$ transition is three order of magnitude smaller in comparison to the ICS of other transitions considered in the present work, therefore, we have not shown it in Fig. 1. The small value of ICS for the above transition may be attributed to the corresponding value of oscillator strength as can be seen in Table 2. Further, the relative values of ICSs for various transitions can be understood on the basis of their oscillator strengths. A transition having a greater value of oscillator strength has larger cross sections over entire range of the incident electron energy. The nearly identical shape of all the cross sections from ground state represents the decreasing of cross section with increasing electron energy.

4   Conclusions

In the present work, we have applied RDW theory to study dipole allowed transitions from the $6s^2 6p$ $^2P_{1/2,3/2}$ states to the $6s^2 6d$ $^2D_{3/2,5/2}$, $6s6p^2$ $^2D_{3/2,5/2}$ and $6s^2 7s$ $^2S_{1/2}$ states. We found that Pb$^+$ is a strong as well as the challenging candidate to exhibit the correlation effects between valence and core subshells. The accuracy of the atomic wavefunctions is established by comparing our calculated oscillator strengths with other available theoretical and experimental results. Further theoretical and experimental investigations are required to resolve the discrepancy among various results for oscillator strengths. We compared our ICS results with the recent measurements of Gomonai et al. [9] available for one transition $6s^2 6p\ ^2P_{1/2} \rightarrow 6s^2 6d\ ^2D_{3/2}$ only. We find good agreement with the measured cross section at 100 eV as expected. We hope that our results will be useful in encouraging more investigations for Pb^+ for better understanding of spectroscopic properties and collision dynamics of this cosmic abundant element.

References

1. Cardelli, J.A., Federman, S.R., Lambert, D.L., Theodosiou, C.E.: Astrophys. J. **416**, L41–L44 (1993)
2. Welty, D.E., Hobbs, L.M., Lauroesch, J.T., Morton, D.C., York, D.G.: Astrophys. J. **449**, L135–L138 (1995)
3. Heidarian, N., Irving, R.E., Ritchey, A.M., Federman, S.R., Ellis, D.G., Cheng, S., Curtis, L. J., Furman, W.A.: Astrophys. J. **808**, 112 (2015)
4. Alonso-Medina, A.: Phys. Scr. **55**, 49–53 (1997)
5. Colón, C., Alonso-Medina, A.: Can. J. Phys. **79**, 999 (2001)
6. Alonso-Medina, A., Colon, C., Herran-Martinez, C.: J. Quant. Spectrosc. Radiat. Transf. **68**, 351 (2001)
7. Quinet, P., Biemont, E., Palmeri, P., Xu, H.L.: J. Phys. B: At. Mol. Opt. Phys. **40**, 1705–1712 (2007)
8. Safronova, U.I., Safronova, M.S., Safronova, W.R.: Phys. Rev. A **71**, 052506 (2005)
9. Gomonai, A.N., Hutych, Y.I., Gomonai, A.I.: Eur. Phys. J. D **71**, 31 (2017)
10. Milisavljević, S., Rabasović, M.S., Šević, D., Pejčev, V., Filipović, D.M., Sharma, L., Srivastava, R., Stauffer, A.D., Marinković, B.P.: Phys. Rev. A **75**, 052713 (2007)
11. Jönsson, P., Gaigalas, G., Bieron, C., Froese Fischer, C., Grant, I.P.: Comput. Phys. Commun. **184**, 2197 (2013)
12. Sharma, L., Surzhykov, A., Srivastava, R., Fritzsche, S.: Phys. Rev. A **83**, 062701 (2011)
13. NIST Atomic Spectra Database, www.nist.gov/pml/data/asd.cfm
14. Migdalek, J.: J. Quant. Spectrosc. Radiat. Transfer **16**, 265 (1976)
15. Kunisz, M.D., Migdalek, J.: Acta Phys. Pol., A **45**, 715 (1974)

Low-Temperature Scattering with the R-Matrix Method: The Morse Potential

Tom Rivlin[1]([✉]), Laura K. McKemmish[1,2], and Jonathan Tennyson[1]

[1] Department of Physics and Astronomy, University College London, London
WC1E 6BT, UK
{t.rivlin,j.tennyson}@ucl.ac.uk
[2] School of Chemistry, University of New South Wales, Kensington, Sydney,
Australia

Abstract. Experiments are starting to probe collisions and chemical
reactions between atoms and molecules at ultralow temperatures. We
have developed a new theoretical procedure for studying these collisions
using the R-matrix method. Here, this method is tested for the atom—
atom collisions described by a Morse potential. Analytic solutions for
continuum states of the Morse potential are derived and compared with
numerical results computed using an R-matrix method, where the inner
region wavefunctions are obtained using a standard nuclear motion algorithm. Results are given for eigenphases and scattering lengths. Excellent
agreement is obtained in all cases. Progress in developing a general procedure for treating ultralow energy reactive and non-reactive collisions
is discussed.

1 Introduction

The ability to perform very low-energy collisions between heavy particles is leading to a quiet revolution at the border between atomic physics and experimental
quantum chemistry [1]. Studies of reactive and non-reactive collisions at temperatures very significantly below 1 K are starting to probe processes which are
not easily resolved at higher temperatures. These experiments study chemical
reactions and scattering at the quantum scattering limit where, asymptotically,
only a few partial waves contribute [2].

To address these problems theoretically requires the development of new computational techniques. Recently, we proposed adapting R-matrix theory to the
study of ultralow energy reactive and non-reactive, heavy-particle collisions [3].
R-matrix theory involves the division of space into an inner region encompassing the whole collision complex and an outer region where species involved in
the scattering can be separately identified. Procedures based on the computable
R-matrix method have proved outstandingly successful for the study of electron
collisions with atoms and molecules [4,5], and are increasingly being adopted
in other areas [6]. In the computable R-matrix method, the Schrödinger equa-

© Springer Nature Singapore Pte Ltd. 2019
P. C. Deshmukh et al. (eds.), *Quantum Collisions and Confinement of Atomic
and Molecular Species, and Photons,* Springer Proceedings in Physics 230,
https://doi.org/10.1007/978-981-13-9969-5_25

tion for the restricted inner region is solved once and for all for each scattering symmetry, independent of the precise scattering energy. For heavy-particle scattering, this procedure is particularly appropriate for reactive or non-reactive collisions which occur over deep potential energy wells. Thus, for example, H + H_2 collisions do not occur over a deep well as the H_3 system is only weakly bound [7], while collisions between H^+ + H_2 occur over the deep well of the H_3^+ potential energy surface [8].

Strongly bound systems with deep potential energy wells support many bound states. Even the very lowest continuum states which are associated with ultralow-energy scattering feel the effect of these many bound states which lie below them in energy. The result is that even the lowest scattering state has a complicated wavefunction which couples many channels which are asymptotically closed. It is well known that this situation leads to a plethora of quasibound states, or resonances, in the near-dissociation region [9–13]. Use of the R-matrix method allows the region of the deep potential well to be treated using variational nuclear motion programs which are capable of giving highly accurate results for energy-independent problems with complicated wavefunctions [14]. It is then only necessary to treat a few partial waves in the energy-dependent outer region. In this region it may be necessary to propagate solutions to very large interparticle separations [15] and to scan over the many energies necessary to characterise narrow resonances.

At present we are in the process of developing a heavy particle R-matrix scattering code, RmatReact, based on the use of a variety of variational nuclear motion codes in the inner region [16–20]. Doing this involves developing computational procedures which extend methods of the solutions into the continuum [12, 13, 21]. In particular, the problem must be solved within a finite region and, critically, use basis functions which give reliable amplitudes at the R-matrix boundary. These amplitudes, and the associated inner region energies, are used to construct the scattering energy-dependent R-matrix which links the inner and outer regions [3].

In this paper, we report on tests we have performed using our methodology for the Morse oscillator potential. Section 2 gives an overview of the general theory while Sect. 3 demonstrates that the scattering problem can be solved analytically for a Morse oscillator potential. This allows the rigorous assessment of our numerical procedures, which are discussed in Sect. 4. Results are given in Sect. 5, and conclusions and some pointers to our future work are given in the final Section.

2   Theory: The RmatReact Method

The theory behind the RmatReact method has been discussed extensively [3–5], and much of this explanation derives from those discussions. The general principle behind the method is the partitioning of space into an inner and outer region, dependent on the reaction coordinate, as discussed above.

In the case of two atoms colliding, there is only one reaction coordinate: the internuclear distance r. A point $r = a_0$ is defined such that any internuclear distance lower than that is the inner region and any distance larger is the outer region.

Within the inner region, the system is treated as a bound diatom, and the eigenenergies and eigenfunctions of the radial Schrödinger equation with the Morse potential can be determined using software built for nuclear motion calculations. Because the eigenfunctions and values refer to the bound states, they are independent of scattering energy. Likewise, in the outer region, the system is treated as a pair of weakly interacting, unbound atoms. Each atom will have associated atomic channels describing its quantum state.

The inner region was solved in this work using a discrete variable representation (DVR) [22] grid method based on the Lobatto shape functions, which have the property of always having a point defined on both boundaries of the grid. Manolopoulos [23] and Manolopoulos and Wyatt [24] pioneered the use of these functions for scattering problems. Lobatto shape functions [25] can be used to obtain simple expressions for the components of the Hamiltonian matrix, making it computationally efficient to diagonalise whilst avoiding much of the expensive integration usually involved in constructing a Hamiltonian matrix. Once the inner region has been solved to obtain a diagonalised Hamiltonian matrix, a matrix known as the R-matrix, can be constructed on the boundary a_0. The R-matrix is constructed from the scattering energy, E, the bound eigenenergies, and the values of the eigenfunctions on the boundary a_0, known as the *surface amplitudes*.

For a given angular momentum quantum number J, if the mth surface amplitude associated with the ith atomic channel is defined as $w_{im}^J(a_0)$, the mth eigenenergy is defined as E_m^J, and the scattering wavefunction for atomic channel i is defined to be $F_i^J(r, E)$, then the R-matrix has two equivalent definitions at a_0:

$$F_i^J(a_0, E) = \sum_{j=1}^{N_{ch}^J} a_0 R_{ij}^J(a_0, E) \frac{dF_j^J(r, E)}{dr}\bigg|_{r=a_0}, \qquad (1)$$

$$R_{ij}^J(a_0, E) = \frac{\hbar^2}{2\mu a_0} \sum_{m=1}^{N} \frac{w_{im}^J(a_0) w_{jm}^J(a_0)}{E_m^J - E}, \qquad (2)$$

where the sum in (1) is over the N_{ch}^J atomic channels for a given value of J considered in the scattering event, and the sum in (2) is over the N solutions to the Schrödinger equation within an atomic channel.

In the $J = 0$, single channel case considered in this work, $i = j$ and (1) reduces down to a single term, $N_{ch} = 1$, and the single R-matrix element is defined as $R(a_0, E)$. Furthermore, $w_{im}^J(a_0)$ becomes a single surface amplitude, and the sum is over the N surface amplitudes.

As (1) suggests, the R-matrix can be thought of as the 'log-derivative' of the channel function $F^J(r, E)$, which is an outer region function. However (2) shows that the R-matrix can be constructed as a sum over the eigenfunctions and energies of the inner region. The fact that these inner and outer region definitions of the R-matrix are equivalent is what gives the R-matrix method its value: information about the energy-independent inner region provide the starting point for obtaining scattering information.

In the outer region, it is assumed that the potential is small and slowly varying, compared to the deep wells of the inner region. As such, it is possible to use methods which iteratively solve the Schrödinger equations over finite distances to *propagate* the R-matrix from the boundary at a_0 to an asymptotic distance a_p. At a_p, the potential is assumed to be zero. In this work, the propagation method due to Burke [4] and Walker and Light [26] is used.

For the single channel, the propagation algorithm takes as its input the R-matrix element at the inner region boundary, $R(a_0, E)$, and produces the R-matrix element at the asymptotic distance, $R(a_p, E)$. To produce the value of the R-matrix at the outer region point a_s, $R_s(a_s, E)$, the iteration equation takes as its input the value of the R-matrix at a_{s-1}, $R_{s-1}(a_{s-1}, E)$, and has the form:

$$R_s = \frac{-1}{a_s \lambda_s} \left(\frac{1}{\tan(\lambda_s \Delta a)} + \frac{2}{\sin(2\lambda_s \Delta a)} (a_{s-1} R_{s-1} \lambda_s \tan(\lambda_s \Delta a) - 1)^{-1} \right), \quad (3)$$

where $\Delta a = a_s - a_{s-1}$, and

$$\lambda_s^2 = \frac{2\mu}{\hbar^2} (E - V(a_s)). \quad (4)$$

At the asymptotic distance a_p, the R-matrix is used to construct the K-matrix using the equation [4]:

$$K_{ij}^J(k) = - \left(\frac{s^J(kr) - R_{ij}^J(a_p, E) kr s^{J'}(kr)}{c^J(kr) + R_{ij}^J(a_p, E) kr c^{J'}(kr)} \right), \quad (5)$$

where

$$s^\nu(x) = x j^\nu(x) \quad (6a)$$
$$c^\nu(x) = -x y^\nu(x), \quad (6b)$$

where $j^\nu(x)$ is the Spherical Bessel Function of the First Kind, $y^\nu(x)$ is the Spherical Bessel Function of the Second Kind, and $s^{J'}(x)$ and $c^{J'}(x)$ are the derivatives with respect to x of $s^J(x)$ and $c^J(x)$ respectively.

In the single channel, (5) reduces to

$$K(k) = \frac{R(a_p, E) kr - \tan kr}{1 + R(a_p, E) kr \tan kr}. \quad (7)$$

The inner, outer and asymptotic regions are illustrated in Fig. 1.

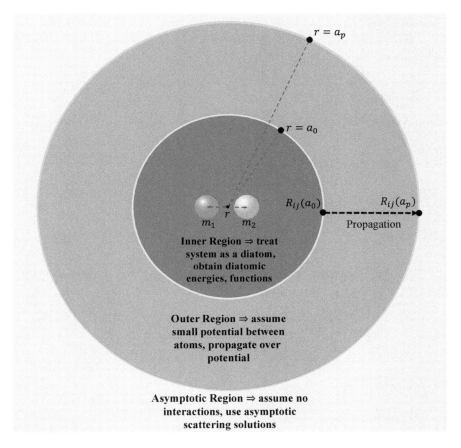

Fig. 1. Schematic outlining the partitioning of space into an inner, outer and asymptotic region in the R-matrix method

3   Analytic Scattering in Morse Oscillators

3.1   Morse Oscillator Solutions

When there is no angular momentum and hence no centrifugal term, the Morse potential for a diatom as a function of the internuclear distance r has the algebraic form:

$$V(r) = D_e \left(\left(1 - e^{-a_{\mathrm{Morse}}(r-r_e)} \right)^2 - 1 \right), \tag{8}$$

where D_e is the well depth (assuming the zero of potential energy is placed at the dissociation energy), r_e is the equilibrium position, or position of the well minimum, and a_{Morse} is a scaling parameter (the so-called Morse parameter) affecting the shape of the well.

The analytic eigenfunctions and eigenenergies of the Schrödinger equation with a Morse potential are well known. For the radial time-independent Schrödinger equation

$$\left(-\frac{\hbar^2}{2m}\frac{d^2}{dr^2} + D_e e^{-2a_{\mathrm{Morse}}(r-r_e)} - 2D_e e^{-a_{\mathrm{Morse}}(r-r_e)}\right)\Psi_n = E_n\Psi_n, \quad (9)$$

the bound eigenenergies E_n and eigenfunctions Ψ_n are given by Morse [27]:

$$E_n^{\mathrm{Morse}} = -D_e + 2a_{\mathrm{Morse}}\sqrt{\frac{D_e\hbar^2}{2\mu}}\left(n+\frac{1}{2}\right) - \frac{1}{4D_e}\left(2a_{\mathrm{Morse}}\sqrt{\frac{D_e\hbar^2}{2\mu}}\left(n+\frac{1}{2}\right)\right)^2,$$
$$(10)$$

and

$$\Psi_n^{\mathrm{Morse}} = N_n z^{(1/(a_{\mathrm{Morse}}r_0))-n-1/2}\exp\left(\frac{-z}{2}\right)L_n^{(2/(a_{\mathrm{Morse}}r_0))-2n-1)}(z), \quad (11)$$

where $L_n^{(\alpha)}(z)$ is the nth associated Laguerre polynomial, and N_n is a normalising factor given by

$$N_n = \left(\frac{\left(\frac{2}{a_{\mathrm{Morse}}r_0}-2n-1\right)a_{\mathrm{Morse}}\Gamma\left(n+1\right)}{\Gamma\left(\frac{2}{a_{\mathrm{Morse}}r_0}-n\right)}\right)^{\frac{1}{2}}, \quad (12)$$

where $\Gamma(x)$ is the standard Gamma function.

3.2 Scattering Observables

It is possible to derive analytic scattering observables for a quantum scattering event involving the Morse oscillator potential energy curve because the time-independent Schrödinger equation with a Morse potential is analytically soluble.

Similar to (9), for a scattering event between particles with reduced mass μ with energy E interacting over a Morse potential, the radial wavefunction $\Psi(r)$ is given by the time-independent radial Schrödinger equation:

$$\left(-\frac{\hbar^2}{2\mu}\frac{d^2}{dr^2} + D_e e^{-2a_{\mathrm{Morse}}(r-r_e)} - 2D_e e^{-a_{\mathrm{Morse}}(r-r_e)}\right)\psi = E\psi. \quad (13)$$

By defining

$$k = \sqrt{\frac{2\mu E}{\hbar^2}}, \quad (14)$$

one constraint that may be placed on $\psi(r)$ is that in the no-potential limit it must behave like the wavefunction of a free particle, i.e. it must be a plane wave. Likewise, this means that in the infinite distance limit where the potential's strength tends to zero, the wavefunction must be sinusoidal such that

$$\lim_{r\to\infty}\psi(r) = \sin(kr + \delta(k)), \quad (15)$$

where $\delta(k)$ is defined to be the phase shift (also known as the eigenphase) induced in the particle by its interaction with the potential.

Furthermore, by defining

$$r_0 = \sqrt{\frac{\hbar^2}{2\mu D_e}}, \tag{16}$$

$$z(r) = \frac{2}{a_{\text{Morse}} r_0} e^{-a(r - r_e)}, \tag{17}$$

and

$$\Phi(z) = z^{\frac{1}{2}} \psi(z), \tag{18}$$

then it can be shown [28] that (13) can be re-written as

$$\frac{d^2 \Phi}{dz^2} + \left(-\frac{1}{4} + \frac{1}{a_{\text{Morse}} r_0 z} + \frac{\frac{1}{4} + \left(\frac{k}{a_{\text{Morse}}} \right)^2}{z^2} \right) \Phi(z) = 0. \tag{19}$$

In this form, the equation is equivalent to the well-known Whittaker equation, whose solutions are the Whittaker functions. There are two linearly independent solutions to (19):

$$\psi_{\pm}(z) = e^{-z/2} z^{\pm ik/a_{\text{Morse}}} {}_1F_1 \left(\frac{1}{2} - \frac{1}{a_{\text{Morse}} r_0} \pm \frac{ik}{a_{\text{Morse}}}, 1 \pm \frac{2ik}{a_{\text{Morse}}}; z \right), \tag{20}$$

where ${}_1F_1(x, y; z)$ is the Kummer confluent hypergeometric function of the first kind, and the $\psi_{\pm}(z)$ functions represent incoming and outgoing waves.

Using the results for the analytic scattering wavefunctions of the Morse potential in (20), it is possible to construct an analytic equation for the eigenphase $\delta(k)$ associated with scattering with the Morse potential. The eigenphase of the scattering event is desired because it can be used to generate other observables such as the cross section and scattering length.

The derivation below follows that of Rawitscher et al. [28] and Selg [29,30].

The general solution $\psi(r)$ to (13) can be written in terms of the two solutions to (19), which are given by (20), such that:

$$\psi(r) = C_+ \psi_+(r) + C_- \psi_-(r), \tag{21}$$

where $C_{\pm}$ are two constants.

There are two boundary conditions on $\psi(r)$ that can be used to obtain an expression for the eigenphase. Firstly, the asymptotic radial function must vanish at $r = 0$, such that $\psi(0) = 0$. This fact can be used to express one of the $C_{\pm}$ coefficients in terms of the other. Secondly the $r \to \infty$ asymptotic limit is given by (15). As $r \to \infty$, $z \to 0$. This means that due to a property of the Kummer confluent hypergeometric functions, both hypergeometric functions tend to 1 as $r \to \infty$.

The S-matrix can be defined in the $r \to \infty$ limit as the negative of the ratio of the coefficients of the outgoing plane wave component of the asymptotic radial wavefunction to the incoming plane wave component [4].

Then, by defining z_0 such that

$$z(r=0) = z_0 = \frac{2}{a_{\text{Morse}} r_0} e^{a r_e}, \tag{22}$$

the following expression can be obtained:

$$\left(\frac{z}{z_0} \right)^{\pm \frac{ik}{a_{\text{Morse}}}} = e^{\mp ikr}. \tag{23}$$

Using the boundary conditions and (23), one can obtain an expression for the ratio of the coefficients of $\psi_{\pm}$ in this limit, and hence one can obtain an analytic expression for the S-matrix:

$$S(k) = \lim_{r \to \infty} \frac{C_+}{C_-} = \frac{{}_1F_1\left(\frac{1}{2} - \frac{1}{a_{\text{Morse}} r_0} + \frac{ik}{a_{\text{Morse}}}, 1 + \frac{2ik}{a_{\text{Morse}}}; z_0 \right)}{{}_1F_1\left(\frac{1}{2} - \frac{1}{a_{\text{Morse}} r_0} - \frac{ik}{a_{\text{Morse}}}, 1 - \frac{2ik}{a_{\text{Morse}}}; z_0 \right)}. \tag{24}$$

Besides (15), another way of defining the eigenphase is as the argument of the S-matrix, such that:

$$S(k) = e^{2i\delta(k)}. \tag{25}$$

Note that the factor of 2 in the exponent is arbitrary, and other authors define it differently, depending on whether the eigenphase is defined as the argument of the S-matrix (as in [29]), or as the arctangent of the K-matrix, which is equivalent to defining the eigenphase to be half of the argument of the S-matrix (as in this work, and [28]).

The analytic expression for the eigenphase is then given by:

$$\delta(k) = \frac{1}{2} \arg \left(\frac{{}_1F_1\left(\frac{1}{2} - \frac{1}{a_{\text{Morse}} r_0} + \frac{ik}{a_{\text{Morse}}}, 1 + \frac{2ik}{a_{\text{Morse}}}; z_0 \right)}{{}_1F_1\left(\frac{1}{2} - \frac{1}{a_{\text{Morse}} r_0} - \frac{ik}{a_{\text{Morse}}}, 1 - \frac{2ik}{a_{\text{Morse}}}; z_0 \right)} \right). \tag{26}$$

Once the eigenphase has been obtained for a given Morse potential, then many scattering observables can be derived, including the K-matrix, and the T-matrix (also known as the transition matrix):

$$K(k) = \tan \delta(k), \tag{27}$$

$$S(k) = \frac{1 - iK(k)}{1 + iK(k)}, \tag{28}$$

$$T(k) = S(k) - 1. \tag{29}$$

Note that other authors use different definitions of the T-matrix such as the negative of its definition given here.

The total cross section at a given energy, $\sigma_{\text{tot}}(k)$, which is the integral of the differential cross section over all solid angles, can be obtained from the eigenphase:

$$\sigma_{\text{tot}}(k) = \frac{4\pi}{k^2} \sin^2(\delta(k)). \tag{30}$$

Finally the scattering length, A, and the effective range, r_{eff}, are characteristic length scales associated with low-energy scattering. A is defined as the limit

$$A = \lim_{k \to 0} \left(\frac{-\tan(\delta(k))}{k} \right), \tag{31}$$

for the $J = 0$, s-wave (lowest energy) eigenphase [4]. The scattering length can be thought of as the low-energy $k \to 0$ limit of the gradient of the eigenphase. The effective range can be analytically determined through an integral over all spaces of the difference between the zero-energy scattering wavefunction, and the zero-energy potential-free scattering wavefunction [31]. It can be thought of as a length parameter which measures the overall effect the potential has on the scattering event, since it is defined by the difference between scattering in the cases with and without a potential. As such, calling it the effective *range* of the potential is natural.

One way of obtaining these two quantities from the eigenphase is by taking a Taylor expansion of the eigenphase close to zero scattering energy [4]:

$$k \cot \delta(k) = \frac{-1}{A} + \frac{1}{2} r_{\text{eff}} k^2 + O(k^4). \tag{32}$$

4 Method

4.1 Potentials Investigated

The main Morse potential used in this work is presented in Fig. 2. This Morse potential uses parameters with reduced mass of $\mu = 33.71525621$ Da (and a value of $\hbar$ obtained from Mohr et al. [32]). The value for μ was chosen for numerical convenience when testing the algorithm, as it meant that $\hbar^2/2\mu$ had a value of 0.5 to seven decimal places in the units of cm^{-1} and Å used in this work. The specific value used for a_{Morse} was chosen such that the ground state eigenenergy was 90 cm^{-1} to six decimal places, for ease of comparison. The values of D_e and r_e used in this work were chosen in analogy with the Ar_2 dimer, which is currently being used to investigate the application of this method to more sophisticated potentials. The analytic eigenenergies were generated from these parameters and (10).

Other Morse potentials were tested, notably several obtained from [33] for actual diatoms: LiH, H_2, HCl, and CO. Figure 3 shows one of these potentials: LiH. In this paper, we present only results for the Morse potential shown in Fig. 2. Similar numerical behaviour was observed for all of the potentials tested, however.

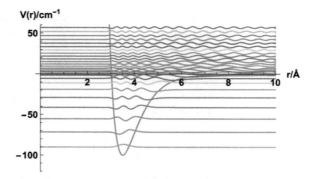

Fig. 2. A Morse oscillator potential energy curve for an Ar_2-like potential with $D_e = 100$ cm^{-1}, $r_e = 3.5$ Å, $a_{Morse} = 1.451455517$ Å^{-1}. Wavefunctions of the vibrational bound states are also shown at their associated eigenenergies, along with the continuum states between 0 and 60 cm^{-1}. The bound and continuum states were generated by solving the Schrödinger equation with $\mu = 33.71525621$ Da with an R-matrix method with a boundary of 10 Å

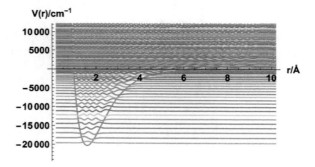

Fig. 3. Morse oscillator potential and states for LiH. Parameters used are $D_e = 20287.62581$ cm^{-1}, $r_e = 1.5956$ Å, $a_{Morse} = 1.128$ Å^{-1} [33]. The states were generated by solving the Schrödinger equation with $\mu = 0.8801221$ Da [33] with an R-matrix method with a boundary of 10 Å

4.2   Numerical Details

The R-matrix method was used to generate scattering results, including the eigenphase and the scattering length, for the single channel, $J = 0$ Morse oscillator potential. These results are compared with the analytic results quoted above.

In the construction of the R-matrix, the inner region bound system was solved numerically to generate the bound eigenenergies and radial eigenfunctions of two particles interacting over a Morse potential well. To generate the numeric results, $N = 200$ grid points and eigenfunctions were used to obtain the inner region eigenenergies (and amplitudes) using the Lobatto shape functions DVR method outlined in Sect. 2. The inner region was defined to range from $r_{\min} = 0.01$ Å to $a_0 = 10.0$ Å.

The R-matrix was then constructed on the boundary and propagated to an asymptotic radius. For the results presented in the following, the propagation was performed from $a_0 = 10.0$ Å to $a_p = 25.0$ Å, with $N_{\mathrm{prop}} = 2500$ iterations of the propagation equation over a uniform grid. The propagated R-matrix was then used to construct the eigenphase for the $J = 0$ Morse scattering event.

To explore the low-energy behaviour of the numeric method, the analytic and numeric eigenphases were used to generate the scattering length and effective range. This was done by fitting the low-energy plot to the form given in (32) using *Mathematica*'s FindFit function over the lower scattering energy range $k = 0.0004$ Å to $k = 0.001$ Å. (This is equivalent to $E = 8.0 \times 10^{-8}$ cm^{-1} to $E = 5.0 \times 10^{-7}$ cm^{-1} for this system.)

5   Results

5.1   Comparison Between Analytic and Numerical RmatReact Results

The numerical and analytic results for the eigenenergies are presented in Table 1. For low-lying states whose wavefunctions are essentially completely contained in the inner region, the agreement between the two methods is excellent. The final two states are more diffuse, as seen in Fig. 2, and hence they are more likely to have significant amplitude outside the inner region. Due to this, the inner region solution energies lies slightly below the true answer.

Table 1. Comparison of the analytic and numeric bound eigenenergies of the Morse diatomic system for vibrational energy levels n = 0–9. The relative error refers to the difference between each level's numeric and analytic values, divided by the analytic value (analytic minus numeric, divided by analytic)

n	Analytic/cm^{-1}	R-matrix/cm^{-1}	Relative error
0	-90.000000	-90.000000	1.73×10^{-12}
1	-71.580042	-71.580042	4.59×10^{-11}
2	-55.266807	-55.266807	1.30×10^{-11}
3	-41.060295	-41.060295	1.82×10^{-11}
4	-28.960506	-28.960506	5.73×10^{-12}
5	-18.967441	-18.967441	1.33×10^{-12}
6	-11.081099	-11.081099	3.25×10^{-12}
7	-5.3014807	-5.3014807	-6.42×10^{-12}
8	-1.6285853	-1.6286033	-0.000011
9	-0.062413189	-0.094633937	-0.516

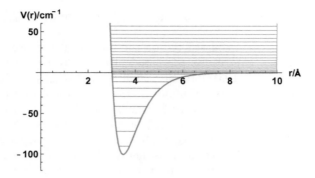

Fig. 4. The same Morse oscillator potential as in Fig. 2. Energy levels of the continuum states generated by the R-matrix below 60 cm^{-1} are coloured differently to the vibrational bound states in order to distinguish the states close to dissociation from the states just above dissociation. The R-matrix inner region boundary, $a_0 = 10$ Å, is also highlighted

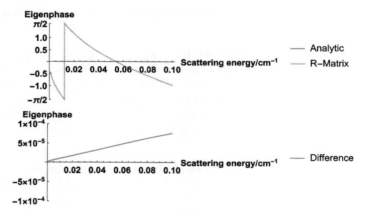

Fig. 5. Upper plot: eigenphase (in radians) for a scattering event for the Morse potential of Fig. 2 calculated both analytically and using R-matrix methodology. The two lines overlap. Lower plot: difference (analytic − R-matrix) in eigenphase (in radians) between the two methods

Figure 4 compares the RmatReact numerical eigenphase to the analytic solution for the eigenphase given by (26) over the scattering energy range of 0.001 to 0.1 cm^{-1} (0.00144 to 0.144 K). The root mean square difference between the analytic and numeric results is approximately 4.6×10^{-5} radians, which is small (Fig. 5).

Analytic and numerical results for scattering length and effective range are presented in Table 2. Again the results given by the two methods are very similar.

Table 2. Table of comparisons for the analytic and numeric scattering length and effective range. The relative error refers to the difference between each quantity's numeric and analytic values divided by the analytic value (analytic minus numeric, divided by analytic)

	Analytic/ Å	R-matrix/ Å	Relative error
Scattering length	10.166078	10.166133	-5.34×10^{-6}
Effective range	1.6537298	1.6667562	-0.00788

Table 3. Table of numerical parameters

Symbol	Definition	Units
N	Number of inner region states and grid points	Unitless
N_{prop}	Number of propagation points	Unitless
$r_{\min}$	Start of inner region	Å
a_0	End of inner region and start of propagation	Å
a_p	End of propagation	Å
Δr	$\frac{a_0 - r_{\min}}{N-1}$ Average inner region grid spacing	Å
Δr_{prop}	$\frac{a_p - a_0}{N-1}$ Average propagator grid spacing	Å

5.2   Numerical Parameters

To investigate the accuracy of the R-matrix method in comparison to the analytic results, the numerical parameters used in the algorithm were varied and the resultant error was plotted. The seven numerical parameters which the method relies on are summarised in Table 3.

To encapsulate all of the information in the lower plot of Fig. 4 in one number, the error metric used was the root mean square deviation (RMSD) between the eigenphase, $\delta(E)$ calculated using the R-matrix method ($\delta_{\mathrm{num}}(E)$) and the analytic eigenphase ($\delta_{\mathrm{ana}}(E)$). The eigenphase was calculated for 100 equally spaced scattering energy values between 0.001 and 0.1 cm^{-1}. The error characteristic, the RMSD, was then calculated using:

$$\delta_{\mathrm{RMSD}} = \sum_{i=1}^{100} \sqrt{\frac{(\delta_{\mathrm{ana}}(E_i) - \delta_{\mathrm{num}}(E_i))^2}{100}}. \tag{33}$$

A version of this error metric which involved (numerically) integrating the squared difference over the energy range was tested, and found to give the same results as merely sampling over 100 equally spaced points in the energy range. Plotting δ_{RMSD} as a function of different error parameters facilitated the assessment of the numerical stability of the method. These plots can be found in Fig. 6. For all of the plots in Fig. 6, $r_{\min}$ was kept constant at 0.01 Å.

When varying a_0, any a_0 value above approximately 9 Å appears to produce converged results where the error changes very little. This is likely because a

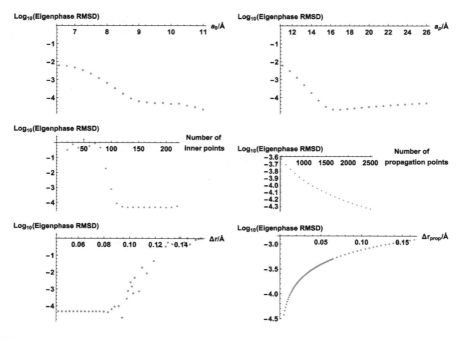

Fig. 6. Top left: The log of the RMSD of the eigenphase plotted against a_0 between 6.5 Å and 11 Å. The other parameters were held constant at $N = 200$, $N_{\text{prop}} = 2500$, $r_{\text{prop}} = 25$ Å. Top right: The log of the RMSD of the eigenphase plotted against a_p between 11 Å and 26 Å. The other parameters were held constant at $N = 200$, $N_{\text{prop}} = 2500$, $a_0 = 10$ Å. Middle left: The log of the RMSD of the eigenphase plotted against N between 20 and 220. The other parameters were held constant at $N_{\text{prop}} = 2500$, $a_0 = 10$ Å, $r_{\text{prop}} = 25$ Å. Middle right: The log of the RMSD of the eigenphase plotted against N_{prop} between 500 and 2500. The other parameters were held constant at $N = 200$, $a_0 = 10$ Å, $r_{\text{prop}} = 25$ Å. Bottom left: The log of the RMSD of the eigenphase plotted against Δr between 0.0445982 Å and 0.156094 Å. N was allowed to vary between 223 and 63 to vary Δr. The other parameters were held constant at $a_0 = 10$ Å, $N_{\text{prop}} = 2500$, $r_{\text{prop}} = 25$. Bottom right: The log of the RMSD of the eigenphase plotted against Δr_{prop} between 0.005 Å and 0.164835 Å. N_{prop} was allowed to vary between 3000 and 90 to vary Δr_{prop}. The other parameters were held constant at $N = 200$, $a_0 = 10$ Å, $r_{\text{prop}} = 25$

value of a_0 which is too small cannot accurately 'capture' all of the bound states of the potential well. Since the final bound state is of the order 10^{-2} cm^{-1} in depth, $V(a_0)$ must be approximately of that order for the state to be found by the method.

When varying a_p, any value above 16 Å appears to produce converged results; however, the error increases slightly as a_p is extended beyond 16 Å. This is likely due to Δr_{prop} increasing as N_{prop} is held constant, which decreases the accuracy of the approximations made in the propagator method.

When varying N and Δr (where Δr is increased by decreasing N and vice versa), there is a clear point where increasing N further has no effect, but where decreasing N even slightly significantly increases the error. This suggests that the method is converging on a solution once the grid spacing is sufficiently small, as is common in numerical integration techniques. This further suggests that this solution's RMSD from the analytic solution is approximately 10^{-4}.

Finally, when varying N_{prop} and Δr_{prop}, the method appears to produce results with very low error for all values of N_{prop} and Δr_{prop} tested, with only slight variation in the error recorded. This suggests that it is possible to propagate the R-matrix using very few, very wide steps and still produce accurate results. However, this may be a consequence of using as the test potential the Morse oscillator potential, since it decreases exponentially with distance and thus varies very little in the outer region. More relativistic potentials are longer range and multipolar in nature at large r, so narrower steps may be needed in the propagation.

6   Conclusions and Outlook

We clearly demonstrate that we can obtain excellent results using our R-matrix implementation for low-energy scattering within a Morse oscillator potential. Asymptotically this potential decays exponentially, which makes it unlike physical potentials, which have a much longer range. Physical potentials decay as r^{-n}, where n is a positive integer.

The next step is to implement DVR shape functions into variational nuclear motion codes to facilitate the calculation of boundary amplitudes within these codes. This been done for the general diatomic code Duo [16] and triatomic code DVR3D [17]. The diatomic problems for which tests have been run so far all involve a single asymptotic channel, which makes R-matrix propagation straightforward. In general, this will not be true and it will be necessary to consider multichannel problems. To address this issue we have successfully performed propagations with a general code originally designed for electron—atom problems [34]. This code now needs generalising to provide automated resonance fitting [35,36] and bound state finding [37] features.

We intend to use this new methodology on physical problems, and to create a generalisation of the R-matrix formalism to allow the explicit treatment of reactive processes. We have conducted preliminary tests similar to the ones presented here on more accurate Ar–Ar potentials with multipolar long-range expansions, for which the leading term is $n = 6$, and also obtained excellent results. All of these results will be reported elsewhere (10.1080/00268976.2019.1615143).

Acknowledgments. This project has received funding from the European Union's Horizon 2020 research and innovation programme under the Marie Sklodowska-Curie grant agreement No 701962 and from the EPSRC.

References

1. Stuhl, B.K., Hummon, M.T., Ye, J.: Annu. Rev. Phys. Chem. **65**, 501 (2014). https://doi.org/10.1146/annurev-physchem-040513-103744
2. Quemener, G., Julienne, P.S.: Chem. Rev. **112**, 4949 (2012)
3. Tennyson, J., McKemmish, L.K., Rivlin, T.: Faraday Discuss. **195**, 31 (2016). https://doi.org/10.1039/c6fd00110f
4. Burke, P.G.: R-Matrix Theory of Atomic Collisions: Application to Atomic, Molecular and Optical Processes, vol. 61. Springer Science & Business Media (2011)
5. Tennyson, J.: Phys. Rep. **491**, 29 (2010)
6. Descouvemont, P., Baye, D.: Rep. Prog. Phys. **73**, 036301 (2010)
7. Tennyson, J.: Chem. Phys. Lett. **86**, 181 (1982)
8. Polyansky, O.L., Alijah, A., Zobov, N.F., Mizus, I.I., Ovsyannikov, R., Tennyson, J., Szidarovszky, T., Császár, A.G.: Phil. Trans. R. Soc. Lond. A **370**, 5014 (2012)
9. Carrington, A., McNab, I.R.: Acc. Chem. Res. **22**, 218 (1989). https://doi.org/10.1021/ar00162a004
10. Mayle, M., Quemener, G., Ruzic, B.P., Bohn, J.L.: Phys. Rev. A **87**, 012709 (2013). https://doi.org/10.1103/PhysRevA.87.012709
11. Zobov, N.F., Shirin, S.V., Lodi, L., Silva, B.C., Tennyson, J., Császár, A.G., Polyansky, O.L.: Chem. Phys. Lett. **507**, 48 (2011)
12. Szidarovszky, T., Csaszar, A.G.: Mol. Phys. **111**, 2131 (2013). https://doi.org/10.1080/00268976.2013.793831
13. Silva, B.C., Barletta, P., Munro, J.J., Tennyson, J.: J. Chem. Phys. **128**, 244312 (2008)
14. Pavanello, M., Adamowicz, L., Alijah, A., Zobov, N.F., Mizus, I.I., Polyansky, O.L., Tennyson, J., Szidarovszky, T., Császár, A.G., Berg, M., Petrignani, A., Wolf, A.: Phys. Rev. Lett. **108**, 023002 (2012)
15. Lara, M., Jambrina, P.G., Aoiz, F.J., Launay, J.M.: J. Chem. Phys. **143**, 204305 (2015). https://doi.org/10.1063/1.4936144
16. Yurchenko, S.N., Lodi, L., Tennyson, J., Stolyarov, A.V.: Comput. Phys. Commun. **202**, 262 (2016). https://doi.org/10.1016/j.cpc.2015.12.021
17. Tennyson, J., Kostin, M.A., Barletta, P., Harris, G.J., Polyansky, O.L., Ramanlal, J., Zobov, N.F.: Comput. Phys. Commun. **163**, 85 (2004)
18. Kozin, I.N., Law, M.M., Tennyson, J., Hutson, J.M.: Comput. Phys. Commun. **163**, 117 (2004)
19. Yurchenko, S., Jensen, P., Thiel, W.: J. Chem. Phys. **71**, 281 (2004)
20. Yurchenko, S., Jensen, P., Thiel, W.: J. Chem. Phys. **90**, 333 (2010)
21. Mussa, H.Y., Tennyson, J.: Comput. Phys. Commun. **128**, 434 (2000)
22. Light, J.C., Carrington, T.: Adv. Chem. Phys. **114**, 263 (2000). https://doi.org/10.1002/9780470141731.ch4
23. Manolopoulos, D.: In: Numerical Grid Methods and Their Application to Schrödingers Equation (pp. 57–68). Springer (1993)
24. Manolopoulos, D., Wyatt, R.: Chem. Phys. Lett. **152**, 23 (1988)
25. Weisstein, E.W.: Lobatto quadrature From MathWorld—A Wolfram Web Resource. http://mathworld.wolfram.com/LobattoQuadrature.html. Accessed on 22 Nov 16
26. Walker, R.B., Light, J.C.: Ann. Rev. Phys. Chem. **31**, 401 (1980). https://doi.org/10.1146/annurev.pc.31.100180.002153
27. Morse, P.M.: Phys. Rev. **34**, 57 (1929)
28. Rawitscher, G., Merow, C., Nguyen, M., Simbotin, I.: Am. J. Phys. **70**, 935 (2002)

29. Selg, M.: Proc. Estonian Acad. Sci. **65**, 267 (2016)
30. Selg, M.: J. Chem. Phys. **136**, 114113 (2012)
31. Bethe, H.: Phys. Rev. **76**, 38 (1949)
32. Mohr, P.J., Taylor, B.N., Newell, D.B.: J. Phys. Chem. Ref. Data **84**, 1527 (2012)
33. Qiang, W.C., Dong, S.H.: Phys. Letts. A **363**, 169 (2007)
34. Burke, V.M., Noble, C.J.: Computer Phys. Comm. **85**, 471 (1995)
35. Tennyson, J., Noble, C.J.: Comput. Phys. Commun. **33**, 421 (1984)
36. Little, D.A., Tennyson, J., Plummer, M., Sunderland, A.: Comput. Phys. Commun. **215**, 137 (2017). https://doi.org/10.1016/j.cpc.2017.01.005
37. Sarpal, B.K., Branchett, S.E., Tennyson, J., Morgan, L.A.: J. Phys. B: At. Mol. Opt. Phys. **24**, 3685 (1991)

Strong-Field Ionization with Few-Cycle Bessel Pulses: Interplay Between Orbital Angular Momentum and Carrier Envelope Phase

Willi Paufler[1]($\boxtimes$), Birger Böning[1], and Stephan Fritzsche[1,2]

[1] Theoretisch Physikalisches Institut, Friedrich Schiller Universität Jena,
Jena, Germany
willi.paufler@uni-jena.de
[2] Helmholtz Institut, Jena, Germany

Abstract. We study strong-field ionization of a hydrogenic target by few-cycle Bessel pulses. In order to investigate the interplay between the carrier envelope phase (CEP) and the orbital angular momentum of a few-cycle pulse (OAM), we apply a semiclassical two-step model. In particular, we here compute and discuss photoelectron momentum distributions (PEMD) for localized atomic targets. We show how these momentum distributions are affected by the CEP and TAM of the incident pulse. In particular, we find that the OAM affects the PEMD in a similar way as the CEP, depending on the initial position of our target.

1 Introduction

The behavior of atoms in strong laser fields is closely connected to nonlinear processes such as above threshold ionization (ATI) [1,2], high harmonic generation (HHG) [3] or nonsequential double ionization (NSDI) [4]. Different theoretical approaches like the numerical solution of the time dependent Schrödinger equation (TDSE), the strong-field approximation (SFA) [1,5] or semiclassical models are used to explore the electron dynamics at the ultrafast timescale. Among the semiclassical models the two-step model for ionization and measurement of direct electrons [6] and the three-step model [7] for re-collisional phenomena [8–11] (HHG, NSDI) have been applied widely and helped to describe many experimental observations. In the three-step model, for instance, the first step describes the tunnel ionization of an electron due to the supresed atomic potential in a strong laser field. Here, the so-called ADK theory (Ammosov Delone Krainov), which was derived from the work of Perelomov et al. [12,13] and Perelomov and Popov [14], models the ionization rate of the electrons. In the second step, the released electron moves freely along its classical trajectory, while its (in)elastic re-scattering, or recombination, is considered in the third step.

© Springer Nature Singapore Pte Ltd. 2019
P. C. Deshmukh et al. (eds.), *Quantum Collisions and Confinement of Atomic and Molecular Species, and Photons*, Springer Proceedings in Physics 230,
https://doi.org/10.1007/978-981-13-9969-5_26

Today, the strong-field ionization of atoms and molecules is quite well understood for incident plane wave radiation for systems that can be described within the single active electron approximation. Spectra for different polarizations [15,16], for different pulse lengths and the corresponding CEP-effects [1,17,18] or two-color ionization have been studied [5] and pulse shaping effects can nowadays be used to control attosecond dynamics.

During the past few years, twisted light beams gained a lot of attention [19,20]. It is well known that plane wave photons carry spin angular momentum (SAM), but twisted photons carry additional orbital angular momentum (OAM). In contrast to plane waves, twisted light typically is characterized by helical phase fronts and a much more complex spatial structure. For such twisted beams, especially, the intensity profile perpendicular to the propagation direction does not exhibit a uniform distribution but has concentric ring maxima and minima, cf. Fig. 2. For the strong-field ionization by twisted light, therefore, the intensity distribution within the profile of the beam is important, because strong-field ionization only occurs near the intensity maxima, where the intensity is high enough. Along the and near to the beam axis (vortex line), there is no ionization. Moreover, since the field vectors point in different directions at different positions in the beam, see [21], it is expected that differently localized single atomic targets will result in different PEMDs.

In this work, we study theoretically the strong-field ionization of hydrogen by twisted few-cycle near-infrared pulses. We apply a semiclassical two-step model (Sect. 2) to calculate the photoelectron momentum distributions (PEMD). We observe an asymmetry in the PEMD similar to the CEP dependent asymmetry in the photo electron spectrum of strong-field ionization by circularly polarized pulses [1,17] (Sect. 3.1). In contrast to the asymmetries in circularly polarized fields, however, the asymmetries observed in this work do not only depend on the CEP but also on the position of the atomic target in the field and the projection of the total angular momentum (TAM), which is the sum of the orbital angular momentum and spin angular momentum, of the Bessel pulse (Sect. 3.2). We will provide two different explanations, how to understand the interplay of OAM and CEP. Finally, we show that the PEMDs produced by Bessel pulses become symmetric for infinitely extended targets.

2 Model

2.1 Semiclassical Two-Step Model

Here we explain the details of our semiclassical two-step model for strong-field ionization. In this model, the atom is exposed to a strong laser field. This strong external field causes a suppression of the atomic potential. Thus, the potential forms a barrier, where the electron can tunnel through. To describe the tunneling of the electron through the suppressed potential, we use the ADK ionization rate. Tunneling is more likely when the atomic potential is more suppressed. Thus the ionization probability is highest when the external field reaches its maximum. The ADK rate attaches a weight to every possible time of ionization.

At this time, we assume a free electron to be "born" at the place of the atom and neglect the Coulomb potential of the parent ion when the tunnel ionization has occurred. We remark, that this is a rough approximation, since the electron cannot be born at the place of the atom. To be more precise, we should take the tunnel exit and the Coulomb potential into account [22]. The coulomb potential for example would change the angle of peak emission in the PEMD [23,24]. However, in this work we will drop these details because they have no effect on the qualitative comparison of strong-field ionization between plane waves and twisted light. In the two-step model, the component of the electron velocity longitudinal to the instantaneous field direction is set to zero, while the initial velocity transversal to the instantaneous electric field is given by a Gaussian distribution and therefore is not necessarily zero. Therefore, the probability for an electron to tunnel through the barrier at a given time t_0 with absolute initial velocity $v_{0\perp}$ is given by Liu [25]

$$w(t_0, v_{0\perp}) = w_{t_0} w_{v_{0\perp}}, \tag{1}$$

where

$$w_{t_0} = \left(\frac{E(\mathbf{b}, t_0)}{4}\right) \left(\frac{4\kappa^4}{E(\mathbf{b}, t_0)}\right)^{2/\kappa} e^{-\frac{2\kappa^3}{3E(\mathbf{b}, t_0)}} \tag{2}$$

gives the probability for ionization at time t_0 and

$$w_{v_{0\perp}} = \sqrt{\frac{\kappa}{\pi E(\mathbf{b}, t_0)}} e^{-\frac{\kappa v_{0\perp}^2}{E(\mathbf{b}, t_0)}} \tag{3}$$

is the Gaussian distribution of initial velocities. $E(\mathbf{b}, t_0)$ denotes the magnitude of the electric field at time t_0 at impact parameter $\mathbf{b} = (b, \varphi_b, z)$, [cf. Fig. 2], and $\kappa = \sqrt{2I_P}$ with I_P being the ionization potential. Note that it is important for a twisted beam to consider the spatial dependence of the electric field $E(\mathbf{b}, t_0)$ and not only the time dependence $E(t_0)$ because in the case of twisted light beams or pulses the electric field depends on the position of the target.

After the release of the electron into the continuum, we neglect the parent ion so that the electron moves freely on its classical trajectory in the external field. Then, the dynamics of the electron are given by Newton's equation

$$\ddot{\mathbf{r}} = -\mathbf{E}(\mathbf{r}, t). \tag{4}$$

Here we consider the spatial dependence of the electric field explicitly to properly describe the twisted pulse. For a twisted Bessel beam, for instance, the electric field components can be easily obtained from the vector potential in Coulomb gauge [19]

$$\mathbf{A}(\mathbf{r}, t) = \sum_{m_s = 0, \pm 1} \eta_{m_s} A_{m_s}(\mathbf{r}) e^{-i\omega t}, \tag{5}$$

where the coefficients $A_{m_s}(\mathbf{r})$ are given by

$$A_{m_s}(\mathbf{r}) = \sqrt{\frac{\varkappa}{2\pi}}(-i)^{m_s}c_{m_s}J_{m_\gamma-m_s}(\varkappa r_\perp)e^{i(m_\gamma-m_s)\varphi}e^{ik_z z}$$

and $\boldsymbol{\eta}_{m_s}$ are eigenvectors of the z-component of the SAM operator

$$\boldsymbol{\eta}_0 = \begin{pmatrix} 0 \\ 0 \\ 1 \end{pmatrix}, \qquad \boldsymbol{\eta}_{\pm 1} = \frac{\mp 1}{\sqrt{2}}\begin{pmatrix} 1 \\ \pm i \\ 0 \end{pmatrix}.$$

The expansion coefficients can be written as

$$c_0 = \frac{\Lambda}{\sqrt{2}}, \qquad c_{\pm 1} = \frac{1}{2}\left(1 \pm \Lambda \cos\vartheta_k\right).$$

The structure and derivation of the vector potential (5) was explained in detail by Matula et al. [19]. To solve Newton's equation (4) in terms of real trajectories, we need a real valued field. Therefore, we only use the imaginary part of the vector potential, so we can rewrite the vector potential as

$$\mathbf{A}(\mathbf{r},t) = A_r\,\mathbf{e_r} + A_\varphi\,\mathbf{e_\varphi} + A_z\,\mathbf{e_z} \tag{6}$$

with components

$$A_r(\mathbf{r},t) = \sqrt{\frac{1}{4\pi\varkappa}}\left(\frac{m_\gamma}{r}J_{m_\gamma}(\varkappa r)\right.$$
$$\left. + \varkappa\Lambda\cos\vartheta_k\left(J_{m_\gamma-1}(\varkappa r) - \frac{m_\gamma}{\varkappa r}J_{m_\gamma}(\varkappa r)\right)\right)\cos(m_\gamma\varphi + k_z z - \omega t), \tag{7}$$

$$A_\phi(\mathbf{r},t) = -\sqrt{\frac{1}{4\pi\varkappa}}\left(\varkappa\left(J_{m_\gamma-1}(\varkappa r) - \frac{m_\gamma}{\varkappa r}J_{m_\gamma}(\varkappa r)\right)\right.$$
$$\left. + \frac{m_\gamma}{r}\Lambda\cos\vartheta_k J_{m_\gamma}(\varkappa r)\right)\sin(m_\gamma\varphi + k_z z - \omega t), \tag{8}$$

$$A_z(\mathbf{r},t) = \sqrt{\frac{\varkappa}{4\pi}}\Lambda\sin\vartheta_k J_{m_\gamma}(\varkappa r)\sin(m_\gamma\varphi + k_z z - \omega t), \tag{9}$$

where m_γ is the projection of the TAM upon the z-axis (propagation direction), Λ is the helicity, $\varkappa = \sqrt{k^2 - k_z^2}$ and $k = \frac{\omega}{c}$, where c is the speed of light.

To consider the ionization of the atoms by a short pulse, we need to multiply the vector potential above by an envelope function:

$$\mathbf{A}^{pulse}(\mathbf{r},t) = f(t)\mathbf{A}(\mathbf{r},t). \tag{10}$$

Here $f(t)$ is chosen to be a $\sin^2$-envelope

$$f(t) = \sin^2\left(\frac{\omega t}{2n_p}\right), \tag{11}$$

and n_p is the number of cycles. Thus, the pulse duration can be written as $T_p = n_p T$ with the cycle duration $T = \frac{2\pi}{\omega}$. The electric field is then given by

$$\mathbf{E}^{pulse}(\mathbf{r}, t) = -\partial_t \mathbf{A}^{pulse}(\mathbf{r}, t). \qquad (12)$$

To obtain an electron's trajectory we randomly choose an initial time t_0 (time of ionization) and an initial velocity perpendicular to the electric field at the time of ionization, as explained above. We use a 4th-order Runge–Kutta algorithm to solve Newton's equations (4). We start the integration of this equation at the time of ionization and propagate it up to the end of the pulse. After the laser is switched off, the momentum of the electron does not change anymore. Thus, the momentum of the electron at the end of the laser pulse is its final momentum. To gather a distribution of final momenta, which we call PEMD, we evaluate about 10^7 trajectories, and associate each trajectory by its weight given by the ADK rate (1).

3   Results

3.1   Strong-Field Ionization by Few-Cycle Circularly Polarized Pulses

Let us first consider the strong-field ionization with circularly polarized few-cycle pulses to analyze how the CEP modifies the PEMD. We model the pulse by

$$\mathbf{A}(t) = A_0 \, f(t) \left(\cos(\omega t + \varphi_{cep}) \, \mathbf{e_x} + \sin(\omega t + \varphi_{cep}) \, \mathbf{e_y} \right)$$

where φ_{cep} is the carrier envelope phase. Since it is well known that the rescattering of the electron can be neglected in an intense circularly polarized field [26], we can apply our semiclassical two-step model, where we only consider photo electrons which do not interact (rescatter) again with the parent ion, but rather move directly to the detector. With this model we can reproduce the prediction of Cormier and Lambropoulos [27] that the photo electron angular distribution of a few-cycle pulse becomes asymmetric because of the CEP of the pulse, which was experimentally confirmed by Paulus et al. [28] and Milošević et al. [29].

In our simulations we applied an 800 nm two cycle pulse with a peak intensity of 10^{14} W cm^{-2}. For a two cycle pulse with $\varphi_{cep} = 0$ the vector potential (13) is symmetric with respect to the x-axis (see Fig. 1c, blue curve). The corresponding PEMD resulting from strong-field ionization by a 2-cycle pulse is also symmetric with respect to the x-axis, see Fig. 1a. Each PEMD mirrors the negative vector potential of the interacting pulse. If we change the CEP of the interacting pulse to $\varphi_{cep} = \frac{\pi}{2}$, the symmetry of the vector potential changes (see Fig. 1c, red curve) and thus the symmetry of the PEMD (Fig. 1b).

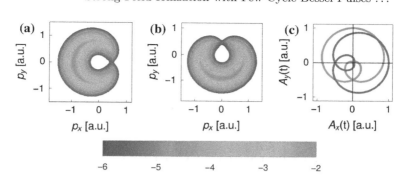

Fig. 1. The momentum distributions of photo electrons for $\lambda = 800$ nm circularly polarized 2 cycle pulses with peak intensity of 10^{14} W cm^{-2}. The pulse propagates in z-direction. The CEP was changed from $\varphi_{cep} = 0$ (**a**) to $\varphi_{cep} = \pi/2$ (**b**). The vector potentials corresponding to (13) are shown in (**c**). The blue curve displays vector potentials for $\varphi_{cep} = 0$ and the red curve for $\varphi_{cep} = \pi/2$, respectively

3.2 Strong-Field Ionization by Few-Cycle Bessel Pulses

Single-Atom Simulations The electric field of a twisted pulse has a more complex structure than a plane wave field. Because of the spatial dependence, the electric field vector points in different directions at different positions or as usually called, impact parameters **b** [30]. In this section we model our target by a single atom which is localized at an impact parameter $\mathbf{b} = (b, \varphi_b, z = 0)$. In order to investigate strong-field ionization, we choose the impact parameter to coincide with the first intensity maximum (see Fig. 2). In these simulations we will vary the angle φ_b. The distance from the beam axis to the first maximum (≈ 600 nm) is large compared to the length of a classical trajectory of an electron (≈ 10 nm), which is accelerated for two cycles in the field. The impact angle of the electron does not change much during the propagation of the electron in the external field. The expression $m_\gamma \varphi$ in (7–9), therefore acts like an additional phase, similar to the carrier envelope phase.

In Fig. 3, we present the PEMDs corresponding to strong-field ionization by a two- cycle Bessel beam with $\Lambda = 1$, $\vartheta_k = 20°$ and $\lambda = 800$ nm. The atom was placed on the first intensity maximum. For each PEMD we analyzed up to 10^7 trajectories. Columns 1, 2 and 3 correspond to $\varphi_b = 0°$, $45°$ and $90°$, respectively, and rows 1, 2 and 3 to TAM of $m_\gamma = 2$, 3, 4. When the TAM projection of the Bessel pulse is increased, the first intensity maximum occurs further away from the beam axis and, hence from the vortex line of zero intensity. The peak intensity on the first maximum decreases, if we increase the projection of the TAM. To compare the PEMDs for beams with different projection of the TAM we normalize each pulse to a peak intensity of 10^{14} W cm^{-2} on its first maximum. In the upper row of Fig. 3 we get similar results to the strong-field ionization with a circularly polarized pulse and different CEP (cf. Fig. 1). As can be seen, the PEMD rotates about $45°$, when we change the impact angle φ_b to $45°$ and from changing this angle φ_b to $90°$ the PEMD rotates about $90°$

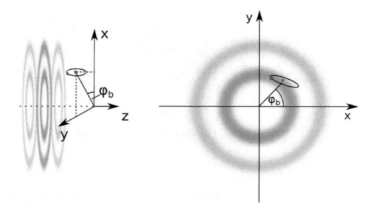

Fig. 2. Scheme of the strong-field ionization of atomic hydrogen with twisted pulses. The red rings represent the intensity profile of the Bessel beam, see also [19]. The pulse propagates in z-direction. A cloud of atoms (or single atom) is centered around an impact parameter $\mathbf{b} = (b, \varphi_b, z = 0)$. The atomic target (cloud) is illustrated by the small atom. To describe the strong-field ionization of atomic hydrogen, an impact parameter $\mathbf{b}$ is chosen for the atom on the first intensity maximum (first ring). On the left we see a side view and right we see a front view, to illustrate better how the impact angle φ_b can be understood

compared to the PEMD with $\varphi_b = 0°$. If we change the projection of the TAM of the pulse from $m_\gamma = 2$ (row 1) to $m_\gamma = 3$ (row 2), the rotation of the PEMD about the propagation axis increases. The vector potential of a Bessel beam is an eigenfunction of the z-component of total angular momentum operator, which is a sum of the projection of the orbital (L_z) and spin angular momentum operator (S_z). However, in general, it is not an eigenfunction of the operators L_z and S_z. Therefore, the beam has no well defined OAM. A Bessel beam can be associated with its dominating OAM, which is given by $m = m_\gamma - \Lambda$. In our case the projection of the TAM $m_\gamma = 2$ and the helicity $\Lambda = 1$, the dominating OAM is $m_\gamma - \Lambda = 1$, in the case of a TAM projection $m_\gamma = 3$, we have $m_\gamma - \Lambda = 2$. The dominating OAM for $m_\gamma = 3$ is twice as large as for $m_\gamma = 2$, thus the angle of rotation of the PEMD is doubled as well. A similar argument holds true for $m_\gamma = 4$.

Another way to explain why the PEMD for $m_\gamma = 3$ is rotated twice as much as for $m_\gamma = 2$ is to analyze the vector potential at each impact parameter. We consider the PEMD for $m_\gamma = 2$, $\varphi_b = 0°$ and $m_\gamma = 2$, $\varphi_b = 90°$ (both Fig. 3). If we examine the term $m_\gamma \varphi_b$, one would intuitively expect a phase shift of

$$(m_\gamma \varphi_b)_{90°} - (m_\gamma \varphi_b)_{0°} = 2 \cdot \frac{\pi}{2} - 2 \cdot 0 = \pi \tag{13}$$

in the vector potential and therefore that the PEMD should rotate about 180°. Instead of a rotation of 180° we observe a rotation about 90° (comp. Fig. 3, first line, left and right column). The x-component of the vector potential at $\varphi_b = 0°$

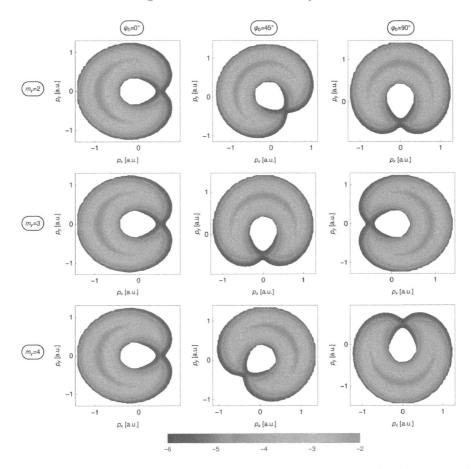

Fig. 3. PEMDs of atomic hydrogen, if ionized with a strong two-cycle 800 nm Bessel pulse with a peak intensity of 10^{14} W cm^{-2}. The pulse propagates in z-direction, and the photo electrons are measured in x-y-direction. Each row corresponds to a different projection of the TAM with $m_\gamma = 2$ (upper row), $m_\gamma = 3$ (middle row) and $m_\gamma = 4$ (lower row). Different columns represent different impact angles. In the first (second, third) column we have $\varphi_b = 0°$ $(45°, 90°)$

can be written as

$$A_x = A_r \cos(\varphi_b) - A_\phi \sin(\varphi_b) = A_r \sim \cos\left(k_z z - \omega t\right). \tag{14}$$

However, the x-component at an impact angle $\varphi_b = 90°$ is similar to

$$A_x = A_r \cos(\varphi_b) - A_\phi \sin(\varphi_b) = -A_\phi \sim \sin\left(\pi + k_z z - \omega t\right). \tag{15}$$

The expressions (14) and (15) differ by a phase shift of $\frac{\pi}{2}$. The same argument can be applied to the y and z component of the vector potential. This phase shift of $\frac{\pi}{2}$ is the reason for rotation of the PEMD about $90°$. If we consider the PEMD for $m_\gamma = 3$ and impact angles $\varphi_b = 0°$ and $\varphi_b = 90°$, respectively, we can calculate a phase shift of π and therefore a rotation of the PEMD about $180°$.

Although for every PEMD the CEP was chosen to be 0 in the twisted pulse, the calculated PEMD exhibit an asymmetry which can be compared to the CEP dependent asymmetries, discussed in Sect. 3.1. Instead of the CEP, only the impact parameter was changed. Because of the dependence of the asymmetry in the PEMD on the impact parameter, we can think of a local phase in a twisted pulse. This local phase is caused due to the OAM of the beam.

Localized Extended Targets If we consider a cloud of atoms in a plane wave field, every single atom is exposed to a field with the same phase. If we expose a cloud of atoms to a short twisted pulse, atoms at different positions in the pulse are exposed to a local field with a different phase. Therefore, in order to obtain the PEMD of an extended target, we need to integrate over different phases. Therefore, an increase of the atomic target reduces the asymmetry due to the averaging over different positions of the atoms in the beam. We model our target by a cloud of atoms. The cloud is described by a Gaussian distribution

$$\rho = \frac{1}{(2\pi\sigma^2)^{3/2}} \exp\left(-\frac{1}{2\sigma^2}((x-b)^2 + y^2 + z^2)\right), \tag{16}$$

where the center of the cloud coincides with the impact parameter $\mathbf{b}$ on the x-axis. We refer the size of the cloud to the full width of half maximum of a Gaussian distribution, which is given by $2.4\,\sigma$.

Figure 4 displays the PEMD for a beam with $m_\gamma = 2$ and $\Lambda = 1$ at impact parameter $\mathbf{b} = (b, \varphi_b = 0, b_z = 0)$, where b is chosen to coincide with the first intensity maximum and for targets of different size: single atom (left), an atomic cloud with diameter $d = 600$ nm (middle) and a cloud with $d = 2400$ nm (right), respectively. The asymmetry in the PEMD vanishes, if we increase the

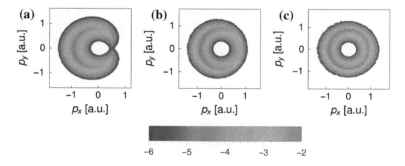

Fig. 4. PEMDs for different sizes of the atomic targets. The spectra were calculated for **a** single atom and clouds with a diameter of **b** 600 nm and **c** 2400 nm. The PEMDs become symmetric for large targets

size of the target. To measure the asymmetry in an experiment, it is necessary to localize the atomic target to if the atoms can be localized to an accuracy of approximately 1 μm scale.

4 Summary and Conclusion

The semiclassical two-step model has been applied for studying the strong-field ionization of atoms by few-cycle, near-infrared twisted pulses. We showed that the PEMD of localized targets is sensitive to the position of the target in the beam. If we ionize an atomic target with a few-cycle circularly polarized pulse, the PEMD strongly depends on the carrier envelope phase. In particular, it is possible to extract the carrier envelope phase from the shape of the PEMD. Similar to the carrier envelope phase in circularly polarized pulses, we found that the PEMD of atoms ionized by strong few-cycle Bessel pulses is affected by the position of the atomic target in the beam. If we change the impact angle φ_b of the target, we also change the local phase of the field that interacts with the atom. This was explained as a position- dependent asymmetry in the vector potential that mirrors in the PEMD. Furthermore, this asymmetry depends on the projection of the TAM of the beam. For higher TAM projection the PEMD is more sensitive to small deviations in the impact angle.

For extended targets with a diameter larger than 2 μm the reported asymmetry in the spectrum of photoelectrons vanishes. Therefore, it is required to localize the atomic targets below this scale to measure these asymmetries in the PEMDs.

Acknowledgements. This work was financially supported within the priority programme QuTiF by German Science Foundation (DFG) under the contract FR 1251/ 17-1.

References

1. Milošević, D.B., Paulus, G.G., Bauer, D., Becker, W.: J. Phys. B: At. Mol. Opt. Phys. **39**(14), R203 (2006). http://stacks.iop.org/0953-4075/39/i=14/a=R01
2. Becker, W., Grasbon, F., Kopold, R., Milošević, D., Paulus, G., Walther, H.: Adv. At. Mol. Opt. Phys. **48**, 35 (2002). https://doi.org/10.1016/S1049-250X(02)80006-4. http://www.sciencedirect.com/science/article/pii/S1049250X02800064
3. Le, A.T., Wei, H., Jin, C., Lin, C.D.: J. Phys. B: At. Mol. Opt. Phys. **49**(5), 053001 (2016). http://stacks.iop.org/0953-4075/49/i=5/a=053001
4. Figueira de Morisson Faria, C., Liu, X., Sanpera, A., Lewenstein, M.: Phys. Rev. A **70**, 043406 (2004). https://doi.org/10.1103/PhysRevA.70.043406
5. Milošević, D.B., Becker, W.: Phys. Rev. A **93**, 063418 (2016). https://doi.org/10.1103/PhysRevA.93.063418
6. Shvetsov-Shilovski, N.I., Lein, M., Madsen, L.B., Räsänen, E., Lemell, C., Burgdörfer, J., Arbó, D.G., Tokesi, K.: Phys. Rev. A **94**, 013415 (2016). https://doi.org/10.1103/PhysRevA.94.013415
7. Corkum, P.B., Krausz, F.: Nat. Phys. **3**(6), 381 (2007). https://doi.org/10.1038/nphys620

8. Paulus, G.G., Becker, W., Nicklich, W., Walther, H.: J. Phys. B: At. Mol. Opt. Phys. **27**(21), L703 (1994). http://stacks.iop.org/0953-4075/27/i=21/a=003

9. Becker, W., Goreslavski, S.P., Milošević, D.B., Paulus, G.G.: J. Phys. B: At. Mol. Opt. Phys. **47**(20), 204022 (2014). http://stacks.iop.org/0953-4075/47/i=20/a=204022

10. Lewenstein, M., Balcou, P., Ivanov, M.Y., L'Huillier, A., Corkum, P.B.: Phys. Rev. A **49**, 2117 (1994). https://doi.org/10.1103/PhysRevA.49.2117

11. Faria, C.F.D.M., Liu, X., Becker, W.: J. Mod. Opt. **53**(1–2), 193 (2006). https://doi.org/10.1080/09500340500227869

12. Perelomov, A., Popov, V., Terent'ev, M.: JEPT **23**, 924 (1966)

13. Perelomov, A., Popov, V., Terent'ev, M.: JEPT **24**, 207 (1967)

14. Perelomov, A., Popov, V.: JEPT **25**, 336 (1967)

15. He, P.L., Takemoto, N., He, F.: Phys. Rev. A **91**, 063413 (2015). https://doi.org/10.1103/PhysRevA.91.063413

16. Wollenhaupt, M., Krug, M., Köhler, J., Bayer, T., Sarpe-Tudoran, C., Baumert, T.: Appl. Phys. B **95**(2), 245 (2009). https://doi.org/10.1007/s00340-009-3431-1

17. Milošević, D.B., Paulus, G.G., Becker, W.: Laser Phys. **13**(7), 948 (2003). https://www.researchgate.net/publication/261173771_Above-threshold_ionization_with_few-cycle_laser_pulses_and_the_relevance_of_the_absolute_phase

18. Wittmann, T., Horvath, B., Helml, W., Schätzel, M.G., Gu, X., Cavalieri, A.L., Paulus, G.G., Kienberger, R.: Nat. Phys. **5**(5), 357 (2009). https://doi.org/10.1038/nphys1250

19. Matula, O., Hayrapetyan, A.G., Serbo, V.G., Surzhykov, A., Fritzsche, S.: J. Phys. B: At. Mol. Opt. Phys. **46**(20), 205002 (2013). http://stacks.iop.org/0953-4075/46/i=20/a=205002

20. Volke-Sepulveda, K., Garcs-Chvez, V., Chvez-Cerda, S., Arlt, J., Dholakia, K.: J. Opt. B: Quant. Semiclass. Opt. **4**(2), S82 (2002). http://stacks.iop.org/1464-4266/4/i=2/a=373

21. Quinteiro, G.F., Reiter, D.E., Kuhn, T.: Phys. Rev. A **91**, 033808 (2015). https://doi.org/10.1103/PhysRevA.91.033808

22. Pfeiffer, A.N., Cirelli, C., Smolarski, M., Dimitrovski, D., Abu-samha, M., Madsen, L.B., Keller, U.: Nat. Phys. **8**(1), 76 (2011). https://doi.org/10.1038/nphys2125

23. Doblhoff-Dier, K., Dimitriou, K.I., Staudte, A., Gräfe, S.: Phys. Rev. A **88**, 033411 (2013). https://doi.org/10.1103/PhysRevA.88.033411

24. Popruzhenko, S.V., Paulus, G.G., Bauer, D.: Phys. Rev. A **77**, 053409 (2008). https://doi.org/10.1103/PhysRevA.77.053409

25. Liu, J.: Classical Trajectory Perspective of Atomic Ionization in Strong Laser Fields (Springer, 2014). https://doi.org/10.1007/978-3-642-40549-5

26. Paulus, G.G., Nicklich, W., Xu, H., Lambropoulos, P., Walther, H.: Phys. Rev. Lett. **72**, 2851 (1994). https://doi.org/10.1103/PhysRevLett.72.2851

27. Cormier, E., Lambropoulos, P.: Eur. Phys. J. D. At. Mol. Opt. Phys. **2**(1), 15 (1998). https://doi.org/10.1007/s100530050104

28. Paulus, G.G., Grasbon, F., Walther, H., Villoresi, P., Nisoli, M., Stagira, S., Priori, E., Silvestri, S.D.: Nature **414**(6860), 182 (2001). https://doi.org/10.1038/35102520

29. Milošević, D.B., Paulus, G.G., Becker, W.: Opt. Expr. **11**(12), 1418 (2003). https://doi.org/10.1364/OE.11.001418. http://www.opticsexpress.org/abstract.cfm?URI=oe-11-12-1418

30. Surzhykov, A., Seipt, D., Fritzsche, S.: Phys. Rev. A **94**, 033420 (2016). https://doi.org/10.1103/PhysRevA.94.033420

.

CPSIA information can be obtained
at www.ICGtesting.com
Printed in the USA
LVHW080126121119
636876LV00002BA/132/P